Responsive Photonic Nanostructures
Smart Nanoscale Optical Materials

RSC Smart Materials

Series Editor:
Hans-Jörg Schneider, *Saarland University, Germany*
Mohsen Shahinpoor, *University of Maine, USA*

Titles in this Series:
1: Janus Particle Synthesis, Self-Assembly and Applications
2: Smart Materials for Drug Delivery: Volume 1
3: Smart Materials for Drug Delivery: Volume 2
4: Materials Design Inspired by Nature
5: Responsive Photonic Nanostructures: Smart Nanoscale Optical Materials

How to obtain future titles on publication:
A standing order plan is available for this series. A standing order will bring
delivery of each new volume immediately on publication.

For further information please contact:
Book Sales Department, Royal Society of Chemistry, Thomas Graham House,
Science Park, Milton Road, Cambridge, CB4 0WF, UK
Telephone: +44 (0)1223 420066, Fax: +44 (0)1223 420247
Email: booksales@rsc.org
Visit our website at www.rsc.org/books

Responsive Photonic Nanostructures
Smart Nanoscale Optical Materials

Edited by

Yadong Yin
University of California, Riverside, California, USA
Email: yadongy@ucr.edu

RSC Publishing

RSC Smart Materials No. 5

ISBN: 978-1-84973-653-4
ISSN: 2046-0066

A catalogue record for this book is available from the British Library

Published by The Royal Society of Chemistry,
Thomas Graham House, Science Park, Milton Road,
Cambridge CB4 0WF, UK

Registered Charity Number 207890

For further information see our web site at www.rsc.org

Printed in the United Kingdom by CPI Group (UK) Ltd, Croydon, CR0 4YY

Preface

Among many types of smart materials, responsive photonic bandgap materials, or more commonly known as responsive photonic crystals, which can change their color in response to external stimuli, have attracted much attention due to their important uses in areas such as color displays, biological and chemical sensors, inks and paints, and many active components in optical devices. The unique colors originating from the interaction of light with periodically arranged structures of dielectric materials are often called structural colors, which are iridescent and metallic, cannot be mimicked by chemical dyes or pigments, and they are free from photobleaching unlike traditional pigments or dyes. Many interesting applications have been proposed for responsive photonic crystal structures. For example, they may be used as optical switches for full automation of optical circuits when significant improvements towards the quality of colloidal crystals and their response time are realized. Military vehicles covered with such materials may be able to dynamically change their colors and patterns to match their surroundings. Such materials might also be embedded in banknotes or other security documents for anti-counterfeiting purposes. The hidden information cannot be revealed until an external stimulus such as a pressure or temperature change is applied. The photonic effect can also be used as a mechanism to develop chemical and biological sensors for detecting target analytes by outputting optical signals. These types of crystals may also find great use as active color units in the fabrication of flexible display media, including both active video displays and rewritable paper that can be reused many times.

Compared to photonic crystals prepared by microfabrication methods, self-assembled photonic crystals, in particular colloidal crystals, can be produced at much lower costs and with higher efficiencies owing to the parallel nature of the self-assembly processes. It is also more convenient to modify the

RSC Smart Materials No. 5
Responsive Photonic Nanostructures: Smart Nanoscale Optical Materials
Edited by Yadong Yin
© The Royal Society of Chemistry 2013
Published by the Royal Society of Chemistry, www.rsc.org

building blocks before or after the formation of crystal structures to enable responsiveness to a given stimulus. As a result, the majority of research on responsive photonic nanostructures has been focused on constructing the photonic crystal structures and incorporating stimulus-responsive materials into the self-assembled photonic crystal structures. In principle, the stimulus can be any means that can effectively induce changes in the refractive indices of the building blocks or the surrounding matrix, and changes in the lattice constants and/or spatial symmetry of the crystalline arrays. While various responsive mechanisms have been developed, such as mechanical stretching, solvent swelling, and temperature-dependent phase change, the research activities in the field have been focused on broadening the tunability of the photonic properties, enhancing the response rate to the external stimuli, improving the reversibility, and integrating into existing photonic devices.

This book highlights several recent areas of progress in the self-assembled responsive photonic nanostructures based on a number of different tuning mechanisms. Among all photonic crystal structures, one-dimensional Bragg reflectors that consist of alternative multilayers of two materials with different dielectric constants are regarded as the simplest type of photonic nano-structures. Calvo and Míguez first discuss recent progresses in the development of such materials for potential applications in sensing owning to their ability to respond to changes in the surrounding environment with a modification of their optical properties, generally caused by a variation of either their refractive index, the thickness of the constituent layers, or both. Self-assembled opals of close-packed colloidal crystals from monodisperse colloidal particles have predominantly served as the starting frameworks for constructing responsive photonic nanostructures. Stimulus-responsive materials can be incorporated into the periodic structures either as the initial building blocks or as the surrounding matrix so that the photonic properties can be tuned. Such colloidal crystals may also be used as the templates to fabricate inverse opals. Various versions of tunable opals and inverse opals have been developed that can respond to a wide range of external stimuli such as mechanical stretching, humidity, light, and temperature change, as reviewed separately by Fudouzi, Gu and Stein and coworkers. Since the opal structures themselves are relatively weak due to the fragile contact points between spheres within the structure, many structurally deformable photonic structures have been made from close-packed or nonclose-packed colloidal crystal arrays encapsulated within a hydrogel or polymer matrix that fills the void space surrounding the colloidal crystal, as discussed by Kanai and Takeoka. Through the infiltration of a defect layer of liquid crystals into photonic structures, the optical properties can be reversely manipulated by the external electric fields to realize the electroch-romatic effect. The relevant research has been summarized by Ozaki and coworkers. Yin and coworkers also highlight recently developed magnetically responsive photonic nanostructures with widely, rapidly and reversely tunable structural colors across the entire visible and near-IR range, which utilize the magnetic field as the convenient stimulus to tune the optical properties by affecting the lattice constant, the orientation, and the structures of the colloidal assemblies. We hope this book will serve as a useful reference to researchers interested in smart nanoscale optical materials, in particular, responsive photonic nanostructures.

Yadong Yin
University of California, Riverside

Contents

RSC Smart Materials No. 5
Responsive Photonic Nanostructures: Smart Nanoscale Optical Materials
Edited by Yadong Yin
© The Royal Society of Chemistry 2013
Published by the Royal Society of Chemistry, www.rsc.org

CHAPTER 1

Responsive Bragg Reflectors

MAURICIO E. CALVO AND HERNÁN MÍGUEZ*

Multifunctional Optical Materials Group, Instituto de Ciencia de Materiales de Sevilla, Consejo Superior de Investigaciones Científicas-Universidad de Sevilla, Américo Vespucio 49, 41092 Sevilla, Spain
*Email: h.miguez@csic.es

1.1 Introduction

Multilayers have been a common subject of study of materials and optical scientists for many decades. The possibility to attain color from the stacking of films of transparent materials and, furthermore, the fine control over light transmission and reflection they offer, have attracted the attention of both scientists and technologists. Indeed, the industrial development of these materials has led to the realization of a myriad of passive optical elements that are commonly found in all kind of spectrophotometers or optical characterization setups. From the manufacturing perspective, most efforts have been put in the preparation of thin films as stable as possible against changes in the surrounding environment. This has been mainly motivated by their use as filters of a range of selected optical frequencies, which would not be constant if their structure varies in the presence of moisture or as a consequence of temperature variations. This feature implied that the constituent layers could not present accessible porosity in which condensation of vapor could take place. Also, unless pore sizes are in the nanoscale range, the presence of voids could easily lead to diffuse scattering that will deteriorate its optical quality.

There is currently a boost in the development of film deposition techniques that permit a strict control to be achieved over porosity at the mesoscale, thus preventing diffuse optical scattering phenomena. While porosity in general

RSC Smart Materials No. 5
Responsive Photonic Nanostructures: Smart Nanoscale Optical Materials
Edited by Yadong Yin
© The Royal Society of Chemistry 2013
Published by the Royal Society of Chemistry, www.rsc.org

endows the film with the potential of hosting guest compounds in the interstitial space, such as potential functional groups or analytes transported from gas or liquid phase, fine tuning of the pore-size distribution yields command over the kinetics of vapor sorption in the layers or molecular-size-selective detection. Actually, many of the porous materials that can be prepared as thin films have already been incorporated in a multilayer structure with the aim of taking advantage of the interplay between the responsive properties that porosity provides. The same aim has been reached by employing a different approach based on the multilayer integration of polymeric films whose thickness and refractive-index change as a function of the species present in their surroundings. In this chapter we will review the main properties that make all these layers interesting building blocks to build responsive optical materials as well as the main synthetic procedures and representative applications that are being explored in this emerging field.

1.2   Fundamentals: Optical Properties of Multilayers

The *optical thickness* of a film is defined as the product of its geometrical thickness times its refractive index. This parameter determines the range of wavelengths for which optical interference effects are observed when a white light beam impinges on the slab surface. Both transmitted and reflected light will present spectral intensity fluctuations whose frequency will depend on the value of the optical thickness relative to the incident wavelength. For a dielectric film, reflectance maxima are expected when half an integer number of wavelengths, respectively, "fit" in the optical thickness of the film. The intensity of these maxima depends on the dielectric constant contrast between the film and the surrounding media, which typically are the air above the film and the substrate supporting it. Optical interferometry of dense thin films is commonly put into practice to prepare antireflection coatings to reduce light insertion losses in sunglasses or devices such as solar cells. For the case of porous films, the possibility was soon realized of making use of the sensitivity of the pattern of lobes observed in transmittance and reflectance spectra to the refractive index of the film for detection and recognition of specific targeted compounds, provided adequate functionalization of the pore walls was achieved. In fact, the first optically responsive films were developed by anchoring antigens to the inner walls of porous silicon films and exposing them to the corresponding antibodies, which gave rise to an increase of the average refractive index of the film that resulted in a redshift of the monitored optical features. The group of Sailor largely contributed in the 1990s and afterwards to the development of porous silicon structures for sensing of different sorts of species based on this approach. An example of this responsive behavior of a porous silicon film is shown in Figure 1.1. Today, there exist many different types of materials that could be shaped as porous films and thus employed as the basis for a responsive interferometric sensing device.

In the last decade, the possibility to stack porous films of different composition and structure preserving the accessibility of the network of

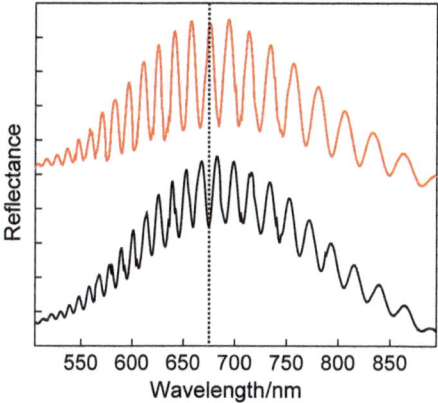

Figure 1.1 Reflectance spectrum from a single layer of porous silicon before (red line) and after (black line) the introduction of a molecular compound in the pores of the structure.
Image reproduced/Adapted from ref. 8.

interstices, *i.e.* achieving continuity of the void space present in the different slabs, has been thoroughly explored. If a rational design of the stack is followed, multilayers in which very intense reflectance peaks (or transmittance dips) are observed as a result of the interference of beams reflected and refracted at each interface. An alternative approach to the construction of responsive optical multilayers without employing porous materials is based on the utilization of block copolymers or hydrogel films as building blocks. These materials may allow the diffusion of certain species through them when put into contact with a solution or vapor containing them. The optical thickness is in this case varied both as a result of swelling, which occurs without loss of the mechanical stability of the films, and refractive-index change. The group of Thomas has greatly contributed with the integration of such thin polymeric coatings in different sorts of photonic multilayered structures.

In what follows we will describe the main optical features of multilayered systems, providing a guide to easily predict their main reflection and transmission properties as well as to understand the possibilities offered by resonant optical modes designed within the structure as far as the responsive character of the material is concerned.

1.2.1 Reflection and Transmission

The main optical feature of periodic multilayers is their capability to reflect, and consequently transmit, light of selected wavelength ranges. At any given incident direction, for those wavelengths for which the beams reflected at each interface present in the stack interfere constructively, an intense specular reflectance peak will be attained. Some simple equations allow one to predict many of the qualitative, as well as some of the quantitative, changes observed in the optical properties of a responsive Bragg mirror when exposed to changes in its environment. In the case in which we have two different types of films alternately distributed in the stack, of thicknesses t_1 and t_2 and refractive indices n_1 and n_2, respectively, the spectral position of the specular reflectance maxima, $\lambda_B(\theta)$, at any given incident angle with respect to the surface normal, θ, can be anticipated by the formula:

$$\lambda_B(\theta) = 2d\sqrt{n^2 - (\sin\theta)^2} \tag{1.1}$$

where $d = t_1 + t_2$ is the unit cell size and n is the average refractive index of the multilayer, which is in turn given by:

$$n = \frac{t_1 n_1 + t_2 n_2}{d} \tag{1.2}$$

The unit cell of a periodic multilayer is defined as the structural unit that repeats itself along the direction perpendicular to the substrate. Expression (1.1) results from the combination of Snell's and Bragg's laws. In fact, periodic multilayers displaying strong optical reflections are known as *Bragg reflectors*, *Bragg stacks*, *distributed Bragg mirrors*, although they can also labeled as *interference filters* or *rugate filters*. For some specific cases, relevant features such as the reflectance peak intensity, R, and spectral width, $\Delta\lambda_0$, can be approximated by relatively simple formulas. In the case in which the layers are designed so that $n_1 t_1 = n_2 t_2 = \lambda_B/4$, reflectance is given by:[1]

$$R = \left[\frac{1 - \left(\frac{n_s}{n_0}\right)\left(\frac{n_2}{n_1}\right)^{2N}}{1 + \left(\frac{n_s}{n_0}\right)\left(\frac{n_2}{n_1}\right)^{2N}} \right]^2 \tag{1.3}$$

where n_0 and n_s are the refractive indices of the incoming and outgoing media and N is the total number of periods in the stack. In all cases, the spectral width of the reflectance peak is given by:

$$\Delta\lambda = \frac{4\lambda_B}{\pi} \sin^{-1}\left[\frac{n_2 - n_1}{n_2 + n_1}\right] \tag{1.4}$$

For the case in which $n_2 \approx n_1$, as it occurs for polymeric multilayers, eqn (1.4) becomes:

$$\Delta\lambda = \frac{2\lambda_B}{\pi}\frac{\Delta n}{\bar{n}} \tag{1.5}$$

where $\Delta n = n_2 - n_1$ and $\bar{n} = (n_2 + n_1)/2$.

If a more detailed analysis of the interplay between the externally induced changes of optical thickness and the performance of the Bragg mirror is pursued, then a simulation and fitting of the full spectrum obtained experimentally is required. This can be done by employing a vector wave calculation based on a transfer matrix approach. A full description of this method can be found elsewhere.[1] In Figure 1.2a, we plot the evolution of the reflectance spectrum of a multilayer as the number of unit cells increases. In this case the

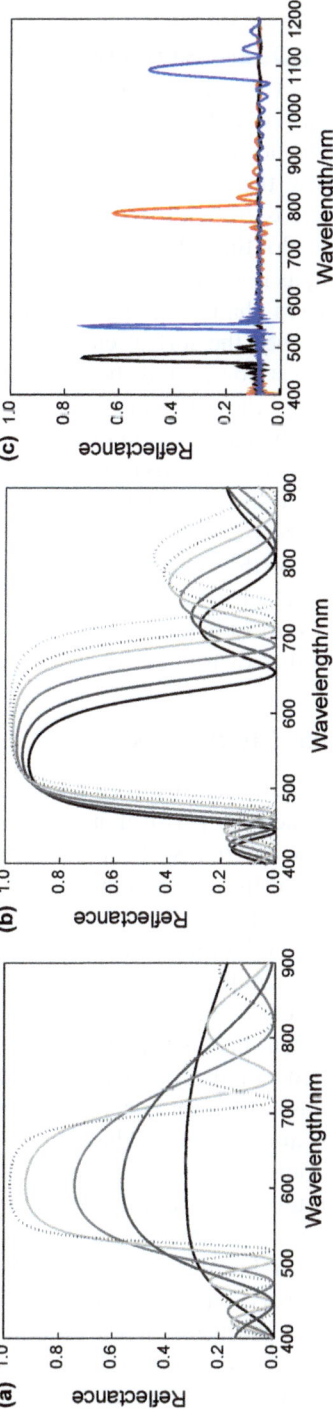

Figure 1.2 (a) Theoretical reflectance spectra of the evolution of a Bragg stack composed by a different number of unit cells, N. $N = 2$ (solid black line), $N = 3$ (solid dark gray line), $N = 4$ (solid gray line), $N = 6$ (solid light gray line) and $N = 8$ (dotted black line). (b) Theoretical reflectance spectra of an 8-unit-cell Bragg stack series in which the refractive index of the denser layer is 1.71 (solid black line), 1.75 (solid dark gray line), 1.82 (solid gray line), 1.89 (solid light gray), 1.93 (dotted gray line) and 2.00 (dotted light gray line). (c) Theoretical reflectance of a 32-layer Bragg stack ($n_{high} = 1.52$; $n_{low} = 1.47$). Thicknesses were fixed at 80 nm for the n_{low} layer and 80 nm (black line), 180 nm (red line) and 280 nm (blue line) for the n_{high} layer. (The imaginary part of the refractive index was neglected in all calculations.)

unit cell is made of a layer of thickness $t_1 = 120$ nm and refractive index $n_1 = 1.23$ and another one of $t_2 = 1.70$ and $n_2 = 90$ nm. These are typical refractive index values of porous TiO_2 and SiO_2, respectively. As explained above, the primary maximum results from the interference of beams partially reflected at each interface present in the multilayer, while the secondary lobes arise from the interference of beams reflected at the top and bottom faces of the structure. In Figure 1.2(b) we show the variation of the reflectance as the refractive index of one of these layers is gradually increased from $n_{1,i} = 1.71$ to $n_{1,f} = 2.00$. Also, in Figure 1.2(c), we plot the reflectance spectra for a multilayer with a low refractive-index contrast, as it is typical of polymeric Bragg mirrors, and its evolution as one of the layers changes its width from $t_{1,i} = 80$ nm to $t_{1,f} = 280$ nm. These shifts exemplify well the sort of changes expected when a layer sequentially incorporates a guest compound, be it by infiltration of its void network (Figure 1.2(b)) or by swelling (Figure 1.2(c)). Sensitivity (how much the optical properties change when a variation of the environment, such as the concentration of a targeted species, occurs) and resolution (minimum external change that gives rise to a distinguishable change in the optical properties) of a responsive Bragg mirror can then be tailored by choosing the right thickness and refractive index of the constituent layers. In this sense, a compromise must be found, since the observation of finer peaks implies the realization of a low refractive-index contrast in the multilayer, as it can be readily inferred from eqn (1.4), what diminishes their intensity according to eqn (1.3).

1.2.2   Field Distribution Inside Multilayers

The particular way in which field intensity is distributed within a multilayer can also be taken advantage of to enhance the performance of responsive Bragg mirrors. In this section we analyze the different sorts of resonant photon modes that can be found inside the stack by means of the method of transfer matrix. In Figure 1.3 we plot the reflectance and transmittance spectra of two types of multilayered structures, one periodic and another one in which the periodicity is intentionally disrupted by a thicker middle layer. Also, we draw the spatial distribution of the squared electric field within the stack in Figures 1.3(c) and 1.3(d), in which resonances can be identified as bright spots. Regions in which the field intensity is strongly depleted can also be easily spotted. These images illustrate that a designer field intensity pattern can be attained by adequate choice of the layers' refractive index and thickness. This allows in turn tailoring the light emission and absorption from imbibed species. It is well known that an increase of the local density of states gives rise to faster emission rates and luminescence intensity. For this effect to occur, the physical volume occupied by the resonant photon modes must be of the order of their wavelength. However, in the case of Bragg mirrors, since light is confined in only one dimension and due to the extended character of the multilayer, the cavity volume is always larger than the resonant wavelengths. Nevertheless, the angular distribution of the output power is largely modified and depends strongly on the optical parameters of the structure. Strong *directional*

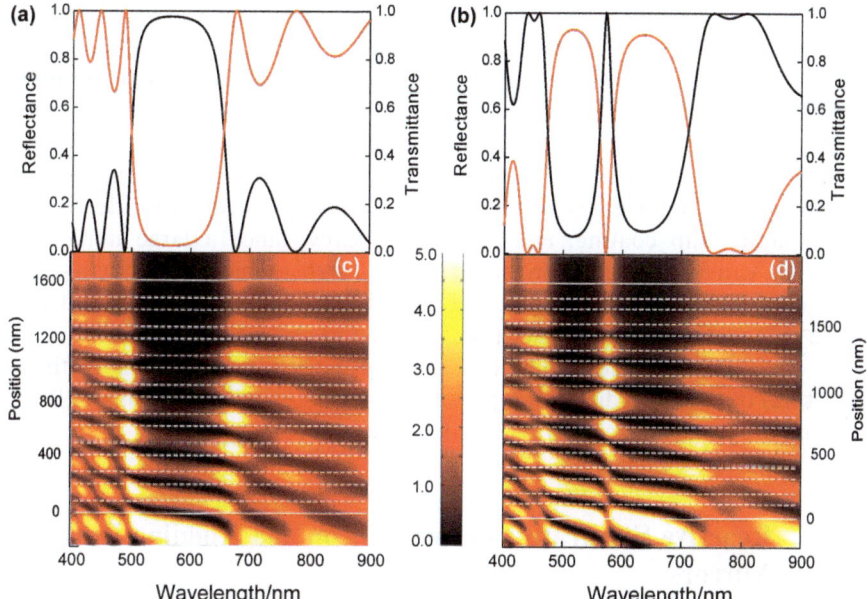

Figure 1.3 Theoretical spectra reflectance (black line) and transmittance (red line) of (a) a Bragg stack built with 8 unit cells ($n_1 = 1.23$; $n_2 = 1.70$; $d_1 = 120$ nm; $d_2 = 80$ nm) and (b) symmetry-disrupted structure with a thicker middle layer. (c) and (d) Calculated spatial distribution of the electric field along a cross section of both types of structures is plotted as a function of the incident wavelength. Horizontal white dashed lines indicate the position of the interfaces between the two types of layers present in the multilayer.

reinforcement or depletion of luminescence can thus be observed. Regarding absorption, its magnitude will also depend on light–matter interaction, which will be much stronger for resonant wavelengths within the multilayer. Hence, by locating nanomaterials or molecules at positions for which a reinforcement of the optical field is devised, and with absorption bands that overlap the spectral position of those resonances, it is possible to tune the resulting absorption spectrum with great precision. The strong dependence of the resonant modes that originate such directional enhancement or suppression of light emission or absorption on the refractive index of the layers provides an excellent means to obtain a fast and abrupt optical response to the presence of a guest compound inside them. Different examples illustrating this point are given in the last section of this chapter.

1.3 Methods and Materials

Both *top-down* and *bottom-up* strategies have been employed to achieve a controlled spatial modulation of the refractive index. These preparation routes will determine not only the microstructure, but also the physicochemical characteristics of the layers. *Top-down* techniques lead to porous Bragg mirrors by engraving a reactive substrate through a strict control of the porosity at the tens of nanometers length scale. Hitherto, only porous Bragg mirrors obtained from silicon wafers and aluminum films have been realized by a top-down approach, in both cases based on the electrochemical etching in highly acidic media. *Bottom-up* strategies are based on the alternating deposition of layers with different refractive index. These types of deposition are carried out by employing a wide diversity of techniques, such as physical vapor deposition, spin coating, dip coating, *etc*. More specifically, since a large number of materials can be processed in liquid media (such as metal and metal-oxide colloidal particles, or polymers), wet deposition methods are preferred over those that make use of condensation from the gas phase.

In what follows, the materials chemistry aspects (preparation methods, structure) of the main different types of multilayers, in which responsive Bragg mirrors properties have been reported, are described.

1.3.1 Reactive Substrates: Porous Silicon and Alumina Bragg Mirrors

The term reactive substrate herein refers to a wafer made of a material through which an electrical current can be applied to diminish its free enthalpy, while a chemical agent etches its surface. Aluminum, doped silicon and, more recently, titanium[2] seem to be the most suitable materials that can be treated using this methodology, leading to nanometer tubular-shaped pores of the corresponding metal oxide that grow perpendicular to the substrate.

Silicon was the first material shaped as a porous 1DPC. Its development was in part possible due to the knowledge acquired in the late 1950s when it was demonstrated that crystalline silicon wafers doped n or p could be made porous by passing an anodic current through them in the presence of a hydrofluoric acid solution.[3,4] Although the full mechanism of oxidation has not been totally elucidated yet, it is well established that it involves four electrons and the coordination of a fluoride ion to a silicon atom, located in the surface and that has previously lost a proton. The hydrolysis of the fluoride-silicon coordinated structure produces a tetravalent complex of silicon from the bulk. Typical layer porosity achieved with these methods ranges between 40% and 80%. The periodic modulation of the refractive index along one direction is in this case achieved by controlled variation of the current during the process of biased etching described above. Since no change in the morphology of the pores occurs when the current is modified, continuous silicon columns are obtained at the end of the process, providing pore connectivity (at least) in the etching direction. Pore size can be tuned from 1 nanometer to a few micrometers by changing the current intensity, the fluoride concentration and the doping level of the active substrate.[5–7] In Figure 1.4(a) we show an image of the cross section of a porous silicon 1DPC, where the periodic modulation of porosity can be readily observed. The preferential pore growth direction is that perpendicular to the substrate, which favors their tubular shape. The group of Sailor has pioneered the research with this type of porous lattices and has made significant advances in the control of the photonic properties of these stacks.[8,9]

In the case of aluminum, in the middle of the last decade it was demonstrated that long pores could be generated in alumina from a mirror polished aluminum foil using hard anodization techniques.[10] Two steps are applied to generate the array of pores. The first step generates the stem pores at a desired pore density, which is inversely proportional to the square of the anodizing potential in the equilibrium condition. The second step generates the columns from the stem pores. The value of voltage applied (in the range of tens of volts) and the nature of the acid determine the pore size of the Al_2O_3 tubes, while the individual slab thickness is related to the duration of these steps. As in the case of silicon, 1DPCs made in this way present a homogeneous composition and the one-dimensional modulation of the refractive index is obtained through the controlled variation of the anodic current.[11,12] Porosities obtained are typically controlled on the order of a few tens of per cent, thus leading to low refractive-index contrast between adjacent layers. Due to this, and in order to attain intense Bragg peaks, a large number of unit cells (more than 30) is needed, which implies that high aspect ratio alumina pores are generated.

1.3.2 Nanoparticle Multilayers

Nanoparticles can also be used to attain porous 1DPC with responsive properties. In this case, the mesoporosity of the layers arises as a result of the incomplete space filling attained when colloidal size particles are packed. Different aggregation states may lead to different pore-size distributions and overall porosity, and in consequence different refractive indices. Hence, it is possible to create Bragg mirrors by alternating layers of different refractive index made of particles of the same composition, which has been shown to add functionalities to the optical material.[13] However, the most common procedure is the alternate deposition of layers of different nanoparticles. In principle, all types of nanoparticles that can be suspended to form a colloid may be integrated in a Bragg stack by this simple method. Wet deposition methods are inexpensive and give rise to uniform coating of substrates with areas about a few tens of square centimeters. Control over the thickness of the layers is achieved through the variation of the parameters that determine the deposition dynamics, which strongly depends on the method of choice. *Layer-by-layer*, for instance, in which electrostatic interactions between particles with different surface charge favors the assembly the whole structure and provides mechanical stability to the ensemble. This approach, developed by the groups of Rubner and Cohen, yields Bragg mirrors made of layers of nanoparticles of SiO_2 and TiO_2 combined with polyanions and polycations, respectively, which allows the surface charge of the layers to be controlled.[14] Once the polymers are removed by thermal treatment, the average porosity of these systems is around 45%. In many cases, the attraction of the particles deposited in a film is greater than the solvation interactions that occur when dispersed in a liquid medium. The origin of this attraction is not well understood yet, and it may be a combination of Van der

Waals and hydration forces, the latter resulting from the presence of hydrogen-bonded water layers surrounding the nanoparticle network. This opens the door to stacking colloidal particles of the same surface charge using similar solvents, as has been repeatedly demonstrated by spin- or dip-coating techniques. As in the previous example, SiO_2 and TiO_2 have been preferred over other materials because of their high refractive-index contrast, but particles of very different composition, such as SnO_2, ZrO_2, In_2O_3, lanthanide fluorides or metals have been integrated in porous Bragg mirrors to take advantage of other optical functionalities, such as, for example, the possibility to absorb or emit light. Figure 1.4(b) displays a cross section of a porous SiO_2/TiO_2 1DPC in which a monolayer of gold nanoparticles (bright circular spots in the SEM image) has been precisely located in the center of an optical resonator.[15] In this particular case, advantage was taken of the effect of the interplay between the absorption caused by localized surface plasmons in the metal and the responsive resonant mode of the porous cavity to devise a multilayer that shows abrupt changes of its absorptance spectrum when infiltrated with different liquids.

1.3.3   Supramolecularly Templated Multilayers

Another way to produce responsive Bragg mirrors is the alternated deposition of mesoporous SiO_2 and TiO_2 films templated by a block copolymer. The added value of these two-scale ordered systems is the opportunity that they present to control the geometry and the distribution of the pores in each layer. In this case, a subwavelength order is a consequence of the self-assembly of a block copolymer included in the same liquid media where a soluble metal oxide precursor is included. Careful control of the hydrolysis and condensation reactions of the metal-oxide precursor accompanied by the layer deposition conditions leads to the formation of an inorganic network of well-defined geometry, since it is directed by the block copolymer assembly. The removal of the block copolymer can be made by extraction or by thermal treatment, the latter providing also the crystallization of the inorganic walls. The stacking of meso-ordered layers is made using the same wet methods described above for nanoparticles. Consolidation of each hybrid inorganic–organic layer without removing the block copolymer allows the deposition of a new layer on top while preventing the infiltration of previously formed layers. Once the final number of layers in the stack is reached, the polymer is removed at once and a porous Bragg stack is developed. The resulting narrow pore-size distribution brings the possibility to use them as base materials for optical sensing in which the probe will induce a change in the optical response depending on its size.[16] In addition, different types of radicals can be anchored to the metal-oxide walls of these porous 1DPC, which provides added functionality to these versatile materials. On the other hand, these materials present the capability to be integrated with dense metal-oxide layers. This characteristic is very desirable when it comes to building different architectures where porous layer is located at a desired position.[17] In Figure 1.4(c), an image of a cross section of a porous meso-structured optical resonator built within a dense 1DPC is shown.

1.3.4 Glancing-Angle Deposition: Columnar Structures

Glancing-angle deposition (GLAD) combines the traditional thin-film vacuum deposition with a particular geometry in which the substrate is tilted with respect to the line connecting the target and the substrate.[18] In this way, the vapor-phase compound arrives obliquely to the substrate onto which it will be deposited. Due to this particular geometry, and the lack of surface diffusion of the deposited species, the microstructure of the deposited material is that of tilted pillars. The ability to control the column orientation leads to interesting applications due to the highly accessible and tunable interconnected pore network of the entire structure. The versatility of GLAD to create photonic crystal structures in different dimensions and the precision it offers to control different types of columnar structures have been thoroughly studied by the group of Brett, who pioneered both the technique and the development of the first photonic structures based on them.[19] One of the main advantages of GLAD is that it allows uniform porous layers with homogeneous optical properties over large areas to be obtained, which makes it compatible with industrial processes. Since GLAD allows control of the porosity of the layers and thus refractive index by means of the angle of deposition, multilayers made of a single material can be prepared by changing, periodically, the tilt angle between the material source (the so-called *target*) and the substrate. The aim of using only one material is to preserve its intensive bulk properties, bringing different functionalities together.[20] In Figure 1.4(d) we show a cross section of a

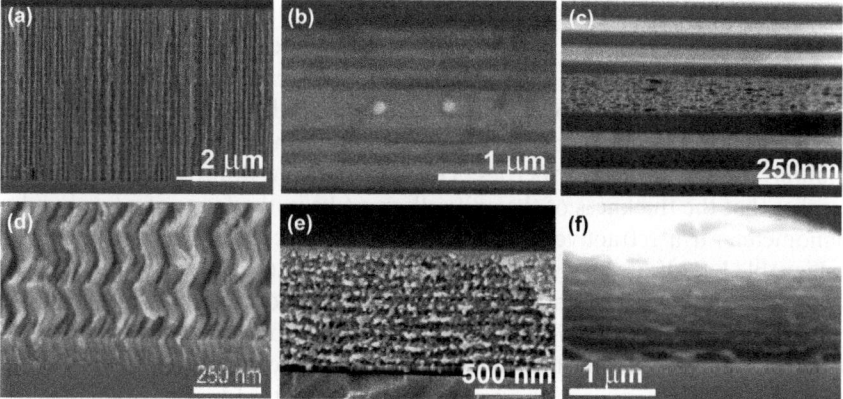

Figure 1.4 Cross-sectional scanning electron microscopy images of Bragg mirrors made of (a) silicon (b) nanoparticle-based TiO_2/SiO_2 multilayers with Au nanoparticles integrated in a thicker middle layer (backscattered electron image) (c) Supramolecularly templated porous layer integrated between two dense TiO_2/SiO_2 Bragg mirrors (d) TiO_2/SiO_2 Bragg stack deposited by GLAD (e) Polystyrene multilayer structure obtained by collective osmotic shock. (f) dry polystyrene-*b*-quaternized poly(2-vinyl pyridine) (PS-*b*-QP2VP) lamellar film deposited on a silicon wafer.
Images reproduced/adapted from ref. 39 (a), ref. 15 (b), ref. 21 (d) and ref. 27 (f).

columnar TiO_2/SiO_2 photonic crystal prepared by GLAD in the labs of the group of González-Elipe.[21]

1.3.5 Polymeric Multilayers

Alternate stacking of layers of different polymers also gives rise to a spatial modulation of the refractive index to construct a one-dimensional photonic crystal. In this case, the higher refractive-index contrast that can be achieved is lower than for the case of inorganic multilayers. In consequence, a high number of layers (more than 10 cell units) is required to reach a significant peak reflectance intensity, although it should be mentioned that, very recently, high dielectric contrast has been achieved in multilayers prepared by vacuum deposition of 4,4'-Bis[4-(diphenylamino)styryl]biphenyl (BDAVBi) and 2-(4-biphenyl)-5-(4-tert-butylphenyl)-1,3,4-oxadiazole (PBD) polymers.[22] *A priori*, polymeric Bragg structures do not present an open and interconnected porosity. Therefore, their responsive optical behavior is related to the capability of the layers to swell under the presence of certain solvents or vapours. Polymer layers are commonly deposited by spin coating from liquid solution. Cross-linking or total swelling of the recently deposited polymer layer is needed in order to prevent the dissolution of the whole structure when a drop of fresh polymer solution takes contact with the previously deposited film.[23] As a consequence, it is difficult to build responsive polymeric structures only with polymers. Zhai and coworkers have found a way to overcome this limitation by using polyelectrolyte layers that provide both porosity and swelling capability to the total ensemble.[24]

Diblock copolymers can also be used to create all-polymeric responsive Bragg mirrors. These dual compounds can self-assemble spontaneously under certain conditions to produce periodic refractive index structures in 1, 2 and 3 dimensions.[25] The lamellar phase of these self-organized materials gives rise to multilayers with 1DPC properties. The tuning of the Bragg peak in the visible implies that the thickness of the unit cells must be in the range of a few hundred nanometers, if a refractive index of around 1.5 is considered. This spacing is achievable by self-assembly of high molecular weight copolymers or by a blend with their respective homopolymers, as Thomas and coworkers demonstrated by using a poly(styrene-b-isoprene) symmetric di-block copolymer.[26] Layers of 100 nm of each block are stacked forming a multilayer. In 2007, the same group reported the first responsive structure based on this block-copolymer self-assembly approach using a hydrophobic–hydrophilic polyelectrolyte block polymer.[27] The responsiveness of this photonic structure is based on the uniaxial expansion of the polyelectrolyte layer when the system is put into contact with different salt solutions. This type of structure is also referred to as a photonic hydrogel, due to its capability to incorporate water into the structure. The structure can be frozen at a desirable thickness by the incorporation of inorganic compound.[28] From a different synthetic approach, block copolymers also offer the possibility to obtain highly ultraviolet-reflecting porous periodic multilayers.[29] In this latter case, a combination of thermally driven self-organization involving phase separation, selective UV degradation and collective osmotic shock, yields a one-dimensional photonic crystal with an interconnected open-pore structure. Figures 1.4(e) and 1.4(f) display cross sections of a swellable polymeric multilayer obtained by spin coating and an ordered porous polystyrene film prepared by collective osmotic shock, respectively.

1.3.6 Hybrid Multilayers

Optical multilayers can also be developed by combining films of very diverse composition and structures, such as polymers, metal oxides, clays, *etc.* Bragg stacks made of organic polymer layers mixed with nanoparticles,[30] or alternated with porous metal oxide layers,[31–33] are examples of hybrid multilayers that display intense responsive optical properties. In these, swelling and pore infiltration occurs simultaneously in the different constituent layers, providing (and preserving after infiltration) enough refractive-index contrast so as to achieve strong reflections with 10 periods.

Other types of compounds that are well integrated into Bragg reflectors are clays. Ozin's group developed a clay Bragg stack based on the alternating layers of laponite (a synthetic philosilicate clay of empirical formula $Na_{0.7}[(Si_8Mg_{5.5}Li_{0.3})O_{20}(OH)_4]$) and metal-oxide porous layers.[34] Laponite layers were successfully deposited from water-based precursor dispersions and templated with polystyrene spheres to allow flowing liquids and gases through them. The importance of the use of clay films as building blocks is the capability of these layered silicates to intercalate, adsorb, and exchange ions within its structure. All these processes modify the optical thickness of clay layers, thus changing the optical properties of the whole ensemble.

1.4 Response to Environmental Changes

The main common effects observed in the response of a Bragg reflector against changes in the environment are described in the first part of this section. As the optical response of layered materials is strongly dependent on the individual response of the constituent layers, a deeper description of the phenomena occurring during infiltration, as well as of the change of the optical parameters it gives rise to, is provided. Also, some representative examples, chosen to illustrate the various flavors of porous and/or swellable Bragg reflectors available and the main features of their response, are presented in the last part of this section.

1.4.1 Effect of Infiltration: Refractive-Index Changes and Swelling

The intense optical effects observed in Bragg mirrors are highly sensitive to the optical thickness of the individual layers present in the stack. In the case of the responsive Bragg mirrors herein described, such optical thickness can be modified either by infiltration of guest compounds in the pore network, which would result in an increase of the refractive index, or by diffusion of species in the polymeric layer, which would cause both swelling, and thus modification of the geometrical thickness, and variation of the refractive index. It is precisely this dependence that is behind the use of these multilayered materials as responsive optical mirrors. The magnitude and specificity of the changes observed will depend, in the former case, on the sorption properties of the pore network, which will in turn be a function of the pore-size distribution and the chemical nature of the pore walls, or in the capability to admit targeted species within the polymeric network for the latter.

From the formulas (1.1)–(1.4), some of the main properties of responsive Bragg mirrors can be readily derived. First, incorporation of a compound into the full structure will cause a spectral redshift of the reflectance peak and the magnitude of this displacement will be directly proportional to the infiltrated amount of such compound. This will be the case no matter whether the effect of the guest compound was to increase the average refractive index or the film thickness. Also, stacks with a larger number of layers will present a higher reflectance and thus a higher signal-to-noise ratio, which favors readability. Peak intensity and width will increase as the refractive-index contrast between the two types of constituent layers does, although the growth of the latter will stabilize as a consequence of the compensation effect of the rising average refractive value, as can be deduced from eqn (1.4). So, a compromise must be found between resolution, which is affected by peak width, and sensibility, which depends on the magnitude of the reflectance peak.

However, if the infiltration does not take place simultaneously in the whole structure, information from the optical response can only be extracted from the full detailed analysis of the reflectance or transmittance spectra. This is actually the case in most multilayered structures, in which films of different pore-size distribution and composition are alternated. Several examples demonstrate that a complex dynamics of incorporation of guest compounds occurs, which makes the fitting of the full optical spectra necessary.

In the case of porous films, n_1 and n_2 are the result of averaging the refractive index of the walls and that of the pores, which might be empty or partially filled. In the most general case in which pore filling is not complete, the average refractive index of the ith layer, n_i, can be obtained from the equation derived by Bruggeman for a three-component dielectric medium,[35] which is based on an effective medium theory:[36]

$$f_{\text{wall}} \frac{n_{\text{wall}}^2 - n_i^2}{n_{\text{wall}}^2 + 2n_i^2} + f_{\text{solvent}} \frac{n_{\text{ads}}^2 - n_i^2}{n_{\text{ads}}^2 + 2n_i^2} + f_{\text{medium}} \frac{n_{\text{bkg}}^2 - n_i^2}{n_{\text{bkg}}^2 + 2n_i^2} = 0 \qquad (1.6)$$

which is derived considering that our inhomogeneous film as composed by the material of which the pore walls are made of, with refractive index n_{wall}, and the adsorbed species present in the pores, n_{ads}, embedded in an otherwise homogeneous background of $n_{\text{bkg}} = 1$. Here f_{wall}, f_{ads} and f_{bkg} are the volume fractions of the material composing the pore walls, the adsorbed species and the surrounding medium, respectively. Knowing n_{wall} and n_{ads}, and extracting the effective refractive index of the film n_i from the fittings of the specular reflectance of a multilayer measured from the empty sample ($f_{ads} = 0$) and by means of eqn (1.3), we can estimate f_{wall} and thus the total pore volume ($f_{medium} = f_{pore} = 1 - f_{wall}$) of the starting material. Then, from the effective refractive index of the film obtained at different environmental conditions and thus degrees of filling, we can estimate the volume fraction occupied by the adsorbed, and eventually condensed, species, f_{ads}, since we can write $f_{medium} = 1 - f_{wall} - f_{ads}$, leaving f_{ads} as the only unknown parameter. The ratio f_{ads}/f_{pore} is the ratio between the volume occupied by the infiltrated species, V_{ads}, and the originally free pore volume, V_{pore}. The specific spectral and temporal response to a particular event (*i.e.* gradual raise of gas pressure or increase of concentration of a target compound in solution) will be determined by the amount and rate of incorporation of the adsorbed species in the Bragg mirrors. Some examples are provided in Section 1.4.2. It should be noticed that, conversely, this tool has been used to analyze the pore-size distribution of mesostructured films by relating the changes of adsorbed vapour volume with the size of the pore employing the Kelvin equation.

For the case of swelling films, such as those made of hydrogel, the materials can be considered homogeneous and eqn (1.3) does not apply any longer, but the effect of the concomitant change in the thickness of the constituent slabs must be taken into consideration.

1.4.2 Examples of Optical Response

The specificity of the response of Bragg reflectors to the presence of targeted compounds, as well as its magnitude, speed, *etc.* depend on the physical or chemical mechanisms involved (diffusion, adsorption, condensation, swelling, *etc.*), the nature of the building blocks and the presence of functional groups in the structure. An exhaustive enumeration and description of all responsive layered optical materials would be too long and beyond the scope of this chapter. The sort of effects observed in the reflectance, transmittance, or, in the case of the integration of optically active materials, luminescence spectra of responsive Bragg reflectors are herein illustrated with a few representative cases.

As was pointed out above, porous mesostructured materials are susceptible of being used as building blocks of responsive Bragg reflectors for their capability to adsorb and eventually condense molecules from their environment. This is a phenomenon that depends on the vapor pressure and therefore, a gradual change of the optical response will occur as this parameter is varied. Several groups have reported this effect, and make use of it to show that these materials can be the base of gas sensors. In Figure 1.5(a), the response of a porous alumina optical resonator, developed by Brett and coworkers using GLAD, to varying humidity conditions is displayed.[19] Large redshifts of the

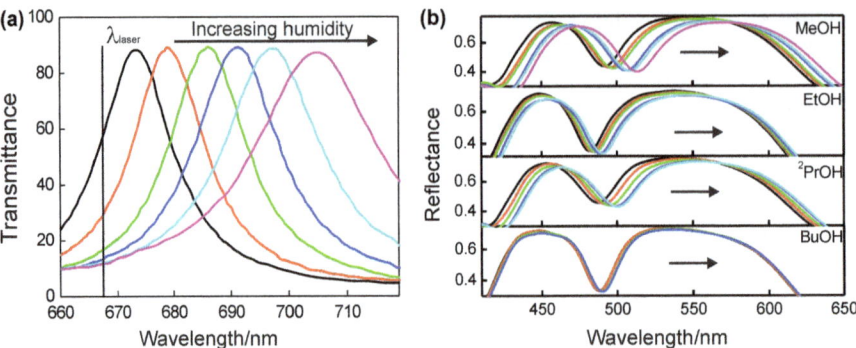

Figure 1.5 (a) Transmittance spectra of a porous 1DPC with an optical resonant mode exposed to increased humidity content. (b) Series of reflectance spectra attained at different pressures of (a) methanol, (b) ethanol, (c) 2-propanol, and (d) butanol from mesostructured optical resonators with pores of average size 9 nm. The direction of increasing pressure is highlighted by an arrow. Images reproduced/Adapted from ref. 19 (a) and ref. 17 (b).

cavity resonance can be observed in the transmittance spectra. In Figure 1.5(b), we plot the reflectance spectra of resonators built by sandwiching meso-structured layers prepared by supramolecular templating between two dense 1DPC as the vapor pressure of different alcohols is varied. From top to bottom, the solvent vaporized in the chamber is methanol, ethanol, isopropanol and butanol. Strict control over pore-size distribution yields layered media capable of responding only when the size of the molecules is such as to allow them to flow through the pore network of the optical cavity, thus adsorption and condensation takes place and, consequently, gives rise to the redshift of the resonance. This latter example shows the added value brought by controlled porosity, since it may allow selective detection depending on the molecule size. The integration of mesostructured layers of this sort in different layered configurations has been demonstrated to give rise to responsive optical materials in which a rich interplay between diffusion, adsorption and condensation phenomena takes place.[37]

Responsive Bragg reflectors have also shown a very clear response to the concentration of specific species in solution. One relevant example of this behavior is provided by the different lamellar Bragg reflectors prepared with self-organized block copolymers. In Figure 1.6(a) the absorbance spectra of a polystyrene-*block*-quaternium poly2vynilpyridine (PS-b-QP2VP) Bragg mirror swollen by contact with different concentrations of NH_4Cl aqueous solution (concentration in mol dm^{-3} is noted between brackets) are plotted. The numbers above the peaks indicate the diffraction order. Please notice that the initially swollen structure decreases its thickness as the concentration of salts in the medium increases. Each one of the swollen states of these Bragg mirrors can be frozen by introduction of SiO_2 particles, which allows removal of the sample from the solution without altering its thickness and thus preserving its color. A series of pictures of PS-b-QP2VP Bragg mirrors treated in this way is shown

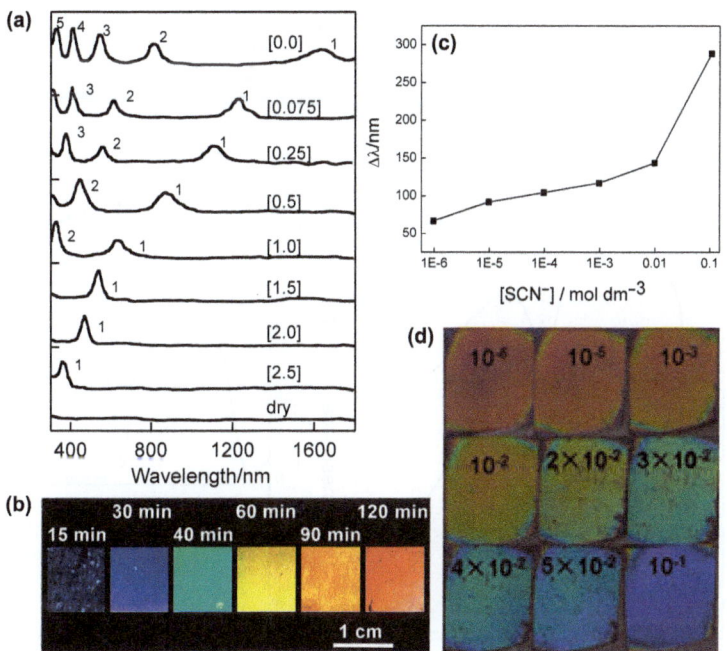

Figure 1.6 (a) Absorbance spectra of polystyrene-*b*-quaternized poly(2-vinyl pyridine) (PS-*b*-QP2VP) Bragg mirrors swollen by contact with different concentrations of NH_4Cl aqueous solution (concentration in mol dm^{-3} is noted between brackets). The numbers above peaks indicate the diffraction order. (b) Optical images of the Bragg mirror system described in (a) after the introduction of SiO_2 particles. (c) Spectral shift in the Bragg peak of a hybrid TiO_2/polydimethylaminoethyl methacrylatecoethylene glycol dimethacrylate (PDMAEMA-co-PEGDMA) systems at different concentrations of SCN^- (d) Optical images of films exposed at different concentrations of SCN (concentrations are indicated over the images). Images reproduced/Adapted from ref. 27 (a), ref. 28 (b) and ref. 32 (c) and (d).

in Figure 1.6(b). Similar effects have also been observed in hybrid structures combining inorganic layers and block polymers. An example of the spectral shift of the Bragg peak measured from a TiO_2/polydimethylaminoethyl methacrylate-co-ethyleneglycol dimethacrylate (PDMAEMA-co-PEGDMA) Bragg mirror at different concentrations of isothiocyanate (SCN^-) groups, as well as pictures showing the observed color change, are displayed in Figures 1.6(c) and 1.6(d).

As has been mentioned before, the intrinsically accessible nature of responsive optical layered media allows both the integration of optically active molecules or nanomaterials inside them, and tailoring of the emission or absorption properties of such guest compounds through the interaction with the confined electromagnetic field. The luminescence of a layer of rare-earth-based nanophosphors (europium-doped yttrium fluoride nanoparticles) integrated in a porous nanoparticle based optical resonator exemplifies well this

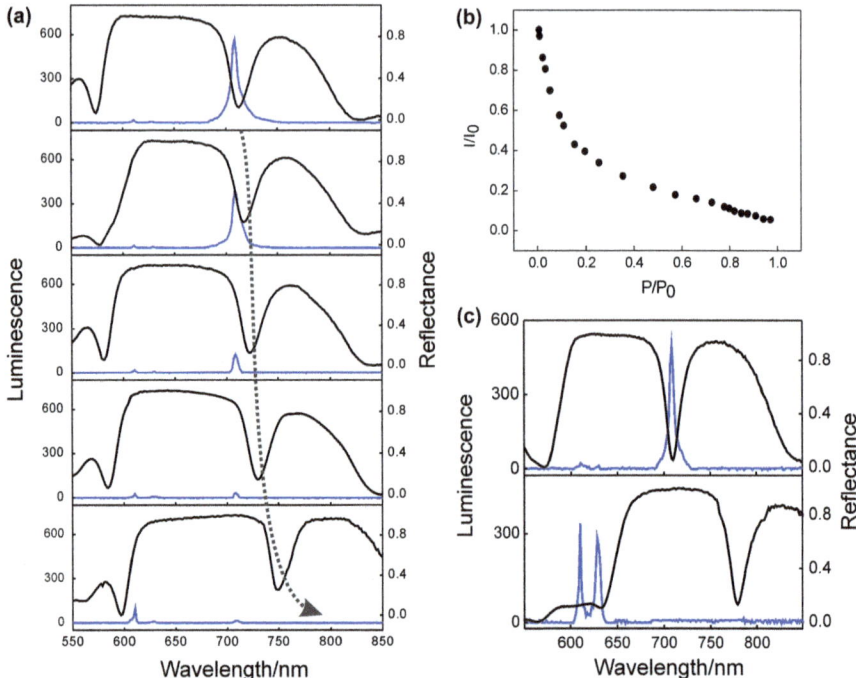

Figure 1.7 (a) Luminescence (blue lines) and reflectance (black lines) spectra obtained from a nanophosphor containing optical resonator built using two Bragg mirrors made of 5 unit cells after being exposed to a gradually increasing normalized pressure (P/P_0) of isopropanol vapor, namely from top to above: $P/P_0 = 0$; $P/P_0 = 0.07$; $P/P_0 = 0.19$; $P/P_0 = 0.66$; $P/P_0 = 1$. Black dotted line indicates the shift in the resonance position as pressure of isopropanol increases in the chamber. (b) Variation of the photo-luminescence intensity of the emission line located at $\lambda = 708$ nm (normalized at $P = 0$) as the pressure of isopropanol (black solid circles) increases in the chamber. (c) Luminescence (blue lines) and reflectance (black lines) spectra obtained from the same system described in (a) before (upper part) and after (lower part) being infiltrated with tetrahydrofurane. Images reproduced/Adapted from ref. 38.

type of effect.[38] Emission (blue lines) and reflectance (black lines) spectra obtained from such a structure when exposed to different pressure of isopropanol are plotted in Figure 1.7(a), while Figure 1.7(b) displays the changes in the emission intensity of the selected line ($\lambda = 708$ nm) as the vapor pressure increases. This same structure shows an abrupt change in the emission properties when it is soaked in tetrahydrofurane, as can be seen in Figure 1.7(c). The Bragg reflector is herein causing the directional reinforcement or suppression of selected emission lines, largely modifying the standard lumi-nescence spectrum of these nanophosphors (all measurements were taken at normal incidence and collection). At the same time, the pore network of the photonic matrix provides the path for fluids to infiltrate the material and give rise to the spectral and intensity changes observed in the emitted light.

1.5 Concluding Remarks

Responsive Bragg reflectors present a great potential for the development of materials that could serve to build new types of nanostructured optical sensors as well as devices in which the emission and absorption of selected infiltrated species could be tailored to measure. Research activity in this emerging field is currently devoted to the exploration of new types of responsive films that could be used as building blocks, their functionalization, the improvement of its optical quality and device integration.

Acknowledgements

The research leading to the author's results has received funding from the European Research Council under the European Union's Seventh Framework Programme (FP7/2007-2013)/ERC grant agreement n$^\cup$ 307081 (POLIGHT), the Spanish Ministry of Economy and Competitiveness under grants MAT2011-23593 and CONSOLIDER HOPE CSD2007-00007, and the Junta de Andalucía under grants FQM3579 and FQM5247.

References

1. P. Yeh, *Optical Waves in Layered Media*, John Wiley & Sons, Hoboken, New Jersey, 2005.
2. L. Zheng, H. Cheng, F. Liang, S. Shu, C. K. Tsang, H. Li, S. T. Lee, Y. Y. Li, *J. Phys. Chem. C*, 2012, **116**, 5509.
3. A. Ulhir, *Bell Syst. Tech. J.*, 1956, **35**, 333.
4. D. R. Turner, *J. Electrochem. Soc.*, 1958, **105**, 402.
5. L. T. Canham, *Appl. Phys. Lett.*, 1990, **57**, 1046.
6. R. Herino, G. Bomchil, K. Barla, C. Bertrand and J. L. Ginoux, *J. Electrochem. Soc.*, 1987, **134**, 1994.
7. G. Vincent, *Appl. Phys. Lett.*, 1994, **64**, 2367.
8. E. J. Anglin, L. Cheng, W. R. Freeman and M. J. Sailor, *Adv. Drug Deliv. Rev.*, 2008, **60**, 1266–1277.
9. M. J. Sailor, *ACS Nano*, 2007, **1**, 248–252 and references therein.
10. W. Lee, R. Ji, U. Gösele and K. Nielsch, *Nature Mater.*, 2006, **5**, 741.
11. X. Hu, Y. J. Pu, Z. Y. Ling and Y. Li, *Opt. Mater.*, 2009, **32**, 382.
12. W. J. Zheng, G. T. Fei, B. Wang, Z. Jin and L. D. Zhang, *Mater. Lett.*, 2009, **63**, 706–708.
13. M. E. Calvo, S. Colodrero, T. C. Rojas, M. Ocaña, J. A. Anta and H. Míguez, *Adv. Func. Mater.*, 2008, **8**, 2708.
14. Z. Wu, D. Lee, M. F. Rubner and R. E. Cohen, *Small*, 2007, **3**, 1445.
15. A. Jiménez-Solano, C. López-López, O. Sánchez-Sobrado, J. M. Luque, M. E. Calvo, C. Fernández-López, A. Sánchez-Iglesias, L. M. Liz-Marzán and H. Míguez, *Langmuir*, 2012, **28**, 9161–9167.
16. N. Hidalgo, M. E. Calvo and H. Miguez, *Small*, 2009, **5**, 2309.

17. N. Hidalgo, M. E. Calvo, M. G. Bellino, G. J. A. A. Soler-Illia and H. Míguez, *Adv. Funct. Mater.*, 2011, **21**, 2534.

18. M. J. Brett and M. M. Hawkeye, *Science*, 2008, **319**, 1192.

19. J. J. Steele, A. C. van Popta, M. M. Hawkeye, J. C. Sit and M. J. Brett, *Sens. Actuators B*, 2006, **120**, 213.

20. M. F. Schubert, J. Q. Xi, J. K. Kim and E. F. Schubert, *Appl. Phys. Lett.*, 2007, **90**, 141115.

21. L. González-García, G. Lozano, A. Barranco, H. Míguez and A. R. González-Elipe, *J. Mater. Chem.*, 2010, **20**, 6408–6412.

22. D. Yokoyama, K. Nakayama, T. Otani and J. Kido, *Adv. Mater.*, 2012, DOI: 10.1002/adma.201202422.

23. J. Bailey and J. S. Sharp, *J. Polym. Sci., Part B: Polym. Phys.*, 2011, **49**, 732–739.

24. L. Zhai, A. J. Nolte, R. E. Cohen and M. F. Rubner, *Macromolecules*, 2004, **37**, 6113–6123.

25. J. Yoon, W. Lee and E. L. Thomas, *MRS Bull.*, 2005, **30**, 721.

26. A. Urbas, Y. Fink and E. L. Thomas, *Macromolecules*, 1999, **32**, 4748.

27. Y. J. Kang, J. Walish and E. L. Thomas, *Nature Mater.*, 2007, **6**, 957.

28. C. Kang, E. Kim, H. Baek, K. Hwang, D. Kwak, Y. Kang and E. L. Thomas, *J. Am. Chem. Soc.*, 2009, **131**, 7539.

29. P. Zavala-Rivera, K. Channon, V. Nguyen, E. Sivaniah, D. Kabra, R. H. Friend, S. K. Nataraj, S. A. Al-Muhtaseb, A. Hexemer, M. E. Calvo and H. Miguez, *Nature Mater.*, 2012, **11**, 53–57.

30. A. Convertino, A. Capobianchi, A. Valentini and E. N. M. Cirillo, *Adv. Mater.*, 2003, **15**, 1103.

31. Z. Wang, J. Zhang, Z. Tian, Z. Wang, Y. Li, S. Liang, L. Cui, L. Zhang, H. Zhang and B. Yang, *Chem. Commun.*, 2010, **46**, 8636–8638.

32. Z. Wang, J. Zhang, J. Li, J. Xie, Y. Li, S. Liang, Z. Tian, C. Li, Z. Wang, T. Wang, H. Zhang and B. Yang, *J. Mater. Chem.*, 2011, **21**, 1264.

33. Z. Wang, J. Zhang, J. Xie, C. Li, Y. Li, S. Liang, Z. Tian, T. Wang, H. Zhang, H. Li, W. Xu and B. Yang, *Adv. Funct. Mater.*, 2010, **20**, 3784–3790.

34. B. V. Lotsch and G. A. Ozin, *ACS Nano*, 2008, **2**, 2065.

35. H. C. van de Hulst, *Light Scattering by Small Particles*, Dover Publications, New York, USA, 1981.

36. N. Nagy, A. Deak, Z. Horvolgyi, M. Fried, A. Agod and I. Barsony, *Langmuir*, 2006, **22**, 8416–8423.

37. M. C. Fuertes, S. Colodrero, G. Lozano, A. R. Gonzalez-Elipe, D. Grosso, C. Boissiere, C. Sanchez, G. J. D. A. Soler-Illia and H. Miguez, *J. Phys. Chem. C*, 2008, **112**, 3157.

38. O. Sanchez-Sobrado, M. E. Calvo, N. Nuñez, M. Ocaña, G. Lozano and H. Míguez, *Nanoscale*, 2010, **2**, 936.

39. I. Rea, G. Oliviero, J. Amato, N. Borbone, G. Piccialli, I. Rendina and L. De Stefano, *J. Phys Chem C.*, 2010, **114**, 2617.

CHAPTER 2

Stop-Bands in Photonic Crystals: From Tuning to Sensing

ZHUOYING XIE, YUANJIN ZHAO, HONGCHEN GU,
BAOFEN YE AND ZHONG-ZE GU*

State Key Laboratory of Bioelectronics, School of Biological Science and
Medical Engineering, Southeast University, Nanjing 210096, China
*Email: gu@seu.edu.cn

2.1 Natural Photonic Crystals and Bioinspiration

2.1.1 Photonic Stop-Band and Structural Color

By coupling light with nanostructures many skills in controlling light have been achieved, most of which are impossible by bulk materials.[1,2] We defined this type of novel material as photonic nanomaterials. Due to their special optical properties and high integratability, photonic nanomaterials have been used to construct new optical devices for display,[3–5] communication,[6,7] sensing[8–11] and so on.[12] Photonic crystals are typical and important photonic nanomaterial. They were highlighted as one of ten great progresses in natural science twice by *Science* in 1998 and 1999. A photonic crystal is a material that possesses a spatially periodical dielectric constant, which can strongly modulate electromagnetic waves and drive a photonic stop-band.[13,14] The propagation of light during the stop-band is forbidden. When wavelength of light is in visible

RSC Smart Materials No. 5
Responsive Photonic Nanostructures: Smart Nanoscale Optical Materials
Edited by Yadong Yin
© The Royal Society of Chemistry 2013
Published by the Royal Society of Chemistry, www.rsc.org

range, the diffracted color will be seen. Because it is driven from physical microstructure, the color is called a structural color.[15]

From the study of prehistoric life, it is apparent that life began to produce all kinds of structural colors using microstructures early on from 5.5 hundred million years ago.[16] Though many of them disappeared during evolution, we still see abundant structural colors formed after long-time survival[17–19] (see Figure 2.1). In 1665, Hooke first presented that such color was originated from microstructure. In 1730, Newton explained the mechanism of structural color from peacock's feathers. The detailed description was finished by Anderson and Richards in 1942. It was attributed to the progress of microscopic observation means such as electron microscopy. In 2001, Andrew Parker from the University of Oxford reported that photonic crystals existed in living beings.[20]

Morpho menelaus   *Morpho sulkowskyi*   *Papilio ulysses telegonus*   *Ornithoptera priamus arruama*   *Charaxes laodice*

Figure 2.1 Typical photonic nanostructures in natural creatures: (a) *Ammonite*, 80 million years old, South Dakota, USA (Reproduced with permission from Ref. 16. Copyright 2000 IOP Publishing Ltd.); (b) Open pennaceous contour feather from the *Eocene Messel Oil Shale* (Reproduced with permission from Ref. 17. Copyright 2009 The Royal Society); (c) *Peacock* (*Indian peafowl*) (Reproduced with permission from Ref. 18. Copyright 2008 IOP Publishing Ltd.); (d) Viewing-angle dependence of the color change in *jewel beetles* (Reproduced with permission from Ref. 18. Copyright 2008 IOP Publishing Ltd.); (e) Gold chrysalis of the butterfly *Euoplea* core (Reproduced with permission from Ref. 16. Copyright 2000 IOP Publishing Ltd.); (f) Typical structural color butterflies. (Photographed by Zhongde Mu.)

Following the upsurge in studies of photonic crystals, scientists found many biological photonic crystals in butterfly, beetle, cuttlefish, sea mouse and so on.[2-23] The microstructures of photonic crystal are very delicately designed by nature and produce fantastic optical effects. People were surprised to find that some of them were what they had designed or manufactured after great efforts, and some of them were impossible to be manufactured by present technologies. Taking photonic crystal fiber for example, this new class of optic fiber is delicately designed and manufactured by periodically packing arrays of holes or solid fibers. It is now finding applications in fiber-optic communications, fiber lasers, nonlinear devices, high-power transmission, highly sensitive gas sensors, and other areas.[24-26] This structural material has existed for hundreds of millions of years, though it was discovered and characterized by Parker *et al.* in the hairs of a sea mouse that inhabits shallow tropical water. Each hair contains thousands of hollow, close-packed and longitudinally oriented cylinders with a diameter of about 230 nm. When the light is diffracted by cylinders collectively, bright structural colors are produced.

2.1.2 Variable Structural Color in Nature

In the biological world, there exists a class of structural color animals that can reversibly change their colors in response to their surrounding environment.[27-31] These animals include *paracheirodon innesi, paradise whiptail, blue damselfish, tortoise beetle, hercules beetle, etc.* Generally, the structural color changes of these animals are based on the spacing variation of the multilayer cell plates. Taking *aracheirodon innesi* for example, it normally displays a structural color of cyan, which is produced by interference of light on periodically stacked microplatelets of skin cells (see Figure 2.2(a)). Under stressful conditions, the structural color of the fish changes rapidly to yellow, triggered by the simultaneous spacing expansion of reflecting plates, to evade its pursuers. Further research suggested that the increase of the spacing among the platelets is caused by the inflow of water into the fish skin cell, which consequently leads to variable structural colors due to the swell-induced spacing expansion mechanism.[32] Similar to this process, the *tortoisebeetle* and *hercules beetle* alter their structural colors by varying the filling amount of water in the cuticle. During this process, not only the thickness but also the average reflective index of their thin skin films are responsible for the structural color variations[33,34] (see Figure 2.2(b)).

2.1.3 General Strategies of Tuning

As described above, there are various kinds of brilliant structural colors existing in Nature, and some of these colors are variable according to the external environment. All these natural photonic nanostructures offer an enormous number of blueprints for the artificial photonic materials designs, fabrications and applications. Bioinspired by the natural creatures with variable structural colors, researchers have developed a lot of

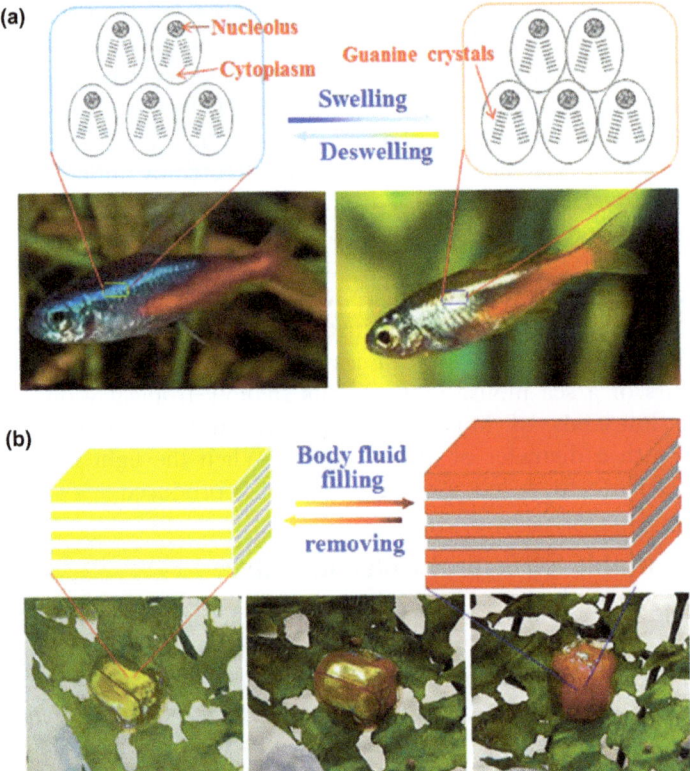

Figure 2.2   Natural creatures with variable structural colors: (a) the *Paracheirodon innesi* normally displays a structural color of cyan, after the swelling of the oval iridophores, its stripe changes to yellow (Reproduced from Ref. 95 with permission of The Royal Society of Chemistry.); (b) *Charidotella egregia* switches its structural color from gold within 1.5 min. This transformation was triggered by touching its wings, plausibly simulating a missed attempt at predation.
(Reproduced with permission from Ref. 34. Copyright 2007 by the American Physical Society.)

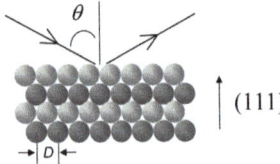

Figure 2.3   Diffraction of periodical packed colloidal photonic crystal.

tunable methods of the photonic materials.[35–38] Here, we focus on the tunable photonic crystals based on colloidal crystals. The colorful appearances of colloidal photonic crystals are ascribed to the diffraction of periodical packed colloidal particles[39,40] (Figure 2.3). The diffraction (or refraction) peak originated from the stop-bands can be estimated by the Bragg–Snell equation:

$$\lambda = 2D \left(n_{\text{eff}}^2 - \sin^2 \theta \right)^{1/2} \tag{2.1}$$

where λ is the wavelength of the reflected light, n_{eff} is the average refractive index of the constituent photonic materials, D is the distance of diffracting plane spacing, and θ is the incidence angle of the light falling on the nanostructures. Based on the equation, there are several protocols for tuning the structural colors of the photonic materials, such as changing the incidence angle θ, the average refractive index n_{eff}, the diffracting plane spacing D, and simultaneously changing these parameters. However, when changing the diffracting plane spacing, the photonic materials should be swollen or shrunk, and this process is usually accompanied by the change of the refractive index. In this situation, the effect of n_{eff} is often overlooked because the change of the diffracting plane spacing could cause a more significant change of stop-band. In the following sections, we will introduce the tuning research of colloidal photonic crystals based on changes of the average refractive index and the diffracting plane spacing, respectively.

2.2 Responsive-Molecules-Based Tunable Photonic Crystals

2.2.1 Photochromic Photonic Crystals

Photochromism dye is a type chemical species that could reversibly transform between two forms by the absorption of electromagnetic radiation.[41] The two forms have different absorption spectra of electromagnetic wave. So the two forms usually show different colors. The photoinduced form change of dyes also influences its refractive index. For the dye, its complex refractive index (n_{cmp}), whose real part is the refractive index (η) and the imaginary part is the extinction coefficient (k), can be expressed by

$$n_{\text{cmp}}^2 = (\eta + ik)^2 = \frac{Ne^2}{\varepsilon_0 m} \sum_k \frac{f_k}{\omega_k^2 - \omega^2 + i\gamma_k \omega} \tag{2.2}$$

where e is the electric charge, m is the electron mass, N is the number of oscillators per unit volume, f_k is the oscillation strength, ω_k is the kth resonant frequency of the dye molecule, and γ_k is the kth damping coefficient due to optical absorption. This indicates that the dramatic change in refractive index when the optical frequency approaches the resonant frequency. Hence, photochromic dye is promising to construct tunable photonic crystal by infiltration.[42]

To induce a large change in the stop-band, one should choose an appropriate photochromic dye, that is, one in which an oscillation frequency of the aggregated state is close to the photonic bandgap, while that of the nonaggregated state is far from the gap. One example of such dye is a spiro derivative,

1′,3′-dihydro-3′,3′-dimethyl-6-nitro-1′-octadecyl-8-docosanoyloxymethylspiro[2H-1-benzopyran-2,2′-[2H] indole] (SP1). Its molecular structures and photochromic reaction is shown in Figure 2.4(a). The dye molecule undergoes photoisomerization from the spiropyran form to the merocyanine form upon illumination with UV light, and forms *J*-aggregates in a thermal process. Its resonant absorption is around 600 nm (Figure 2.4(c)). By Kramers–Kronig analysis,[43] the refractive index of the spiropyran before illumination is almost constant (∼1.7) in the visible region, but dramatically changes after illumination. It decreases below 600 nm, while it increases above 600 nm, and reaches a maximum of 3.0 at 626 nm. As a result, the reflection peak of the dyed opal photonic crystal with a peak at 725 nm shifted to longer wavelength by 37 nm during the illumination, while that of the dyed opal with peak at 488 nm shifted to shorter wavelength by 6 nm (Figure 2.4(d)). This means the stop-band of

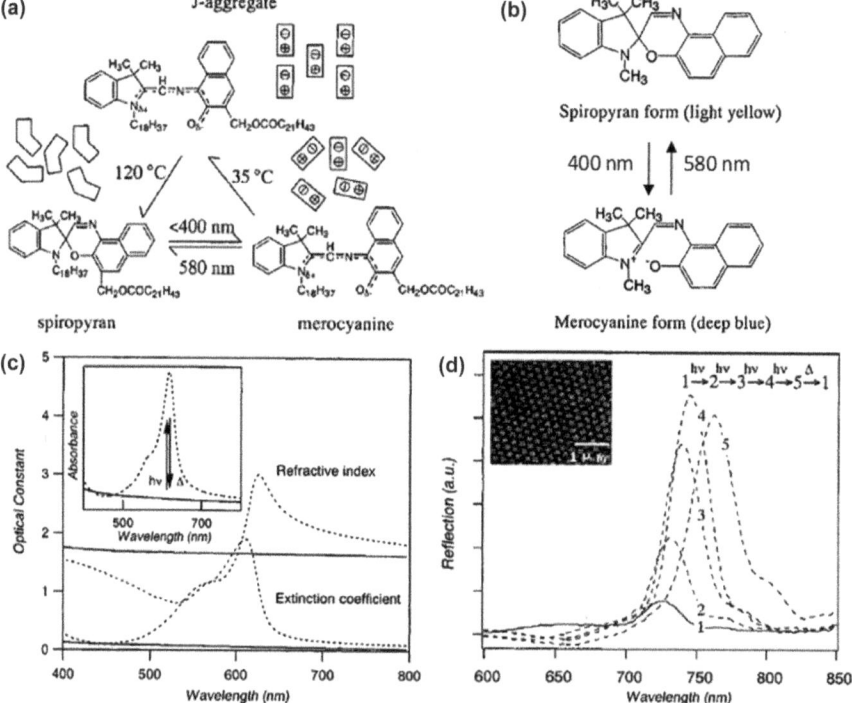

Figure 2.4   (a) Photochromic reaction of a spiro dye SP1; (b) Molecular structures and the photoreactions of SP2; (c) Optical constants of SP1 calculated from transmittance spectra by KK analysis. The solid and the dashed lines are curves for the optical constants before and after illumination with UV light, respectively. Insert is the absorption spectra of the dye film; (d) Photoinduced changes in the 0° reflection spectra of the dye-filled opal. Spectra 1–5 are measured after being irradiated for 0, 1, 3, 10, and 60 min. The (111) face of the dyed opal is shown in the inset.
(Reprinted with permission from Ref. 42. Copyright 2000 America Chemical Society.)

such photonic crystals can be continuously tuned in certain regions by UV light. The direction and extent of the stop-band shift depend on the change of dye's refractive index in the region.

The photochromic dye mentioned above needs thermal treatment to recover the aggregation state. For all-light optical devices, the photoreversible tunable photonic crystal will possess more advantages. To achieve this by some approach, a photochromic dye with reversible photoreactions, 1,3-dihydro-1,3,3-trimethyspiro-[2H-indol-1,3′-[3H]-naphth[2,1–b][1,4] oxazine] (SP2), is a good substitute for infiltration[44] (Figure 2.4(b)). For the dyed opal with a reflection peak at 723 nm, irradiation with UV light shifted the stop-band by 15 nm to longer wavelength. The state after irradiation can be completely recovered to the original state by irradiation at 600 nm. So, all of the changes in the reflection spectra were photoreversible. Also, during the cycles of alternate irradiation by UV and visible light, the dyed opal also demonstrated excellent stability and repeatability.

2.2.2 Liquid-Crystal Photonic Crystals

2.2.2.1 Shift and Switch of Stop-Band

Liquid crystals refer to a state of matter that has properties between those of a conventional liquid and those of a solid crystal.[45] They may flow like a liquid, but their molecules may be oriented in a crystal-like way. External perturbation can cause significant changes in the arrangement of the liquid-crystal system.[46] As a result, their macroscopic properties, for example refractive indices, can be changed by electric fields, magnetic fields, temperature, surface morphology, light irradiation, *etc.* Due to the successful application in electronic displays,[47] the study of liquid-crystal infiltrated photonic crystals rapidly expanded after Busch and John proposed the first marriage of liquid crystals and photonic crystals in 1999.[48,49]

This kind of material has played an important role in tunable photonic crystalconstruction.[50–53] Generally, the liquid crystals were infiltrated into the free voids of the opal or inverse opal films. The inverse opal composites possesses a larger volume fraction of variable refractive index, which induce changes in the refractive index, and hence produces a wider range of tunable stop-band. Sato and coworkers have developed a tunable photonic crystal film by filling a nematic liquid crystal, 4-pentyl-40-cyanobiphenyl (5CB), into the voids of an inverse opal film[54–56] (Figure 2.5(a)). At room temperature, 5CB kept the nematic phase, and its molecules are aligned parallel to the surfaces of the spherical voids of the inverse opal and form a bipolar structure.[57–59] The axes of the bipolar structures are different in each of the void sand the anisotropic refractive indices in the voids were not uniform. Thus, the composite photonic crystal film was opaque overall because the light was scattered. However, when the film was heated to the phase transition temperature (34 °C) of 5CB, a bright structural color appeared. This phenomenon was ascribed to the transition of 5CB from the anisotropic to

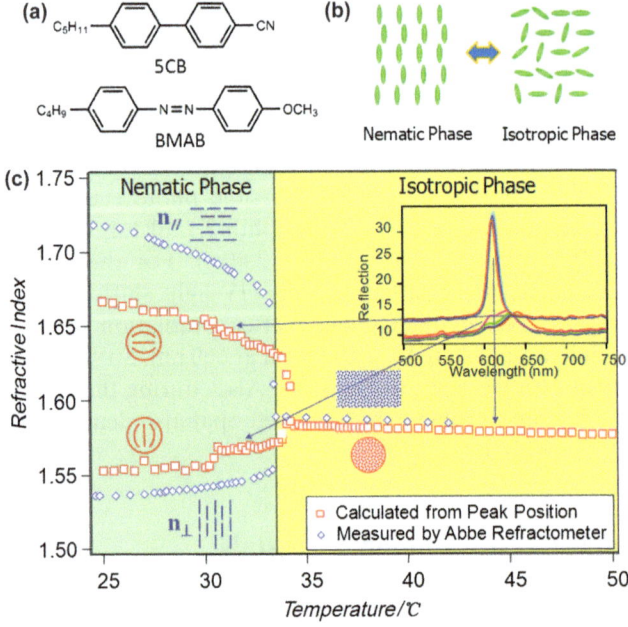

Figure 2.5   The chemical structures of 5CB and BMAB; (b) Transition of nematic liquid crystal from the anisotropic to isotropic phase; (c) Changes of refractive index and stop-band during liquid crystal phase transition. (Reprinted with permission from Ref. 55. Copyright 2004 America Chemical Society.)

isotropic phase (Figure 2.5(b)). In this phase, the refractive indices of every void were uniform. Therefore, the film consisted of periodic refractive indices of liquid crystals in its inverse opal structure, and a stop-band due to the selective Bragg diffraction of light could be observed in the spectrum (Figure 2.5(c)).

In order to accelerate the switching speed, various phototunable photonic crystal films have been prepared by using azobenzene derivatives. The photo-tunable photonic crystal film consists of an inverse opal structure infiltrated by a mixture of 5CB and 4-butyl-4′-methoxyazobenzene (BMAB). When it was irradiated by light, *trans–cis* photoisomerization of the BMAB was induced in the voids.[60] Because the *cis*-form of azobenzene has a "bent" shape, the nematic phase is disorganized by the *cis*-azobenzene, resulting in a phase transition from the nematic phase to the isotropic phase. Hence, as in the case of the thermally induced color change, this change in color can be induced by light. When the composite film was irradiated by UV light (light intensity = ca. 1.5 mW/cm^2) through a photomask, a stop-band appears at the irradiated position. This improvement confers the films a more practical application as switches.

2.2.2.2   *Photoswitch of Birefringence*

To further develop the tunable photonic crystal with special switching capability, an interesting birefringent photonic crystal was constructed by infiltrating the mixture of 5CB and **BMAB** into the stretched poly(methyl-methacrylate) (PMMA) inverse opal films[61] (Figure 2.6). In contrast to the unstretched inverse opal, the liquid crystals could align themselves along the stretching axis and form birefringence when in nematic phase. As a result, a great mismatch of refractive index occurs along the stretching direction, while there is no mismatch along the perpendicular direction. When light illuminated the photonic crystals, a reflection peak was present for polarized light parallel to the stretching direction but not for polarized light perpendicular to the stretching direction. The birefringent behavior of the composite photonic crystal film could also be switched by the light irradiation. Under irradiation with the UV light, a nematic to isotropic phase transition was triggered for the composited liquid crystals. This change caused the film to lose its birefringent properties, and the light reflected from the film became equal for both polarizations. Shown in the spectra, the diffraction peak for parallel polarized light blueshifted and at the same time the diffraction for the perpendicular polarized light appeared. This process can be reversed by irradiating the films with visible light. This material was proposed for polarizing devices in displays or beamsplitters.

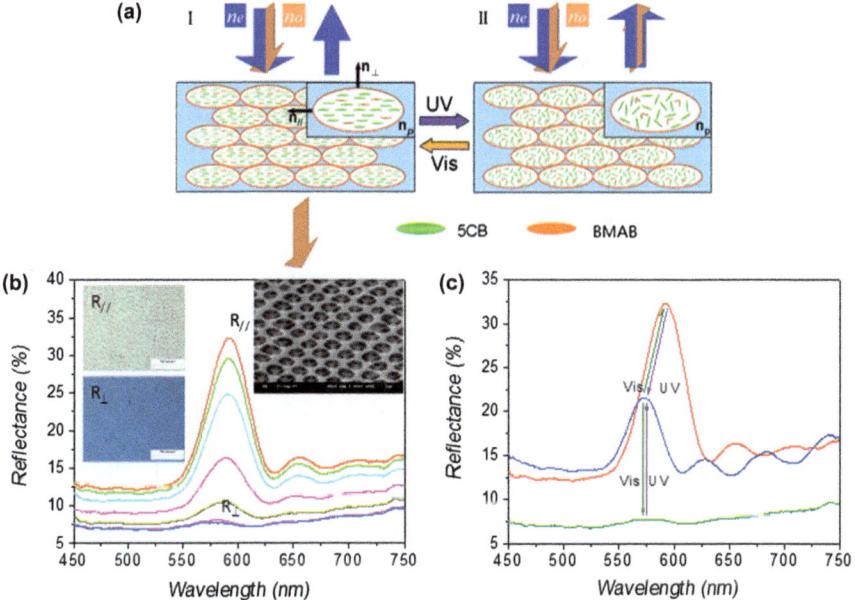

Figure 2.6   (a) Phototunability of a stretched inverse opal infiltrated with BMAB-doped 5CB liquid crystal; (b) Polarized reflection spectra of a stretched PMMA inverse opal infiltrated with liquid crystals. The left inserts are the reflective images under the irradiation of polarized light with polarization parallel ($R_{//}$) and perpendicular ($R_\perp$) to the elongation direction. The right insert is SEM of stretched PMMA inverse opal; (c) Photoinduced changes of the reflection spectra at different polarized directions.
(Reprinted with permission from Ref. 61. Copyright 2008 by Wiley-VCH.)

2.2.2.3   *Reversible Wavelength*

Combining nematic liquid crystal 5CB and azo dye BMAB, a photoswitching photonic device can also be constructed without infiltration in porous photonic crystal. This device is a stack of photonic crystal film, polymer-dispersed liquid-crystal (PDLC) film and holophote in turn[62] (Figure 2.7). For the device, its reflection at the stop-band position can be switched between valley and peak by UV- and visible-light irradiation. When the PDLC film was in the opaque state, a normal reflection peak was observed, which originates from the stop-band of the colloidal crystal. When the device was irradiated with UV light, the PDLC film became transparent and a dramatic intensity increase in the nonstop-band region was observed in reflection spectra. This is because the light that passed through the colloidal crystal film was scattered when the PDLC film was opaque, while it reached the mirror surface when the PDLC film was transparent. Utilizing polystyrene spheres with different diameter,

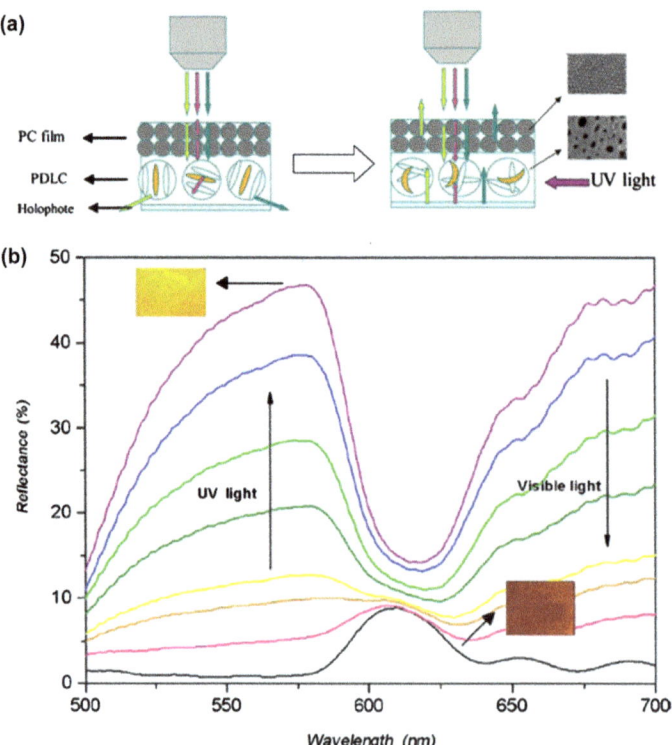

Figure 2.7   (a) Sketch of the photonic device with reversible wavelength choice; (b) Reflection spectrum changes under the irradiation of ultraviolet light. The insets show the colors before and after ultraviolet irradiation. (Reprinted with permission from Ref. 62. Copyright 2007, American Institute of Physics.)

various color and patterns could be exhibited, where light is used as a switch. So this device is not only used for the application of wavelength selection but is also useful for display applications. However, the reversible photoswitch is monostable due to metastable isotropic phase of low molecular weight nematic liquid crystals. For practical applications, a similar bistable photoswitch photonic crystal is realizable by using a polymer liquid crystal to replace the PDLC film.

2.3 Magnetic-Nanoparticles-Based Tunable Colloidal Crystals

2.3.1 Magnetically Tunable Film

The stop-band of photonic crystal depends on the refractive index and spatial periodic structure. The changes in refractive index are usually limited, but changes in the spatial structure induce appreciable tunability in the stop-bands. However, the rates at which such changes can be made are usually too low for practical tunable optical devices. One promising strategy is introducing magnetic properties into the photonic crystals for the intelligent control of the stop-band.[63–70] This type of photonic crystal is composed by monodispersed magnetic colloidal particles suspended in solution. When no external magnetic field is applied, the magnetic colloidal particles display Brownian motion and maintain a disordered status arising from interparticle electrostatic repulsive forces from the charged groups on the surface of the nanoparticles. Under the force of an applied external magnetic field, these spheres respond to the magnetic field and form nonclose-packed chain-like ordered structures, which parallel the direction of the external magnetic field. The ordered particle chains diffract light and exhibit a distinct structural color. The interparticle distance is a result of the balance between the magnetic attractive force and the interparticle electrostatic repulsive force. Hence, the wavelength of the diffracted light mostly depends on the intensity of the magnetic field and the size, surface charge, and magnetic content of particles. Because the particles reach the balance state almost in an instant, the magnetic photonic crystals have a great advantage to construct high-speed tunable photonic crystal devices, such as dynamic displays.

For the purpose of display, a uniform magnetic particles suspension film is expected to fabricate color patterns by a controllable magnetic field. A convenient way is assembling the nonclose-packed magnetic photonic crystal on the surface of a wet hydrogel film. According to this idea, a magnetic-field-responsive colloidal crystal display film was fabricated on the surface of agarose-gel[70] (Figure 2.8). An array of small magnets was used to induce the film to display a specific pattern with different colors. In the film, each small area induced by a single magnet was an independent pixel unit, which was affected by the intensity of the external magnetic field acting on it. Changing the

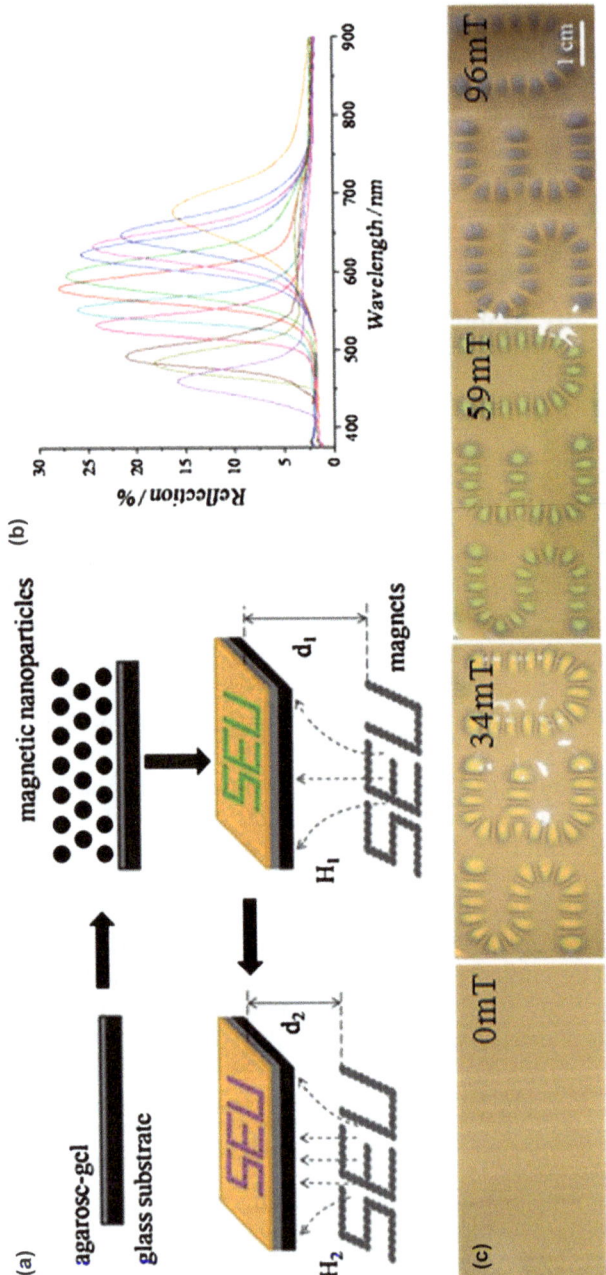

Figure 2.8 (a) Procedure for the fabrication of the magnetically tunable colloidal crystal film on the surface of agarose-gel-coated substrate; (b) Reflection spectra of the magnetically tunable colloidal crystal film in response to varying intensity of the external magnetic field by changing the distance between the film and magnets; (c) Photographs of the magnetically tunable colloidal crystal film under different intensities of the external magnetic field. (Reprinted with permission from Ref. 71. Copyright 2009 by Wiley-VCH.)

distance between the film and magnets, the periodical structures formed by assembly of the magnetic nanospheres were easily regulated due to the alteration of the intensity of the external magnetic field. Therefore, light with different wavelengths was diffracted and various distinct colors can be observed. The tuning range covered almost the whole visible spectrum. Furthermore, by changing the arrangement of the external magnetic array, various patterns were formed as desired. As a result, all patterns with all colors can be conveniently modulated. It opens up a facile way to apply magnetic photonic crystal in displays and also has potential applications in other optical devices.

2.3.2 Magnetochromatic Microcapsules

The magnetic photonic crystals have a wide stop-band tuning range covering the entire visible spectrum as well as a fast and fully reversible response. However, for the construction of display devices, the independently tunable units as pixels are necessary.[71–73] The monodispersed microcapsules encapsulated magnetic colloidal crystals are fine building blocks for fabricating multipixel arrays for displays. The type of microcapsule was composed of a transparent ethoxylated trimethylolpropane triacrylate (ETPTA) hard shell and a magnetochromatic droplet core, which was encapsulated by microfluidic technology[74–76] (Figure 2.9). Tuning the flow rates of various reactants, the

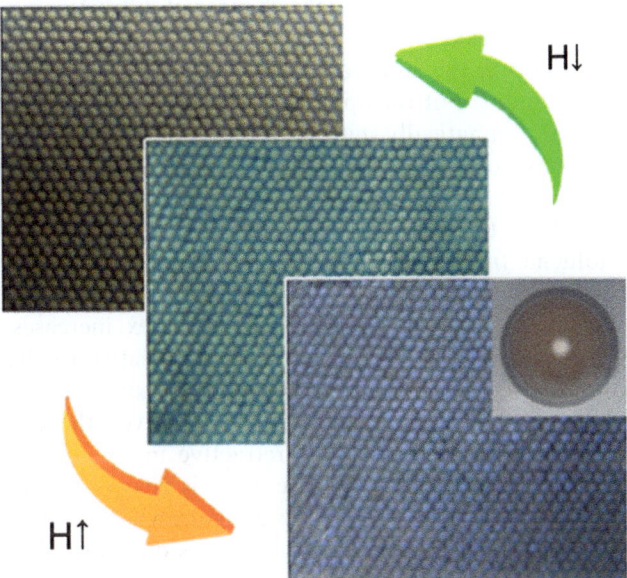

Figure 2.9 Digital photographs of the colors exhibited by the tunable magnetochromatic microcapsules array under different external magnetic stimulation.
(Reprinted with permission from Ref. 76. Copyright 2011 by Wiley-VCH.)

size, shell thickness and shape of capsules can be easily controlled. By arranging the generated microcapsules, a piece of a dynamic switchable multipixel display array was obtained. The exhibited color of the microcapsules could be easily and conveniently controlled with the help of external magnetic stimulation, thus various contents could be displayed *via* the control of each individual microcapsule. Besides, the exhibited color of the magnetochromatic micro-capsules varied a little from different viewing angles and the display itself was extremely stable. It can be foresee that more complex and fine patterns will be achieved by independently controlling electronic magnetic fields in the small scale.[77]

2.4   Smart Photonic Materials for Sensing

2.4.1   Photonic Crystal Films for Label-Free Sensing

2.4.1.1   Solvent Sensor

The original intention of constructing tunable photonic crystals is to controlling the propagation of light by modulating their stop-band. However, everything has two sides. For the particular photonic crystal, the other side of changing the stop-band by external stimuli is sensing external stimuli by the change of stop-band. So, the physical or chemical stimuli can be sensed by ingeniously designing the components and structures of photonic crystal.[78–80] Usually, the stimuli are recognized by the changes of diffraction wavelength or intensity of reflection peak.

As mentioned above, the stop-band of a photonic crystal depends on its average refractive index. For the opal or inverse opal photonic crystals, their reflection peak will dramatically shift when the refractive index of infiltration changes. A demonstration was immersing the porous silica opal in solvents with different refractive indices: methanol ($n = 1.329$), ethanol ($n = 1.360$), 2-propanol ($n = 1.377$), tetrahydrofuran ($n = 1.407$), N,N-dimethyl formamide ($n = 1.431$), toluene ($n = 1.496$) and 1,2-dibromoethane ($n = 1.538$).[81] The stop-band depended on the solvent and shifted by 84 nm from the original 660 nm to 744 nm when the refractive index increases from 1 to 1.538. Hence, the larger shift of the stop-band is sensitive to changes in the surrounding medium and suitable for the precise measurement of the change in refractive index of the surrounding medium. However, the intensity of the stop-band peak is also sensitive to the refractive index of the medium. It decreases significantly when the refractive index is near to that of silica, and disappears at around 1.45. Therefore, it is a big drawback that the refractive index of the surrounding medium in this region is difficult to measure.

2.4.1.2   Gas Sensor

For solvents, their refractive-index differences are big enough to be distinguished by the shift of the stop-band. But for gaseous components, the difference of refractive index is usually small. So, a problem arises in how to amplify the signal? A feasible approach is selecting some materials that can interact with the gas to construct a photonic crystal. An example is platinum-doped tungsten trioxide (Pt-WO$_3$) inverse opal, which presents a specific response to hydrogen gas (H$_2$).[82] Under H$_2$ stimulation, the reflection peak of the photonic crystal rapidly shifted to short wavelength within 3 s as well as showing an intensity decline. The peak could return to its original location when the film was exposed to an oxygen atmosphere. The Pt-WO$_3$ photonic crystal also demonstrated the performance of rapid response time, good recovery, and high-sensitivity gas sensing.

2.4.1.3 Ion Sensor

Heavy metal ions, such as Hg^{2+} and Pb^{2+}, are well-known bioaccumulative, nonbiodegradable, and highly toxic pollutants of the environment and human health. Current techniques for detecting these ions include atomic absorption spectrometry, X-ray fluorescence spectroscopy, and inductively coupled plasma–mass spectrometry (ICP–MS). Although these methods can produce ultrasensitive analytical results, they usually require expensive equipment, complicated and time-consuming experiments, and high levels of operator skill. To overcome these limitations, we developed a robust method for the visual detection of heavy metal ions (such as Hg^{2+} and Pb^{2+}) by using aptamer-functionalized colloidal photonic crystal hydrogel (CPCH) films.[96] The CPCHs were derived from a colloidal crystal array of monodisperse silica nanoparticles, which were polymerized within the poly(acrylamide) hydrogel. The heavy-metal-ion-responsive aptamers were then crosslinked in the hydrogel network. In the case of detection, specific binding of heavy-metal ions and the crosslinked single-stranded aptamers in the hydrogel network caused the hydrogel to shrink, which was detected as a corresponding blueshift in the Bragg diffraction peak position of the CPCHs. The shift value could be used to estimate quantitatively the amount of target ion. It was demonstrated that our CPCH aptasensor could screen a wide concentration range of heavy metal ions with high selectivity and reversibility.

2.4.1.4 Biomolecule Sensor

In label-free bioassays, target molecules are not labeled using fluorescent dyes but detected in their natural forms.[83–85] These types of biosensors could not only eliminate the time-consuming and expensive labeling step for the targets or the reporter molecules, but also allow for quantitative and kinetic measurement of molecular interaction. Optical label-free biosensors based on a photonic crystal rely on its high sensitivity to the refractive index and diffracting plane spacing because the interactions of analytes and probe molecules will result in some physicochemical changes of the photonic crystals.[86,87] These changes of the photonic crystals could be simply detected as the changes of their Bragg diffraction peaks.

We first introduced the inverse opal film into a label-free biosensor by replicating the silica colloidal crystals with polystyrene.[88] Before being used in bioassay, the free-standing macroporous photonic crystal film was first modified with a probe on the pore surface. After the probe was immobilized in the pores, the places where no protein adsorbed were blocked with bovine serum albumin. When used in bioassay, the immobilized probes could selectively capture corresponding antigens. The formation of an immuno-complex in the pore of the photonic crystal film resulted in an increase of average refractive index and this was detected as a corresponding redshift in the diffraction peak position. The shift values induced by the immunoreactions could even be used for quantitatively estimating the amount of the bound antigens (Figure 2.10(a)).

However, for multiplex assay, which simultaneously measures multiple analytes in a single assay, the probes should be encoded to distinguish different binding events in parallel. The most common approach is a planar array, in which the probe molecules are immobilized on a substrate and encoded by the coordinates of their positions. The drawbacks of planar arrays are the limitations in reaction speed, repeatability and flexibility. These problems can be solved by using suspension arrays. One of the key techniques of suspension arrays is encoding. So when the photonic crystal films are performed for label-free multiplex assay, the suspended microcarriers need to be encoded.[89] Photonic crystals could be self-encoded, whose code is the characteristic reflection peak originating from the stop-band. By appro-priately assigning a code region and an analysis region, a series of photonic crystal encoded microcarriers were fabricated using stretched polymer inverse opal films.[90] Because stretching the poly(methylmethacrylate) (PMMA) inverse opal films could shift the reflection spectra of the films to shorter wavelengths, different kinds of encoded microcarriers could be prepared by cutting the films with different stretching degree into microsquares. Every wave band was assigned as one code, and the analytes were detected by peak shifts. Both decoding and bioreaction detection are a measurement of white-light reflection, which makes the detection and the analyzing apparatus extremely simple.

In contrast to these photonic crystal biosensors sensitive to refractive index, spatial structures responsive photonic crystals are more common for biosensors, which have appreciable changes in the stop-bands. This type of photonic biosensors is usually constructed by a soft polymer gel network, which embeds probe molecules in a polymer net matrix. Taking glucose for example, we demonstrated a glucose reporter by immobilizing glucose oxidase onto the backbone of poly(acrylamide-*co*-acrylic acid) hydrogel[91] (Figure 2.10(b)). In the sensing process, the oxidation of glucose substrate was catalyzed by the enzyme. The oxidized product actuated change in the hydrogel volume and consequently the reflection wavelength. Hence, the glucose concentration could be analyzed simultaneously by the peak shift.

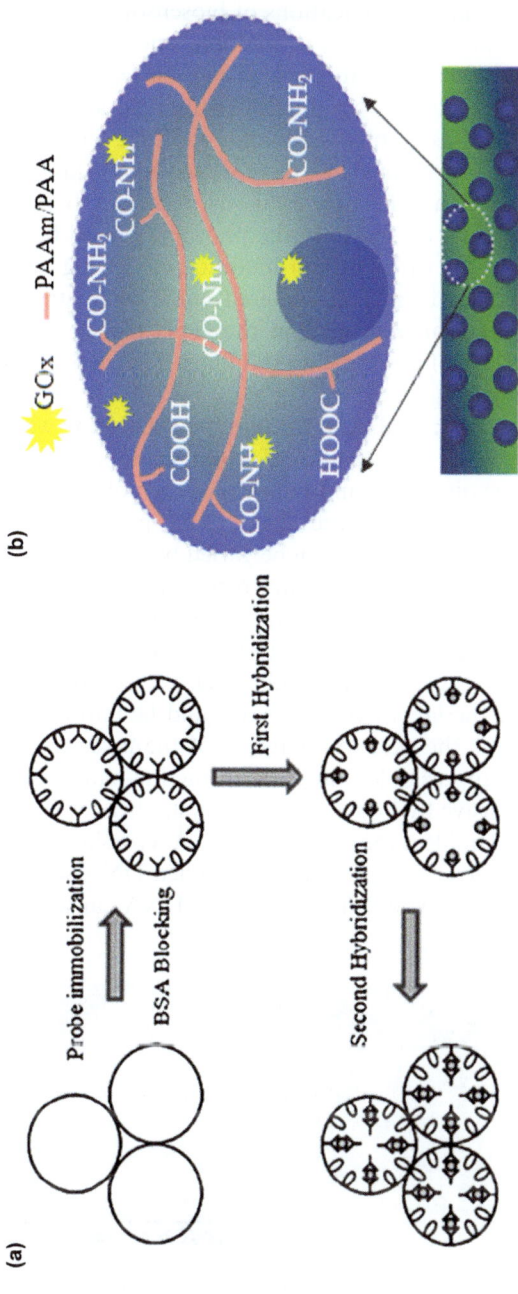

Figure 2.10 (a) Procedure of using inverse opal carriers for label-free multiplex immunoassay; (Reproduced from Ref. 90 with permission of The Royal Society of Chemistry.) (b) Glucose-responsive hydrogel inverse opal film based on the principle of enzyme–catalytic redox reaction.
(Reproduced from Ref. 91 with permission of The Royal Society of Chemistry.)

2.4.2   Photonic Crystal Beads for Bioassays

In the decades of research, the most common and simple form of photonic crystals is a planar shape film in applications of biosensors. However, its stopband is angle dependent and its mobility in solution is not satisfactory. Also, in multiplex bioassays, they were easy to stack during bioreaction and decoding processes, which would result in a wrong test results. An alternative form of photonic crystal is a self-assembled colloidal photonic crystal bead. The ordered colloidal particles formed a three-dimensional photonic crystal structure, giving the bead a reflection peak in a certain bandgap region. Compared with the colloidal crystal film, the photonic crystal beads have the virtue that the reflected color varies little from different viewing angles because monodisperse nanoparticles form face-centered cubic symmetry in colloidal crystals with an hexagonal arrangement in the (111) plane. The beads also overcome the shortcomings of photonic crystal films. They have good mobility and are not prone to stacking.

An inverse opal photonic crystal bead for multiplex detection was formed by droplet crystallization of a suspension of silica nanoparticles and poly-styrenespheres in silicone oil at first and removal of the polystyrene spheres by high-temperature calcination[92] (Figure 2.11). The position of this peak constituted the identifying code, which can be varied by changing the size of the polystyrene spheres used to mold the macroporous. As the inverse opal structure colloidal crystal films, the label-free bioassay was realized based on the shift of their reflection spectra, were resulted from the specific analyte binding caused by the average refractive-index change. Compared to the hard beads, the sensitivity of the label-free bioassay could be improved by using a bioresponsive hydrogel as the skeleton material of the inverse opal photonic crystal beads due to the appreciable change of diffracting-plane spacing. A new kind of molecular-responsive hydrogel photonic crystal bead was developed by polymerization of acrylamide monomer and DNA crosslinker used as the skeleton material of the inverse opal.[93] In the matched DNA solution, specific hybridization of the free DNA and the crosslinker DNA caused the hydrogel

Figure 2.11   (a) SEM images of the inverse opal photonic beads; (b) 3D image of the seven kinds of inverse opal photonic beads in water. Scale bar 500 mm. (Reproduced with permission from Ref. 92. Copyright 2009 by Wiley-VCH.)

shrinking and this was detected as a corresponding blueshift of the stop-band. The shift degree of the stop-band was used to quantitatively estimate the amount of free DNA in the solution. For multiplex detection, the DNA responsive photonic crystal beads were tagged by some quantum dots (QDs) as the coding elements.

When using molecular-imprinted hydrogel as the skeleton material, the prepared imprinted photonic crystal beads could be used for label-free detection of proteins without using immunological antibodies[94] (Figure 2.12). This type of assay was demonstrated by the optical response of the bovine hemoglobin (Hb) imprinted photonic beads incubated in the solutions of bovine Hb at different concentrations. As the biomolecule enters a complementary imprinted nanocavity can form a multitude of simultaneous hydrogen bonds between the oriented amide groups in the nanocavity and the polar surface residues of the biomolecule. These cooperative, multivalent hydrogen bonds lead to a significantly increased selective biomolecule binding affinity. The imprinted photonic crystal beads reported this molecular recognition event through a gradual shift of the position of the Bragg diffraction peak to long

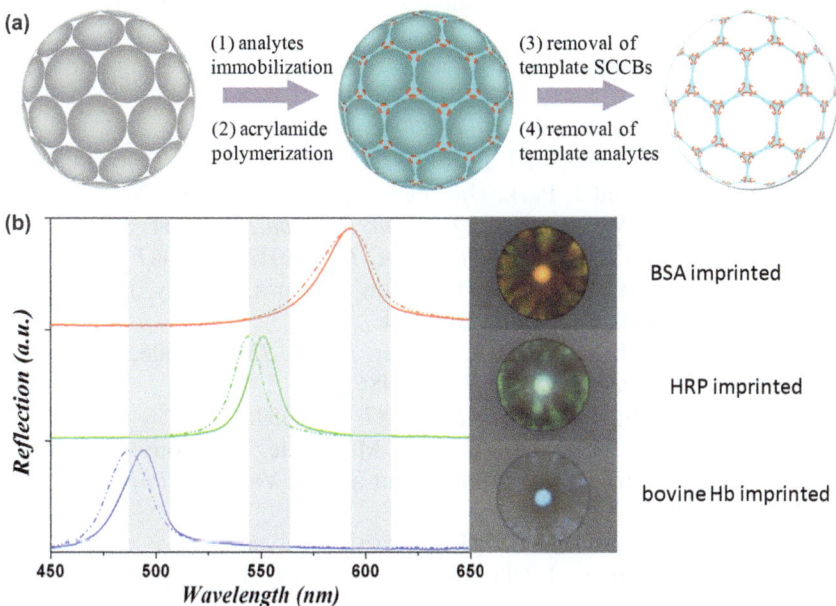

Figure 2.12 (a) Schematic diagram of the preparation method for molecularly imprinted polymer beads with photonic crystal structure; (b) Reflection spectra (left) and bright-field microscopic images (right) of three kinds of MIPBs. The cyan, green, and red MIPBs were imprinted with bovine Hb, HRP, and BSA, respectively. The dashed lines and solid lines are the spectra of the MIPBs before and after multiplex detection. The gray areas are the encoding of the MIPBs.
(Reproduced with permission from Ref. 94. Copyright 2009 by Wiley-VCH.)

wavelengths with an increase of the bovine Hb concentration. Remarkably, a trace amount of bovine Hb (1 ng mL^{-1}) was enough to lead to a red-shift of the diffraction peak. For multiplex label-free bioassay, different spectral ranges encoded photonic beads imprinted with different proteins, respectively, were mixed in a single tube containing analytes. The analyte proteins could be detected simultaneous. A powerful feature of this method is that both the decoding and biomolecule detection are as simple as one-step measurement of the diffraction peak of the inverse opal photonic crystal beads, which simplifies the detection instruments and procedures.

References

1. X. Xie, Y. Liu, M. Zhang, J. Zhou and K. Wong, *Physica E*, 2012, **44**, 1109.
2. G. Freymann, A. Ledermann, M. Thiel, I. Staude, S. Essig, K. Busch and M. Wegener, *Adv. Funct. Mater.*, 2010, **20**, 1038.
3. B. Comiskey, J. Albert, H. Yoshizawa and J. Jacobson, *Nature*, 1998, **394**, 253.
4. A. Arsenault, D. Puzzo, I. Manners and G. Ozin, *Nature Photon.*, 2007, **1**, 468.
5. Q. Sun, Y. Wang, L. Li, D. Wang, T. Zhu, J. Xu, C. Yang and Y. Li, *Nature Photonics*, 2007, **1**, 717.
6. J. Clark and G. Lanzani, *Nature Photon.*, 2010, **4**, 438.
7. M. Wu, A. Solgaard and J. Ford, *J. Lightwave Technol.*, 2006, **24**, 4433.
8. B. Lee, S. Roh and J. Park, *Opt. Fiber Technol.*, 2009, **15**, 209.
9. R. Nair and R. Vijaya, *Prog. Quantum. Electron.*, 2010, **34**, 89.
10. R. Freeman and I. Willner, *Chem. Soc. Rev.*, 2012, **41**, 4067.
11. M. Stewart, C. Anderton, L. Thompson, J. Maria, S. Gray, J. Rogers and R. Nuzzo, *Chem. Rev.*, 2008, **108**, 494.
12. Y. Li, F. Qian, J. Xiang and C. Lieber, *Mater. Today*, 2006, **9**, 18.
13. S. John, *Phys. Rev. Lett.*, 1987, **58**, 2486.
14. E. Yablonovitch, *Phys. Rev. Lett.*, 1987, **58**, 2059.
15. R. McPhedran, N. Nicorovici, D. McKenzie, G. Rouse, L. Botten, V. Welch, A. Parker, M. Wohlgennant and V. Vardeny, *Physica B*, 2003, **338**, 182.
16. A. Parker, *J. Opt. A*, 2000, **2**, R15.
17. J. Vinther, D. Briggs, J. Clarke, G. Mayr and R. Prum, *Biol. Lett.*, 2010, **6**, 128.
18. S. Kinoshita, S. Yoshioka and J. Miyazaki, *Rep. Prog. Phys.*, 2008, **71**, 076401.
19. G. Tayeb, B. Gralak and S. Enoch, *Opt. Photon. News*, 2003, **14**, 38.
20. A. Parker, R. McPhedran, D. McKenzie, L. Botten and N. Nicorovici, *Nature*, 2001, **409**, 36.
21. A. Parker, V. Welch, D. Driver and N. Martini, *Nature*, 2003, **426**, 786.
22. A. Parker, *Phil. Trans. R. Soc. Lond. A*, 2004, **362**, 2709.
23. S. Kinoshita and S. Yoshioka, *ChemPhysChem*, 2005, **6**, 1442.

24. P. Russell, *Science*, 2003, **299**, 358.
25. J. Dudley and J. Taylor, *Nature Photon.*, 2009, **3**, 85.
26. A. Cerqueira, *Rep. Prog. Phys.*, 2010, **73**, 024401.
27. H. Hinton and G. Jarman, *Nature*, 1972, **238**, 160.
28. W. Crookes, L. Ding, Q. Huang, J. Kimbell, J. Horwitz and M. McFall-Ngai, *Science*, 2004, **303**, 235.
29. D. Stuart-Fox and A. Moussalli, *Philos. Trans. R. Soc. B*, 2009, **364**, 463.
30. S. Yoshioka, B. Matsuhana, S. Tanaka, Y. Inouye, N. Oshima and S. Kinoshita, *J. R. Soc. Interface*, 2011, **8**, 56.
31. H. Cong, B. Yu and X. Zhao, *Opt. Exp.*, 2011, **19**, 12799.
32. J. Lythgoe and J. Shand, *J. Exp. Biol.*, 1989, **141**, 313.
33. F. Liu, B. Dong, X. Liu, Y. Zheng and J. Zi, *Opt. Exp.*, 2009, **17**, 16183.
34. J. Vigneron, J. Pasteels, D. Windsor, Z. Vértesy, M. Rassart, T. Seldrum, J. Dumont, O. Deparis, V. Lousse, L. P. Biró, D. Ertz and V. Welch, *Phys. Rev. E*, 2007, **76**, 031907.
35. A. Parker and H. Townley, *Nature Nanotechnol.*, 2007, **2**, 347.
36. M. Kolle, P. Salgard-Cunha, M. Scherer, F. Huang, P. Vukusic, S. Mahajan, J. Baumberg and U. Steiner, *Nature Nanotechnol.*, 2010, **5**, 511.
37. A. Parker, *Philos. Trans. R. Soc. A*, 2009, **364**, 1759.
38. T. Stegmaier, M. Linke and H. Planck, *Philos. Trans. R. Soc. A*, 2009, **367**, 1749.
39. V. Colvin, *MRS Bull.*, 2001, **26**, 637.
40. F. Meseguer, *Colloid. Surface. A*, 2005, **270**, 1.
41. M. Irie, *Chem. Rev.*, 2000, **100**, 1683.
42. Z. Gu, S. Hayami, Q. Meng, T. Iyoda, A. Fujishima and O. Sato, *J. Am. Chem. Soc.*, 2000, **122**, 10730.
43. P. Nilsson, *Appl. Opt.*, 1968, **7**, 435.
44. Z. Gu, T. Iyoda, A. Fujishima and O. Sato, *Adv. Mater.*, 2001, **13**, 1295.
45. S. Chandrasekhar, *Liquid Crystals*, Cambridge University Press, Cambridge, 2nd edn, 1992.
46. P. Gennes and J. Prost, *The Physics of Liquid Crystals*, Clarendon Press, Oxford, 1974.
47. T. Brody, *Inform. Disp.*, 1997, **13**, 28.
48. K. Busch and S. John, *Phys. Rev. Lett.*, 1999, **83**, 967.
49. E. Yablonovitch, *Nature*, 1999, **401**, 539.
50. D. Kang, J. Maclennan, N. Clark, A. Zakhidov and R. Baughman, *Phys. Rev. Lett.*, 2001, **86**, 4052.
51. M. Ozaki, Y. Shimoda, M. Kasano and K. Yoshino, *Adv. Mater.*, 2002, **14**, 514.
52. S. Choi, S. Morris, W. Huck and H. Coles, *Adv. Mater.*, 2009, **21**, 3915.
53. J. Guo, H. Wu, F. Chen, L. Zhang, W. He, H. Yang and J. Wei, *J. Mater. Chem.*, 2010, **20**, 4094.
54. S. Kubo, Z. Gu, K. Takahashi, Y. Ohko, O. Sato and A. Fujishima, *J. Am. Chem. Soc.*, 2002, **124**, 10950.

55. S. Kubo, Z. Gu, K. Takahashi, A. Fujishima, H. Segawa and O. Sato, *J. Am. Chem. Soc.*, 2004, **126**, 8314.

56. S. Kubo, Z. Gu, K. Takahashi, A. Fujishima, H. Segawa and O. Sato, *Chem. Mater.*, 2005, **17**, 2298.

57. G. Springer and D. Higgins, *J. Am. Chem. Soc.*, 2000, **122**, 6801.

58. B. Luther, G. Springer and D. Higgins, *Chem. Mater.*, 2001, **13**, 2281.

59. D. Rudhardt, A. Fernández-Nieves, D. Link and D. Witz, *Appl. Phys. Lett.*, 2003, **82**, 2610.

60. J. Sung, S. Hirano, O. Tsutsumi, A. Kanazawa, T. Shiono and T. Ikeda, *Chem. Mater.*, 2002, **14**, 385.

61. Z. Xie, L. Sun, G. Han and Z. Gu, *Adv. Mater.*, 2008, **20**, 3601.

62. G. Han, Z. Xie, D. Zheng, L. Sun and Z. Gu, *Appl. Phys. Lett.*, 2007, **91**, 141114.

63. X. Xu, G. Friedman, K. Humfeld, S. Majetich and S. Asher, *Adv. Mater.*, 2001, **13**, 1681.

64. X. Xu, G. Friedman, K. Humfeld, S. Majetich and S. Asher, *Chem. Mater.*, 2002, **14**, 1249.

65. C. Reese, A. Mikhonin, M. Kamenjicki, A. Tikhonov and S. Asher, *J. Am. Chem. Soc.*, 2004, **126**, 1493.

66. J. Ge, Y. Hu and Y. Yin, *Angew. Chem. In. Ed.*, 2007, **46**, 7428.

67. J. Ge and Y. Yin, *J. Mater. Chem.*, 2008, **18**, 5041.

68. J. Ge and Y. Yin, *Adv. Mater.*, 2008, **20**, 3485.

69. J. Ge, H. Lee, L. He, J. Kim, Z. Lu, H. Kim, J. Goebl, S. Kwon and Y. Yin, *J. Am. Chem. Soc.*, 2009, **131**, 15687.

70. J. Ge, L. He, J. Goebl and Y. Yin, *J. Am. Chem. Soc.*, 2009, **131**, 3484.

71. C. Zhu, L. Chen, H. Xu and Z. Gu, *Macromol. Rapid Commun.*, 2009, **30**, 1945.

72. H. Kim, J. Ge, J. Kim, S. Choi, H. Lee, H. Lee, W. Park, Y. Yin and S. Kwon, *Nature Photonics*, 2009, **3**, 534.

73. H. Lee, J. Kim, H. Kim, J. Kim and S. Kwon, *Nature Mater.*, 2010, **9**, 745.

74. H. Yang, Y. Han, X. Zhao, K. Nagai and Z. Gu, *Appl. Phys. Lett.*, 2006, **89**, 111121.

75. G. Xiong, G. Han, C. Sun, H. Xu, H. Wei and Z. Z. Gu, *Adv. Funct. Mater.*, 2009, **19**, 1082.

76. C. Zhu, W. Xu, L. Chen, W. Zhang, H. Xu and Z. Gu, *Adv. Funct. Mater.*, 2011, **21**, 2043.

77. C. Sun, Y. Yao, L. Sun and Z. Gu, *Colloid. Surface. A*, 2011, **384**, 720.

78. Y. Zhao, X. Zhao and Z. Gu, *Adv. Funct. Mater.*, 2010, **20**, 2970.

79. R. Nair and R. Vijaya, *Prog. Quantum. Electron.*, 2010, **34**, 89.

80. Y. Yin and J. Ge, *Angew. Chem. Int. Ed.*, 2011, **50**, 2.

81. Z. Gu, R. Horie, S. Kubo, Y. Yamada, A. Fujishima and O. Sato, *Angew. Chem. Int. Ed.*, 2002, **41**, 1154.

82. Z. Xie, H. Xu, F. Rong, L. Sun, S. Zhang and Z. Gu, *Thin Solid Films*, 2012, **520**, 4063.

83. X. Zhao, Z. Liu, H. Yang, K. Nagai, Y. Zhao and Z. Gu, *Chem. Mater.*, 2006, **18**, 2443.

84. X. Zhao, Y. Cao, F. Ito, H. Chen, K. Nagai, Y. Zhao and Z. Gu, *Angew. Chem. Int. Ed.*, 2006, **45**, 6835.

85. Y. Zhao, X. Zhao, C. Sun, J. Li, R. Zhu and Z. Gu, *Anal. Chem.*, 2008, **80**, 1598.

86. Y. Guo, J. Ye, C. Divin, B. Huang, T. Thomas, J. Baker and T. Norris, *Anal. Chem.*, 2010, **82**, 5211.

87. M. Schwartz, S. Alvarez and M. Sailor, *Anal. Chem.*, 2007, **79**, 327.

88. W. Qian, Z. Gu, A. Fujishima and O. Sato, *Langmuir*, 2002, **18**, 4526.

89. Z. Xie, Y. Zhao, L. Sun, X. Zhao, Y. Shao and Z. Gu, *Chem. Commun.*, 2009, **45**, 7012.

90. Z. Liu, Z. Xie, X. Zhao and Z. Gu, *J. Mater. Chem.*, 2008, **18**, 3309.

91. L. Jin, Y. Zhao, X. Liu, Y. Wang, B. Ye, Z. Xie and Z. Gu, *Soft Matter*, 2012, **8**, 4911.

92. Y. Zhao, X. Zhao, J. Hu, M. Xu, W. Zhao, L. Sun, C. Zhu, H. Xu and Z. Gu, *Adv. Mater.*, 2009, **21**, 569.

93. Y. Zhao, X. Zhao, B. Tang, W. Xu, J. Li, J. Hu and Z. Gu, *Adv. Funct. Mater.*, 2010, **20**, 976.

94. Y. Zhao, X. Zhao, J. Hu, J. Li, W. Xu and Z. Gu, *Angew. Chem. Int. Ed.*, 2009, **48**, 7350.

95. Y. Zhao, Z. Xie, H. Gu, C. Zhu and Z. Gu, *Chem. Soc. Rev.*, 2012, **41**, 3297.

96. B. Ye, Y. Zhao, Y. Cheng, T. Li, Z. Xie, X. Zhao and Z. Gu, *Nanoscale*, 2012, **4**, 5998.

CHAPTER 3

Opal Photonic Crystal Films with Tunable Structural Color

HIROSHI FUDOUZI

Applied Photonic Materials Group, Photonic Materials Unit, National Institute for Materials Science 1-2-1 Sengen, Tsukuba 305-0047, Japan
Email: FUDOUZI.Hiroshi@nims.go.jp

3.1   Introduction

Opal is a gemstone with iridescence that exhibits the effect known as color play. Using a scanning electron microscope (SEM), the Austrian mineralogist Sanders discovered that natural opals have an ordered array structure of monodispersed colloidal silica spheres.[1] Subsequently, artificial opals were synthesized using the principles of a chromogenic phenomenon called structural color, which was attracting significant attention. The Stober method produces monodispersed amorphous silica spheres starting from hydrolysis and condensation reactions of tetraethoxysilane.[2] In the early 1970s, Gilson was the first to prepare artificial opals as synthetic gemstones on an industrial scale.[3] The refractive index of artificial opals has a periodic structure owing to the array of colloidal spheres.

Figure 3.1 shows an artificial opal, its microstructure, and its structural color mechanism. The artificial bulk opal shown in Figure 3.1(A) was developed by the Kyocera Company. The slow settling of silica spheres from a suspension results in the formation a microstructure of close-packed silica spheres. An illustration of the traditional bulk opal microstructure can be seen in Figure 3.1(B). Both natural and artificial opals have micrometer-scale domains with polycrystalline microstructures. When the periodic length of the array is

RSC Smart Materials No. 5
Responsive Photonic Nanostructures: Smart Nanoscale Optical Materials
Edited by Yadong Yin
© The Royal Society of Chemistry 2013
Published by the Royal Society of Chemistry, www.rsc.org

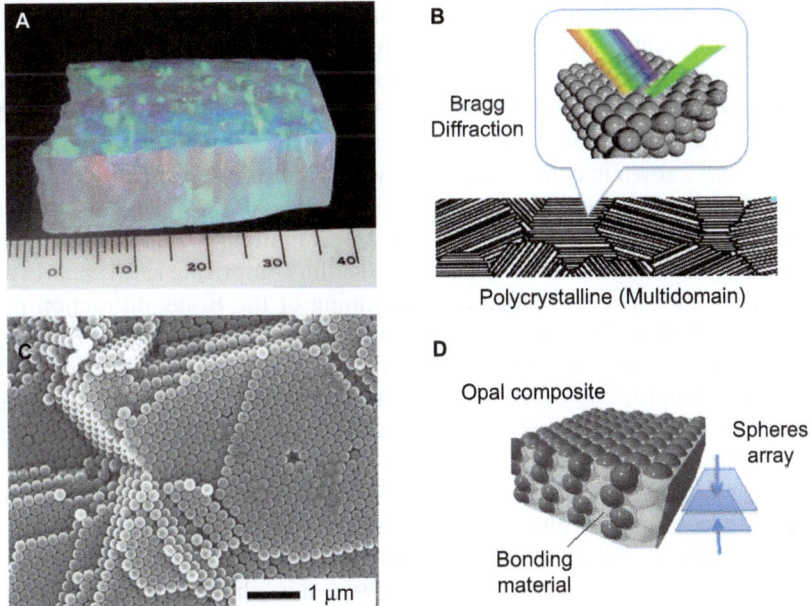

Figure 3.1   Structural color in opals. (A) Kyocera-created bulk silica opal with a
rainbow color for synthetic gems; (B) Bragg diffraction of white light from
an array of spheres and the origin of iridescence (the random orientation
of the multidomain); (C) SEM image of 200 nm diameter cubic close-
packed (ccp) polystyrene spheres; and (D) Opal composite film using a
bonding material for tunable interspacing of the array of spheres.

approximately half that of the wavelength of visible light, a specific wavelength
of light selectively reflects owing to Bragg's law of diffraction. In addition, play
of color arises from the incident angle owing to the random orientation of the
domains. Further details on opal research are described by Marlow *et al.*[4]

In the past decade, artificial opals and related structures have received much
attention as three-dimensional photonic crystals and monochromatic structural
color materials.[5–13] Opal films are fabricated on the basis of a convective self-
assembly process without the need for cleanroom facilities and expensive
equipment.[14,15] High-quality opal films with crystal planes oriented on the
substrate induce monochromatic structural color. A scanning electron
microscopy (SEM) image of close-packed polystyrene (PS) spheres on a silicon
wafer is shown in Figure 3.1(C). Here, monodispersed 200 nm PS spheres form
a periodic 3D nanostructure with cubic close packing (ccp). The ccp (111)
planes are parallel to the substrate. The vertical stacking of the ccp (111)
planes on the substrate results in monochromatic structural color. This color
depends on the refractive index, tilting angle, and distance between the ccp
(111) planes.

An opal composite composed of ordered colloidal spheres filled with a
bonding material can be seen in Figure 3.1(D). The 3D arrayed colloidal

spheres are immobilized or fixed by thermal heating, UV irradiation, or reaction with a chemical. When the interspaces of the ccp (111) plane are changed by a stimulus, the optical properties of the opal composite will respond, and thus, the opal composite acts like a smart material. Similar tunable optical properties obtained by varying the interspacing of ccp (111) planes is well known in colloidal crystals, which are similar nanostructures. The original idea was proposed by Asher and coworkers in the 1990s for nonclose-packed colloidal crystals prepared as a hydrogel material with a tunable Bragg diffraction peak.[16–18] A more rigid composite film comprising silica spheres and methyl acrylate was also fabricated, and tuning of the Bragg diffraction peak was demonstrated for this film.[19] The robust composite film tunes the diffraction to shorter wavelengths with uniaxial stretching, and in contrast, to longer wavelengths by swelling with a monomer. In the 2000s, a new perspective was gained, and these colloidal crystal composites began to be regarded as photonic crystals on the basis of stimulus-responsive tunable stop-bandgap materials.[20,21] These colloidal crystals embedded in hydrogels were recognized as tunable photonic crystals and applied in bandpass filters, dielectric mirrors, and new types of photonic crystal devices. Details on these materials are described in another chapter; this chapter focuses on opal composites.

3.2   Tunable Optical Properties of Opal Composites

Opal photonic crystals are applied in tunable photonic devices, filters, displays, and sensors. Figure 3.2 shows the photonic band structure for a silica opal with cubic close packing calculated using the BandSOLVE simulation engine based on the Plane Wave Expansion algorithm.[22] The stop-bandgap ccp (111) and ccp (100) planes correspond to points L and X in Figure 3.2(A), respectively. In the opal film shown in Figure 3.1(C), the stop-bandgap at point L, *i.e.* the ccp (111) plane, is the most important with respect to the tunable optical properties of the opal photonic crystal. A comparison of the reflectance diffraction from the ccp (111) plane and the photonic band calculation for the Γ–L region is shown in Figure 3.2(B). The measured Bragg peak for the opal film is in agreement with the theoretical calculation. Here, the Bragg diffraction peak position is located at the center of the stop-bandgap at point L. In the case of opal composites with a high-contrast refractive index, the stop-bandgap and diffraction peak will be widened. A wide stop-bandgap is expected to enable new photonic crystal engineering: a perfect dielectric mirror, a resonant cavity, or a waveguide. Yoshino and coworkers demonstrated early examples of the tunable stop-bandgap in photonic crystals composed of opal composites[23,24] using a two-electrode cell-packed opal film filled with a liquid crystal molecule. Their pioneering work demonstrated a tunable stop-bandgap by applying an electric field or increasing temperature. The change in the refractive index of the liquid-crystal molecules caused the tuning of the stop-bandgap. In addition, variable-periodic structures in opal composites can also lead to the tuning of the stop-bandgap.[25]

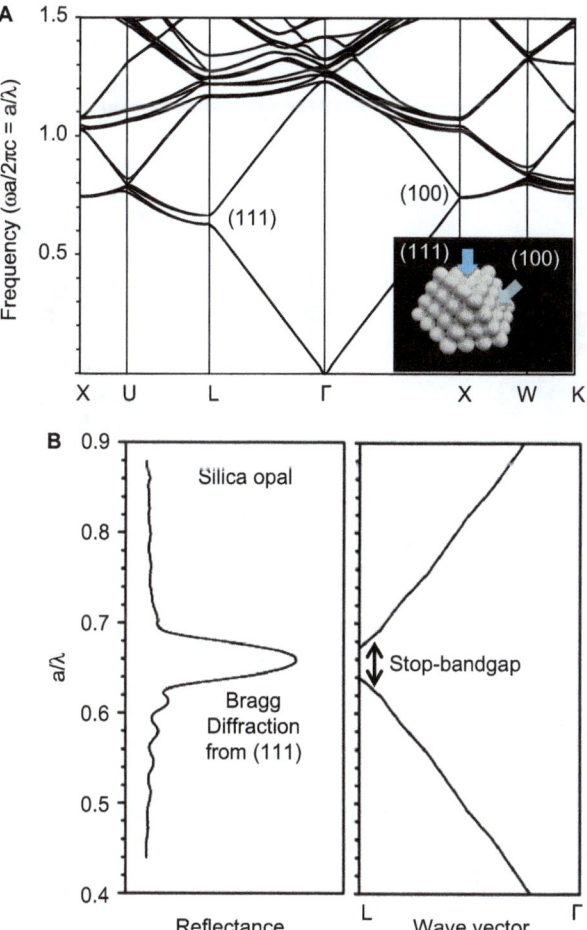

Figure 3.2 Photonic bandgap (PBG) and Bragg diffraction. (A) Photonic bandgap diagram for the ccp structure of silica spheres in air; and (B) Comparison of the Bragg diffraction of a ccp (111) opal film with the theoretical PBG calculation.

Figure 3.3 shows a schematic of the processes for fabricating opal composites. The left scheme depicts the standard approach. First, the opal is crystallized from a colloidal sphere suspension,[14,15] and then, the spaces between the spheres are filled with precursors. The center scheme presents the coassembly procedure, in which the colloidal spheres and precursors are assembled at the same stage. One advantage of the coassembly method is the release of stress during crystallization, which results in the formation of a crack-free opal composite film using a hydrolyzed silicate sol-gel precursor[26] and an aqueous monomer.[27] The right scheme shows the assembly of core–shell colloidal spheres that comprise a hard core of PS spheres covered with a soft shell of elastic material.[28,29]

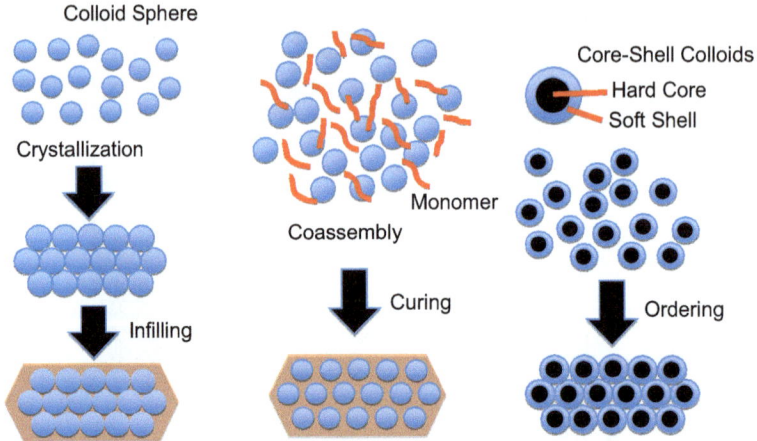

Figure 3.3   Schematic of procedures for the preparation of opal composites. Left: Crystallization and infilling; Center: Coassembly and curing; Right: Core–shell colloids and ordering.

The most widely used processes are the evaporation-driven self-assembly methods shown in Figure 3.4. Figures 3.4(A) and 3.4(B) are widely used for three types of opal composites. Withdrawal[14] and vertical deposition[15] methods are standard procedures. Both methods are based on the evaporative self-assembly of colloidal spheres shown in Figure 3.4(C). Convective flow and capillary forces play important roles in the formation of high-quality colloidal crystal films.[30] The colloidal spheres form an opal film, *i.e.* an ordered structure as a close-packed colloidal crystal.

We have been developing a method for the colloidal crystal growth of opal films immersed in silicone oil, as shown in Figure 3.4(D).[31] This method is one of the evaporative self-assembly techniques for opal films from colloidal sphere suspensions. Understanding the mechanism of the process is important for ensuring the coating of high-quality opal thin films. First, a 3-in silicon wafer was covered with PS colloidal spheres, and then, the concentrated crystallization of close-packed colloidal crystal films was achieved from an aqueous suspension immersed in silicone oil.[32] The arrow indicates the interface between the white and green structural colors. Using an optical microscope and *via* spectroscopy, a phase transition was observed to occur at the interface between the disordered colloidal suspension and the colloidal crystal film. Recently, we have been developing the opal film coating equipment shown in Figure 3.4(E). At the interface, the Bragg diffraction peaks were measured with a miniature fiber probe optical spectrometer. In this instrument, the crystal growth for high-quality and uniform opal films is controlled by a feedback system using the diffraction peak.

Using a completely different approach, Ruhl and Hellmann reported a uniform and rapid fabrication method for rubber opal composites using core–shell colloids without a solvent.[33] This melt shear stress process involves

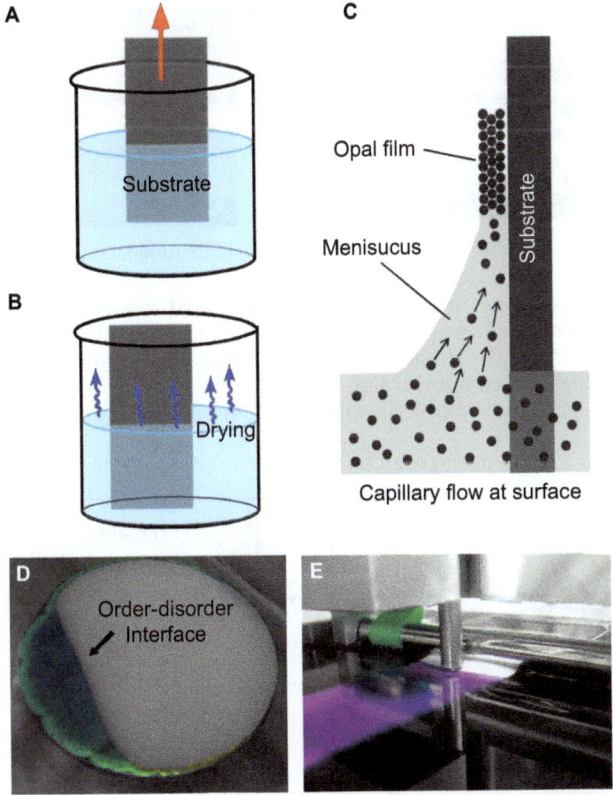

Figure 3.4 Evaporative self-assembly to form opal films. (A) Withdrawal of the substrate from a colloidal suspension; (B) Vertical deposition on the substrate in a colloidal suspension; (C) Self-assembly mechanism *via* capillary flow in a meniscus; and (D) and (E) Crystallization concentrated colloidal suspension under a silicone oil layer. A high-quality opal film forms *via* a disorder–order phase transition at the interface.

rapid ordering *via* axial hot pressing and results in large opal composite films. Figure 3.5 shows the mechanism of the melt flow ordering process (A) and the opal composite film with ordered ccp (111) planes (B). The ccp (111) planes are parallel to the substrate, and the vertical stacking of the array of spheres results in monochromatic structural color.[34] At the present stage, this approach is suitable for industrial-scale manufacturing, such as role-to-role film coating[35] or nozzle fiber injection.[36]

In initial work on the tunable opal photonic crystals shown in Figure 3.6, a bulk opal composite was developed with 280 nm styrene-butadiene polymer spheres *via* a sedimentation process. A 7 mm×7 mm×7 mm cubic specimen was mechanically compressed or elongated in the direction vertical to the ccp (111) plane.[25] Figure 3.6(A) shows photographs of the bulk opal composite under mechanical stress. When the pressure was increased, the color changed from blue-green to orange-red. Figure 3.6(B) shows the reflection spectrum as a

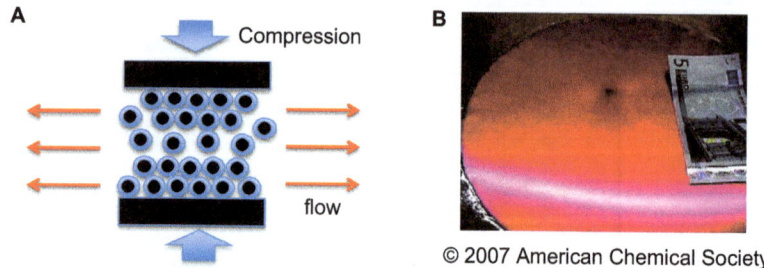

© 2007 American Chemical Society

Figure 3.5 Shear stress ordering of a polymer opal composite using core–shell polymer spheres. (A) Shear flow alignment of the core–shell polymer spheres under viscous melt condition through compression; and (B) Large-scale opal composite film.
Reproduced with permission (Ref. 34).

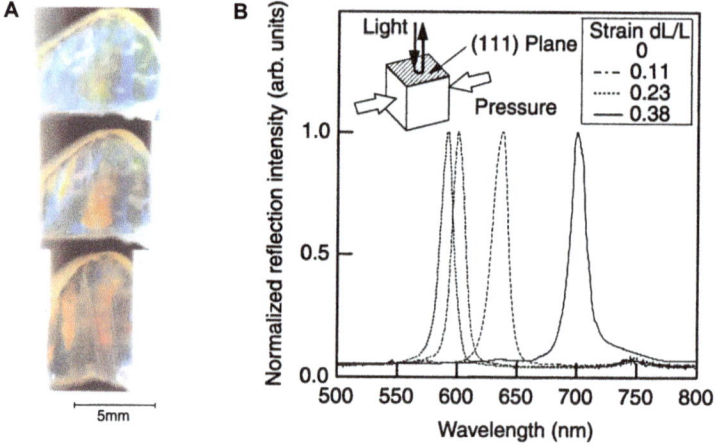

© 1999 The Japan Society of Applied Physics

Figure 3.6 Opal photonic crystal with a tunable stop-bandgap. (A) Structural color change by applying compression on a cubic block; and (B) Bragg diffraction shifting to a longer wavelength under varying strain.
Reproduced with permission (Ref. 25).

function of strain under pressure. The reflection peak shifted to a longer wavelength. The results obtained for this opal composite indicated that tunable photonic crystals can be employed as mechanical sensors. However, this opal composite shows iridescence, and its preparation and microstructure were not described in detail. In addition, the quality of the bulk sample was not sufficiently high for engineering applications, such as sensors, optical devices, and displays. However, during the past decade, several preparation methods have been developed that provide high-quality and well-controlled opal composites.[5,7,8,11,14,15,26,27,32,34,35] Figure 3.7 shows the various opal composites designed as tunable photonic crystals for use in various engineering fields.

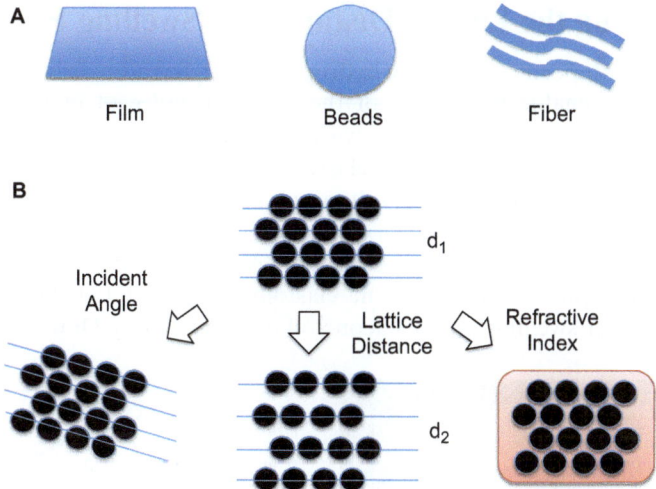

Figure 3.7 Shape and color change factors in opal composites. (A) Opal composite shapes, including films, beads, and fibers; and (B) Three major factors include the incident angle, lattice distance, and refractive index. Tunable structural color means tuning the stop-bandgap or the Bragg diffraction peak.

Figure 3.7(A) classifies the morphology of opal composites as thin films, microbeads, and microfibers.[36] Opal composite films are fabricated on a supporting solid substrate. Microbeads are fabricated on a liquid interface[37] or by a microinjection technique.[38] Recently, color-tunable microbeads composed of opal composites have been applied to chemical/biosensors[39] and magnetic response structural colors.[40]

One of the popular topics in opal composites and related nanostructures is tunable structural color and tuning stop-bandgaps[41–45] Figure 3.7(B) shows the mechanisms for tunable photonic stop-bandgaps and Bragg diffraction wavelengths. When the diffraction wavelength is located in the visible region, *i.e.* between 400 nm and 700 nm, the opal composite shows a bright and monolithic structural color. This structural color depends on the incident angle, lattice distance, and refractive index. The diffraction wavelength, λ, is expressed as below:

$$\lambda = 2d \sqrt{(n_{\text{eff}}^2 - \sin^2 \theta)} \qquad (3.1)$$

Here, d indicates the lattice distance, n_{eff} the average refractive index, and θ the incident angle. As is clear from the equation, structural color depends on the interspacing of the ccp (111) planes, refractive index, and the tilting angle of the ccp (111) planes.

Next, we focus on the tuning of the structural color in opal composites by varying the lattice distance of the ccp (111) planes.

3.3   Tuning the Lattice Distance *via* Swelling Phenomena

Swelling gel composites, including hydrogels,[46] polyferrocebylsilane gels,[47] and silicone gels,[48] enable the tuning of the lattice distance of opal composites. Figure 3.8 shows an opal composite with a lattice distance that is tunable using swelling phenomena.[10] A silicone elastomer is swollen with various hydrophobic solvents. Expanding the volume of a silicone elastomer leads to the enlargement of the lattice distance of the ccp (111) planes. The lattice change was analyzed by transmittance spectroscopy. An opal composite film composed of 202 nm PS spheres and a silicone elastomer (Sylgard 184, Dow Corning) before and after swelling with a silicone oil (0.65 cSt, Toray Dow Corning) can

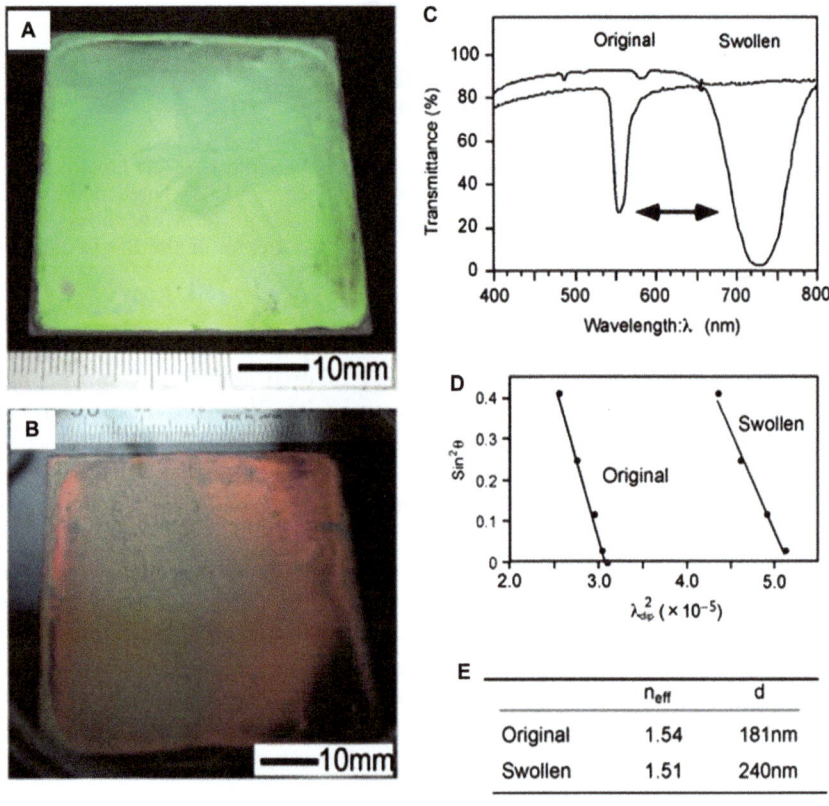

© 2009 The Society of Powder Technology Japan

Figure 3.8   Reversible structural color change in response to swelling and shrinking. (A) Opal composite film composed of 202 nm PS spheres on a 2 in glass plate; (B) After being swollen with 0.65 cSt silicone oil. Color change from green to red due to the expanding lattice distance of the ccp (111) planes; (C) Reversible and repeatable color change over a wide range (>150 nm); (D) Plot of the analysis of the original and swollen films using Bragg's law; and (E) Effective refractive index, n_{eff}, of the original and swollen films and the interspacing of the ccp (111) planes.
Reproduced with permission (Ref. 10).

be seen in Figures 3.8(A) and 3.8(B), respectively. The dry opal composite film was green in color and changed to a reddish color in the swollen condition. Figure 3.8(C) compares the transmission spectroscopy before and after swelling using a volatile silicone oil. The dip position, or stop-bandgap, of the dry sample was located at 550 nm and was redshifted to 725 nm in the swollen condition. The shift of the dip position is reversible, and the original position is perfectly recovered after the solvent is evaporated from the silicone elastomer.

The dry and swollen conditions analyzed in eqn (3.1) are shown in Figure 3.8(D), and the results are tabulated in Figure 3.8(E). The average refractive index is nearly the same before and after swelling with the silicone oil, because the refractive indices of both silicone elastomer and silicone oil are of similar value. In contrast, there is a large change in the interspacing of the array of ccp (111) planes, d, before and after swelling. The expansion of the lattice distance of the ccp (111) planes reached approximately 60 nm, which was an increase of approximately 33%.

3.4 Tuning the Lattice Distance Using Mechanical Deformation

Opal composites are also prospective materials for mechanical sensing applications and provide a platform for optical strain sensors, tensile testing, and color decoration through stretching or compression. The pioneering studies in this area demonstrated that the Bragg diffraction peak shifts to a lower wavelength in response to uniaxial stretching.[19,49] From the viewpoint of a tunable stop-bandgap of a photonic crystal, Yoshino *et al.* developed a rigid colloidal crystal block assembled with polymer spheres that changed color in response to a compressive stress, as shown in Figure 3.6.[25] In this case, the reflection peak of the colloidal crystal block shifted to a higher wavelength as a function of the compression strain. Soft materials, such as nonclose-packed colloidal crystals embedded in hydrogels, have also been reported to deform more easily under compression strain.[20,21] The reflected light wavelength was tuned over a wide range of wavelengths covering the visible spectrum by applying a linear compression strain. Inverse opal structures may also be used as plastic[50] and elastic colloidal[51] crystals. Baumberg and coworkers demonstrated a polymeric opal with strain-sensitive flexible photonic crystal sheets,[52] whereas Ozin and coworkers reported e-ink elastic photonic crystals for color fingerprinting.[53]

In addition, we reported that opal composite films can also be applied as "photonic rubber sheets"[54] using elastic colloidal crystals composed of arrayed polystyrene microspheres and polydimethylsiloxane (PDMS) with tunable stop-bandgaps that respond to mechanical tension. This elastic material has ordinary rubber properties and is dry, soft, flexible, and durable. Figure 3.9 shows that the structural color of a PS colloidal opal composite film coated on a fluoride rubber sheet was changed reversibly by stretching the sheet.[10] Figure 3.9(A) shows the initial condition of the opal composite on the rubber sheet, which had a red structural color. The rubber sheet was then expanded by applying stress in the vertical direction. Figure 3.9(B) shows the opal composite deformed from the

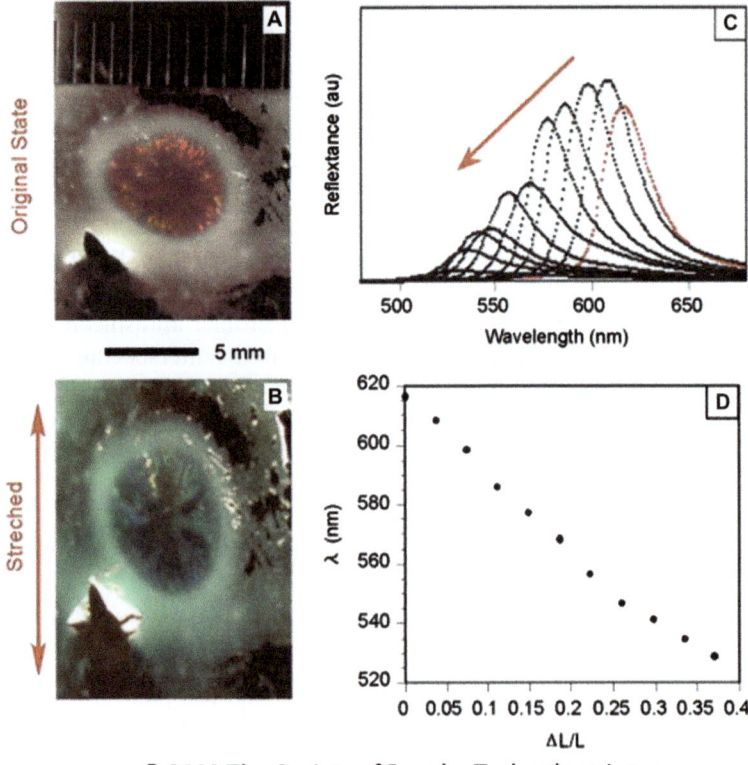

Figure 3.9   Tuning the structural color by stretching an opal composite film. The opal
film was coated on a black fluoride rubber sheet. The original red color
(Photo A) changed to green (Photo B) owing to mechanical elongation.
(C) Bragg diffraction peak shifted beyond the low wavelength; and
(D) Relationship between the peak position and the elongation.
Reproduced with permission (Ref. 10).

horizontal to the vertical ellipsoidal and its structural color change from red to
green. By releasing the applied mechanical stress, the structural color was perfectly
recovered to its initial red state. The change in the structural color due to elastic
deformation was measured as the peak shifting of the Bragg diffraction.
Figure 3.9(C) shows the spectroscopy for different strain conditions on the opal
composite. The peak shifted to a lower wavelength as a function of the expansion
ratio and its intensity decreased. Figure 3.9(D) shows the relationship between the
strain ($\Delta L/L$) and peak position (nm). The peak position decreased as a function
of the elongation. This relationship can be applied to strain-gage indicators.

3.5   Potential Applications Using Structural Color

In this final section, the use of the tunable structural color of opal composite films
for chemical sensing, color display, textile fibers, and strain imaging are discussed.

3.5.1 Chromic Materials for Sensors

Asher and colleagues explored new vistas by using hydrogel colloidal crystals as intelligent sensors that can detect variations in temperature, pH, and ion concentration.[16–18] Subsequently, several types of photonic crystal sensors have been demonstrated in recent years. From the viewpoint of response and sensitivity, an inverse opal structure has some advantages;[9,46,55] a review on inverse opals is described in another chapter written by Takeoka (Chapter 7). Here, we focus on opal photonic crystals. There are also many reports on chromic materials swollen through liquid immersion[47,48] or gas absorption.[56,57]

Figure 3.10 shows two types of opal-based sensing applications—tuning the Bragg diffraction peak and tuning the defect mode in the stop-bandgap. Figure 3.10(A) shows an opal composite film on a Si wafer and measurement using reflectance spectroscopy.[58] The composite film was composed of 175 nm PS arrayed spheres filled with the PDMS elastomer. A drop of isopropanol

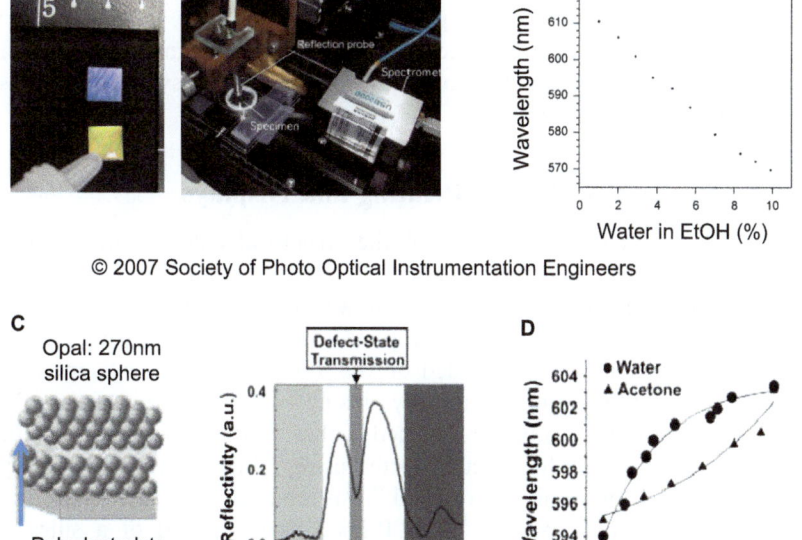

© 2007 Society of Photo Optical Instrumentation Engineers

© 2005 WILEY-VCH Verlag GmbH & Co. KGaA

Figure 3.10 Opal composite film for chromatic sensors. (A) In some solvents, the color change was easily recognized by the naked eye (Left photo). The spectroscopy was measured using a palmtop spectrometer with a reflection fiber probe (Right photo). (B) A small amount of water in isopropanol is measured based on the wavelength of the peak shift; (C) Tuning the planar defect in a photonic stop-bandgap; and (D) Tuning the planar defect and application to vapor/gas sensing. Reproduced with permission (Refs. 58 and 59).

changed its structural color from blue to red. A color change within the visible wavelength is easily recognized by the naked eye. In addition, the color change was measured as a peak shift of the wavelength of the Bragg diffraction. As shown in the photo on the right side, the reflection spectra of the opal composites were obtained using a miniature fiber optic spectrometer (Ocean Optics, USB2000). In this layout, the detector selectively received a nearly regular reflection of the Bragg diffraction from the ccp (111) planes of the specimens. Figure 3.10(B) shows the PDMS elastomer swollen with a mixture of ethanol and water. The peak position shifted to approximately 47 nm for the ethanol–water system with a water content of 0–10%. By measuring the tuning of the Bragg diffraction peak, the water concentration in the ethanol could be determined on the order of 1 vol.%.

With regard to tuning the defect mode in the stop-bandgap,[59] Figure 3.10(C) shows a tuning planar defect photonic crystal composed of a polyelectrolyte layer sandwiched between two layers of a silica opal film. This polyelectrolyte layer produced a planar defect in the stop-bandgap in the transmission spectra shown on the right side. In addition, the polyelectrolyte layer absorbed molecules of vapor. As a result, in the swollen layer, the expanded layer thickness caused a shift in the dip position in the spectrum. By measuring this dip position, the gas pressure of the molecules could be estimated, as shown in Figure 3.10(D). This result demonstrates a simple and easy gas sensor using opal photonic crystal films.

3.5.2 Structural Color for Printing and Displays

Figure 3.11 shows three examples of the structural color of image display materials, including "photonic paper",[48] "Photonic-ink or P-ink",[47] and "M-ink".[60] The photonic paper/ink system was demonstrated for stamping and coloring. Swelling phenomena were applied to color printing using structural color. The structural color depended on the molecular size of silicone oils. In addition, this printing was erasable by rinsing (volatile solvent) or could be fixed by crosslinking using an injected linker.[61] Recently, high-resolution structural color imaging has been reported using inkjet printing technology.[62]

The concept of "photonic ink or P-ink" was reported and then demonstrated as a prototype display device.[63] The P-ink was composed of a silica opal composite with a polyferrocenylsilane (PFS) organic solvent gel and cationic iron sites. This PFS organic gel was swollen by applying an electric field that affects the redox-controlled polarity of the PFS chains. The structural color change was reversible, rapid (subsecond) and covered a broad wavelength range (the entire visible region). As the applied voltage was increased, the coloration changed from green to red, as shown in the middle photos in Figure 3.11. Recently, a P-ink display device was improved using an inverse opal structure.[64]

The concept of M-ink, which is based on magnetic responsive colloidal spheres of Fe_3O_4, has also been proposed.[65] The bottom left photo shown in Figure 3.11 depicts the change in the structural color as an external magnetic

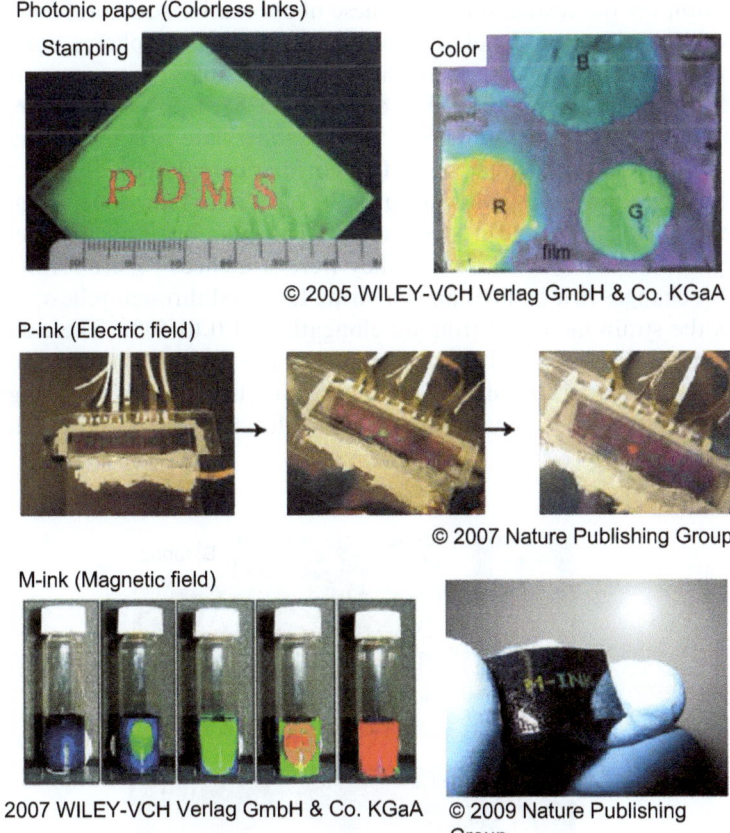

Figure 3.11   Structural color for potential printing and display applications. Photonic paper using swelling liquids, Photonic-ink (P-ink) using applied voltages on a prototype cell, and M-ink using magnetic fields on liquid inks and a flexible sheet.
Reproduced with permission (Refs. 48, 60, 63 and 65).

field is varied. This color change occurs as the distance between the colloidal spheres in the chain is modified in response to the changing magnetic field. (Note that M-ink is not a close-packed colloidal crystal.) Using M-ink on a flexible polymer sheet, a prototype magnetic color recording device was reported, as shown in the bottom right photo. M-ink is a very useful technology for magnetic recording and printing applications.

3.5.3   Color-Tunable Textile Fibers

As mentioned in Figure 3.5, the melt flow ordering process has potential advantages because it is easy to scale-up for industrial mass production. For example, Baumberg and coworkers have been developing polymeric opal fibers.[36] The fiber-shaped opal composite shown in Figure 3.7(A) is a key application for the textile industry. These fibers exhibit structural color on the basis of the self-assembly of submicrometer core–shell spheres possessing a spectrum that is stretch-tunable across the visible region, as shown in Figure 3.12. These polymeric opals consist of a hard polystyrene core coated with a thin polymer layer containing allyl-methacrylate as a grafting agent and a soft polyethylacrylate outer shell. The polymeric opal fibers (Figure 3.12(A)) and a fabric prepared from them (Figure 3.12(B)) were fabricated using existing synthetic fiber manufacturing technology. They exhibited an iridescent color and significant changes in color as they were stretched. Figure 3.12(C) shows the color changes in a 1-mm diameter string from red through yellow/green and blue as the strain increased from an elongation of 0 to 50%. These structural color changes were caused by a decrease in the interplanar distance during stretching. Figure 3.12(D) shows the Bragg diffraction peak of the string. As the string became increasingly stretched, the diffraction peak shifted to lower wavelengths.

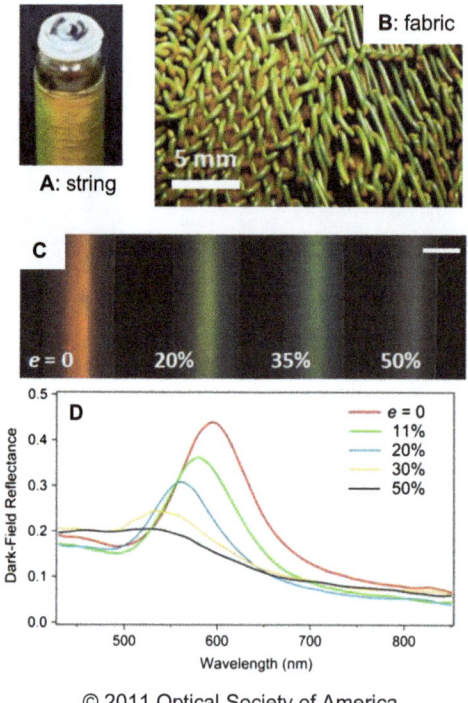

© 2011 Optical Society of America

Figure 3.12 Polymer opal fiber with tunable structural color comprising core–shell spheres. (A) String with long fibers showing iridescent structural color; (B) Knitted into fabrics with a marked stretch-variable structural color effect; (C) Dark-field images of a 1-mm diameter red opal fiber under strains of $e = 0$, 20%, 35%, and 50%; and (D) Corresponding reflectance spectra for the different strains in C.
Reproduced with permission (Ref. 36).

3.5.4 Visualization Technique for the Strain Deformation of Metal Plates

The last application is a simple and low-cost method for visualizing the local strain distribution in deformed metal plates and is shown in Figure 3.13.[66] In this study, aluminum plates were coated with an opal photonic crystal film with tunable structural color. The photonic crystal films consisted of a silicone elastomer that contained an array of submicrometer polystyrene colloidal spheres. When the aluminum sheets were stretched, the change in the spacing of the colloidal spheres in the opal film altered the color of the film. This approach could be useful as a new strain gauge with a visual indicator for detecting mechanical deformation.

Figure 3.13(A) shows an opal composite film on a black, 50-μm supporting polyethylene terephthalate (PET) sheet. From the SEM image, it was determined that the film thickness was 3.5 μm. The opal composite film was red

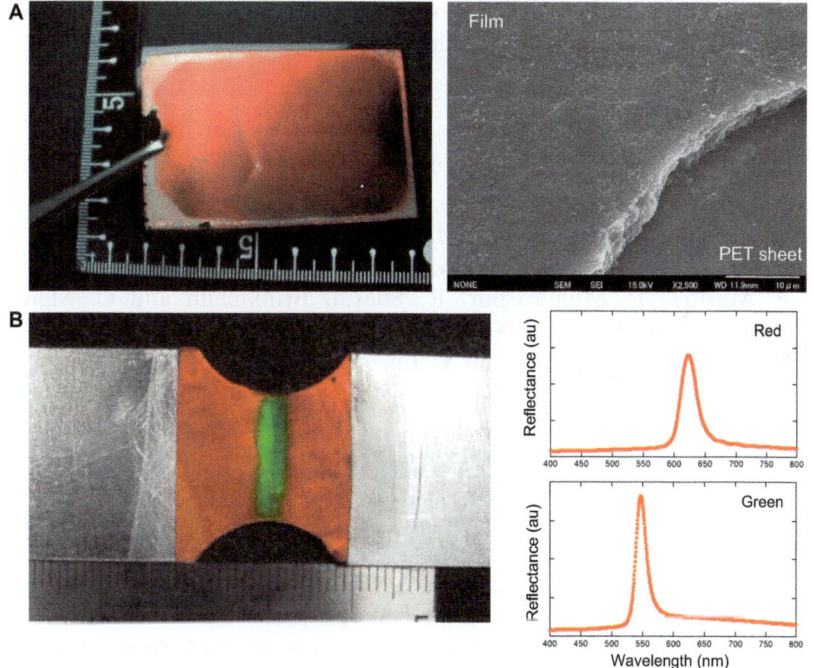

© 2012 Society of Photo Optical Instrumentation Engineers

Figure 3.13 Smart photonic coating for strain imaging of deformed metal plates. (A) Photography of a soft opal film coated on a PET sheet. The structural color is metallic red. The right SEM image shows an edge of the soft opal film with a thickness of 3.5 μm; (B) Change in the structural color in response to uniaxial elongation of the aluminum plate. Reflectance spectroscopy of the undeformed area (red-colored area, $\lambda = 623$ nm) and the deformed area (green-colored area, $\lambda = 548$ nm). Reproduced with permission (Ref. 66).

in color and the Bragg diffraction peak was at 642 nm. Here, the PET sheet plays two functions. It serves as the supporting substrate for the flexible opal composite film and as an absorbing layer for the transmitted light. The sheet is flexible and easy to use as a structural color strain indicator.

For example, to image strain distribution, the PET sheet with the opal composite film was adhered to a target aluminum plate. When the aluminum sheet was stretched with a universal testing machine, the change in the spacing of the colloidal spheres in the opal film altered the color of the film. This approach could also be useful as a new strain gauge with a visual indicator for detecting mechanical deformation. Figure 3.13(B) shows the change in the structural color due to uniaxial elongation of the aluminum plate. The photograph shows the plastic deformation at the neck area, and it can be seen that the strain area is concentrated near the dumbbell's center. Reflectance spectroscopic analysis was performed on both the deformed and undeformed areas, and it was determined that the Bragg diffraction peak shifted from 623 nm to 548 nm owing to elongation of the aluminum plate. This result demonstrates that an image of the local strain in a metal plate can be obtained using a colloidal photonic crystal film.

References

1. J. V. Sanders, *Acta Crystallogr.*, 1968, **A24**, 427.
2. W. Stober, *J. Colloid Interf. Sci.*, 1968, **26**, 62.
3. P. Gilson, *J. Gemmol.*, 1979, **16**, 494.
4. F. Marlow, P. Muldarisnur, R. Sharifi, Brinkmann and C. Mendive, *Angew. Chem. Int. Ed.*, 2009, **48**, 6212.
5. Y. Xia, B. Gates, Y. Yin and Y. Lu, *Adv. Mater.*, 2000, **12**, 693.
6. C. López, *Adv. Mater.*, 2003, **15**, 1679.
7. G. A. Ozin and A. C. Arsenault, *Nanochemistry, 2nd edn*, RCS Publishing, Cambridge, 2008.
8. K. Busch, S. Lölkes, R. B. Wehrspohn, and H. Föll (ed.), *Photonic Crystals: Advances in Design, Fabrication, and Characterization*, Wiley-VCH Verlag GmbH & Co. KGaA, Weinheim, FRG, 2006.
9. M. Harun-ur-rashie, T. Seki and Y. Takeoka, *Chem. Rec.*, 2009, **9**, 87.
10. H. Fudouzi, *Adv. Powder. Tech.*, 2009, **20**, 502.
11. J. Zhang, Z. Sun and B. Yang, *Curr. Opin. Colloid. Interf. Sci.*, 2009, **14**, 103.
12. Y. Zhao, X. Zhao and Z. Gu, *Adv. Funct. Mater.*, 2010, **20**, 2970.
13. S. H. Kim, S. Y. Lee, S. M. Yang and G. R. Yi, *NPG Asia Mater.*, 2011, **3**, 25.
14. P. Jiang, J. F. Bertone, K. S. Hwang and V. L. Colvin, *Chem. Mater.*, 1999, **11**, 2132.
15. Z. Z. Gu, A. Fujishima and O. Sato, *Chem. Mater.*, 2002, **14**, 760.
16. J. M. Weissman, H. B. Sunkara, A. S. Tse and S. A. Asher, *Science*, 1996, **274**, 959.

17. S. A. Asher, J. H. Holtz, J. M. Weissman and G. Pan, *MRS Bull.*, 1998, **11**, 44.
18. J. H. Holtz and S. A. Asher, *Nature*, 1997, **389**, 829.
19. J. M. Jethmalani and W. T. Ford, *Chem. Mater.*, 1996, **8**, 2138.
20. Y. Iwayama, J. Yamanaka, Y. Takiguchi, M. Takasaka, K. Ito, K.T. Shinohara, T. Sawada and M. Yonese, *Langmuir*, 2003, **19**, 977.
21. S. H. Foulger, P. Jiang, A. Lattam, D. W. Smith, J. Ballato, D. E. Dausch, S. Grego and B. R. Stoner, *Adv. Mater.*, 2003, **15**, 685.
22. For example, ftp://kxcad.net/Optical/BeamProp/bandsolve.pdf.
23. K. Yoshino, Y. Shimoda, Y. Kawagishi, K. Nakayama and M. Ozaki, *Appl. Phys. Lett.*, 1999, **75**, 932.
24. M. Ozaki, Y. Shimoda, M. Kasano and K. Yoshino, *Adv. Mater.*, 2002, **14**, 514.
25. K. Yoshino, Y. Kawagishi, M. Ozaki and A. Kose, *Jpn. J. Appl. Phys., Lett.*, 1999, **38**, L786.
26. B. Hatton, L. Mishchenko, S. Davis, K. H. Sandhage and J. Aizenberg, *Proc. Nature Acad. Sci.*, 2010, **107**, 10354.
27. J. Zhou, J. Wang, Y. Huang, G. Liu, L. Wang, S. Chen, X. Li, D. Wang, Y. Song and L. Jiang, *NPG Asia Materials*, 2012, **4**, e21.
28. M. Egen and R. Zentel, *Macromol. Chem. Phys.*, 2004, **205**, 1479.
29. X. Hu, Y. Thomann, R. J. Leyrer and J. Rieger, *Polym. Bull.*, 2006, **57**, 785.
30. N. D. Denkov, O. D. Velev, P. A. Kralchevsky, I. B. Ivanov, H. Yoshimura and K. Nagayama, *Langmuir*, 1992, **8**, 3183.
31. H. Fudouzi, *J. Colloid Interf. Sci.*, 2004, **275**, 277.
32. H. Fudouzi, *Colloids Surf. A*, 2007, **311**, 11.
33. T. Ruhl and G. P. Hellmann, *Macromol. Chem. Phys.*, 2001, **202**, 3502.
34. B. Viel, T. Ruhl and G. P. Hellmann, *Chem. Mater.*, 2007, **19**, 5673.
35. W. Wohlleben, F. W. Bartls, S. Altmann, R. J. Leyrer, C. E. Finlayson, P. Spahn, D. R. E. Snoswell, G. Yates, A. Kontogeorgos, A. I. Haines, G. P. Hellmann and J. J. Baumberg, *Adv. Mater.*, 2011, **20**, 1540.
36. C. E. Finlayson, C. Goddard, E. Papachristodoulou, D. R. E. Snoswell, A. Kontogeorgos, P. Spahn, G. P. Hellmann, O. Hess and J. J. Baumberg, *Opt. Exp.*, 2011, **19**, 3144.
37. O. D. Velev, A. M. Lenhoff and E. W. Kaler, *Science*, 2000, **287**, 2240.
38. T. Kanai, D. Lee, H. C. Shum, R. K. Shah and D. A. Weitz, *Adv. Mater.*, 2010, **22**, 4998.
39. Y. Zhao, X. Zhao, J. Hu, M. Xu, W. Zhao, L. Sun, C. Zhu, H. Xu and Z. Gu, *Adv. Mater.*, 2009, **21**, 569.
40. J. Kim, Y. Song, L. He, H. Kim, H. Lee, W. Par, Y. Yin and S. Kwon, *Small*, 2011, **7**, 1163.
41. C. I. Aguirre, E. Reguera and A. Stein, *Adv. Funct. Mater.*, 2010, **20**, 2565.
42. J. F. Galisteo-Lopez, M. Ibisate, R. Sapienza, L. S. Froufe-Perez, U. Blanco and C. Lopez, *Adv. Mater.*, 2011, **23**, 30.
43. S. Furumi, H. Fudouzi and T. Sawada, *Laser Photon. Rev.*, 2010, **4**, 205.
44. J. Ge and Y. Yin, *Angew. Chem., Int. Ed. Engl.*, 2011, **50**, 1492.

45. Y. Zhao, Z. Xie, H. Gu, C. Zhu and Z. Gu, *Chem. Soc. Rev.*, 2012, **41**, 3297.
46. Y. J. Lee and P. V. Braun, *Adv. Mater.*, 2003, **15**, 563.
47. A. C. Arsemault, H. Miguez, V. Kitaev, G. A. Ozin and I. Manners, *Adv. Mater.*, 2003, **15**, 503.
48. H. Fudouzi and Y. Xia, *Adv. Mater.*, 2003, **15**, 892.
49. S. A. Asher, J. Holtz, L. Liu and Z. Wu, *J. Am. Chem. Soc.*, 1994, **116**, 4997.
50. K. Sumioka, H. Kayashima and T. Tsutsui, *Adv. Mater.*, 2002, **14**, 1284.
51. J. Li, Y. Wu, J. Fu, Y. Cong, J. Peng and Y. Han, *Chem. Phys. Lett.*, 2004, **390**, 285.
52. O. L. J. Pursiainen, J. J. Baumberg, K. Ryan, J. Bauer, H. Winkler, B. Viel and T. Ruhl, *Appl. Phys. Lett.*, 2005, **87**, 101902.
53. A. C. Arsenault, T. J. Clark, G. Von Freymann, L. Cademartiri, R. Sapienza, J. Bertolotti, E. Vekris, S. Wong, V. Kitaev, I. Manners, R. Z. Wang, S. John, D. Wiersma and G. A. Ozin, *Nature Mater.*, 2006, **5**, 179–184.
54. H. Fudouzi and T. Sawada, *Langmuir*, 2006, **22**, 1365.
55. R. A. Barry and P. Wilzius, *Langmuir*, 2006, **22**, 1369.
56. A. C. Arsemault, V. Kitaev, I. Manners, G. A. Ozin, A. Mihi and H. Miguez, *J. Mater. Chem.*, 2005, **15**, 133.
57. E. Tian, J. Wang, Y. Zheng, Y. Song, L. Jiang and D. Zhu, *J. Mater. Chem.*, 2008, **18**, 1116.
58. H. Fudouzi and T. Sawada, *Proc. SPIE*, 2007, **6767**, 676704–4.
59. N. Tereault, A. C. Arsenault, A. Mihi, S. Wong, V. Kitaev, I. Manners, H. Miguez and G. A. Ozin, *Adv. Mater.*, 2005, **17**, 1912.
60. H. Kim, J. Ge, J. Kim, S. Choi, H. Lee, H. Lee, W. Park, Y. Yin and S. Kwon, *Nature Photon.*, 2009, **3**, 535.
61. H. Fudouzi and Y. Xia, *Langmuir*, 2003, **19**, 9653.
62. P. Kang, O. S. Ogunbo and D. Erickson, *Langmuir*, 2011, **27**, 9676.
63. A. C. Arsenault, D. P. Puzzo, I. Manners and G. A. Ozin, *Nature Photon.*, 2007, **1**, 468.
64. D. P. Puzzo, A. C. Arsenault, I. Manners and G. A. Ozin, *Angew. Chem. Int. Ed.*, 2009, **48**, 943.
65. J. Ge, Y. Hu and Y. Yin, *Angew. Chem. Int. Ed.*, 2007, **46**, 7428.
66. H. Fudouzi, T. Sawada, Y. Tanaka, I. Ario, T. Hyakutake and I. Nishizaki, *Proc. SPIE*, 2012, **8345**, 83451S.

CHAPTER 4

Tuning Color and Chroma of Opal and Inverse Opal Structures

DAVID JOSEPHSON AND ANDREAS STEIN*

Department of Chemistry, University of Minnesota, Minneapolis,
MN 55455, USA, Email: josep193@umn.edu
*Email: a-stein@umn.edu

4.1 Introduction

Photonic bandgap materials, first introduced by Yablonovitch[1] and John,[2] have been widely studied in the past two decades. These materials contain periodic arrays of high and low dielectric zones that form photonic Bragg diffraction planes. Diffraction by these planes serves to inhibit the propagation of photons of specific wavelengths through the material. Such materials are commonly called photonic crystals because the periodic structure at optical length scales is analogous to the structure of crystalline materials at atomic length scales. However, whereas crystalline materials diffract X-rays, the dimensions of periodic features in photonic crystals cause the diffraction of longer-wavelength photons in and around the visible range of the electromagnetic spectrum. The spectral position of the diffracted wavelengths corresponds to the energy of the photonic bandgap. If, for a specific geometry, the refractive-index contrast between the low and high dielectric constant phases is below a certain threshold, an incomplete bandgap may be found that is referred to as a photonic "stop-band".

RSC Smart Materials No. 5
Responsive Photonic Nanostructures: Smart Nanoscale Optical Materials
Edited by Yadong Yin
Published by the Royal Society of Chemistry, www.rsc.org

The wavelengths of reflected light caused by this phenomenon can imbue a photonic crystal with color. Because this color effect is based purely on structural characteristics that cause diffraction within the photonic crystal, materials that are colored in this way are said to possess "structural color". Many other, more common, mechanisms for color generation have long been known. Mechanisms such as absorption, birefringence, emission, and numerous others all interact with visible light in different ways to produce color.[3] A majority of these mechanisms arise due to the chemical nature of the materials themselves that can absorb, separate, or emit wavelengths of light based on interaction at an atomic level. The color characteristics of these materials cannot be altered without altering the chemical composition of the material itself. Structural color, however, allows a great deal of control over the color expression of several different material compositions that would otherwise be limited to their natural color state. The degree of tunability that is inherent in structurally colored materials also provides opportunities for new classes of optical sensors and pigments based on dynamic color changes.

Many different examples of structural color are found in nature. For example, the wings of Morpho butterflies,[4] the setae of a species of polychaete worm,[5] and several species of plants[6] possess photonic nanostructures that create structural color and can impart high visibility at specific wavelengths over long distances. Inorganic examples of structural color also exist in minerals such as labradorite,[7] which exhibits optical interference due to the stacking of lamellar layers. Perhaps, the most well-known natural example of a photonic crystal is the opal.[8]

Opals, despite consisting of colorless silica, exhibit intense flashes of color that can span the entire visible spectrum at varying angles of observation. These colors are the result of the face-centered cubic (fcc) array of silica spheres that make up the opal. The periodically varying high and low dielectric zones of silica and air create diffraction planes throughout the material that produce photonic stop-bands at specific wavelengths, depending on the sphere diameter. Because natural opals often contain different planar orientations within numerous crystallite zones, multiple colors can be seen that change as the angle of observation changes. This seemingly random flashing of color is referred to as opalescence.

The fcc structure of natural opals has been replicated using a variety of colloidal materials and assembly methods (Figure 4.1). The simplest method involves a suspension of colloidal spheres that is left to sediment by gravity[9] or centrifuged at a low speed. Over time, the individual particles self-assemble into a predominantly fcc array as they become close-packed.[10–13] This technique is suited for producing relatively large amounts of opaline products in a short period of time. However, natural sedimentation times are typically long, and sedimentation by centrifugation produces colloidal crystals with a relatively large number of defects. To create well-ordered fcc colloidal crystals with fewer defects, other techniques can be used. These techniques include vertical deposition,[14,15] shear-induced order,[16–18] electrophoresis,[19,20] directional growth by charge induction,[21] float packing,[22] and template directed

Figure 4.1   Scanning electron micrographs of a synthetic opal composed of uniformly
sized polymer spheres. Left: The (111) plane of the opal. Note the highly
periodic array of spheres that create Bragg diffraction planes within the
material. Right: An incomplete opal showing the 3D structure of the
material.

Figure 4.2   Scanning electron micrograph of inverse opal zirconia.

growth.[23–25] Most of these techniques are best suited for producing thin opal
films. After their assembly into an opaline structure, the sphere arrays possess a
photonic band structure[26] and are opalescent with color properties similar to
natural opals.

Another type of photonic crystal that is closely related to the opal is the
inverse replica of an opal: an "inverse opal". These structures are created by
using an opal as a template and filling in the interstitial spaces around the opal
spheres. The colloidal spheres are removed and replaced with interconnected
air spheres surrounded by a solid wall where the interstitial spaces were
previously located (Figure 4.2). For the purpose of color generation, these two
types of photonic crystal share similar properties and can be analyzed *via*
similar methods and equations that will be discussed in Section 4.2 for the
generation of their structural color.

Several structural characteristics of opals and inverse opals allow for the dynamic color tuning of these materials. For instance, changes to lattice spacing (d_{hkl}), refractive index of the walls (n_{walls}), and refractive index of the void-filling material (n_{voids}) can all be employed in order to alter the stop-band position of the photonic crystal. The ability to dynamically and reversibly color tune these structures makes them special among colored materials. The inherent structural color also does not undergo photobleaching like organic dyes, so that longer-lasting and more consistent colors can be achieved. The characteristics of structural color are governed by several factors, but the color and chroma of a photonic material depend primarily on stop-band position and stop-band shape, which will be discussed in detail in the following section. The manifestation of this color mechanism within opal and inverse opal photonic crystals and their applications will be the focus of this chapter, followed by a discussion on the inverse opal structure and its beneficial features for sensors and pigmenting applications. The chapter will conclude with a discussion on methods to improve optical signals and color manifestation.

4.2   Stop-Band Features in Opal and Inverse Opal Photonic Crystal

4.2.1   Stop-Band Position

The position of the photonic stop-band in both opals and inverse opals can be estimated on the basis of combining Bragg's law of diffraction with Snell's law of refraction. However, it is more accurately described using dynamical diffraction theory.[27–29] While simple calculations using Bragg's and Snell's laws can provide close estimates for the bandgap position and are often adequate for designing photonic-crystal-based pigments or sensors with specific colors, dynamical diffraction theory accounts for interactions between incident and diffracted wave scattering and corrects for these events. The modified Bragg's law for first order diffraction in a photonic crystal is shown in eqn (4.1),

$$\lambda_B = 2n_{avg}d_{hkl}\sin\theta \qquad (4.1)$$

where λ_B is the bandgap position, d_{hkl} is the spacing between lattice planes, n_{avg} is the average refractive index of the material, and θ is the angle of measurement/observation from normal. The value of n_{avg} can be calculated by multiplying the refractive index (n) of each dielectric material with its respective volume fraction (φ) in the chosen photonic crystal (eqn (4.2)).

$$n_{avg} = \sum n_i \varphi_i \qquad (4.2)$$

The majority of materials discussed in this chapter exhibit an fcc lattice structure. The lattice spacing term (d_{hkl}) can be calculated for an fcc lattice from eqns (4.3) and (4.4),

$$d_{hkl} = \frac{a}{\sqrt{h^2 + k^2 + l^2}} \qquad (4.3)$$

$$a = D\sqrt{2} \tag{4.4}$$

where a is the unit cell parameter and h, k, and l are the Miller indices of the chosen diffraction plane. The stop-band with the highest intensity for an fcc lattice corresponds to the (111) set of diffraction planes. The main color expression for these fcc materials will be governed by the position of that stop-band. For most fcc photonic crystals with stop-bands in the visible region, the only diffraction peaks expressed in the visible spectrum according to UV-vis spectra are the (111) and (200) with the main reflection coming from the (111) diffraction planes, and the (200) stop-band being close enough to the (111) as to be mostly indistinguishable. However, for materials with a (111) stop-band that is sufficiently redshifted into the longer-wave visible region, it should be noted that other diffraction peaks, such as the (220) peak, can enter the visible region and influence color properties (Figure 4.3).

Another property of structurally colored materials is their color dependence on the incident angle of light. Equation (4.1) shows that by altering the incident angle of light, or the angle of observation relative to the incident light, the stop-band wavelength is affected and the perceived color changes. This effect is lessened by creating photonic powders that possess randomized crystal domain orientations and small domain sizes. The Bragg equation can be simplified for these powders by assuming the angle dependence is averaged to normal (90°) due to the myriad of planar orientations exhibiting first order diffraction (eqn (4.5)).

$$\lambda_{\text{powder}} = 2d_{hkl}n_{\text{avg}} \tag{4.5}$$

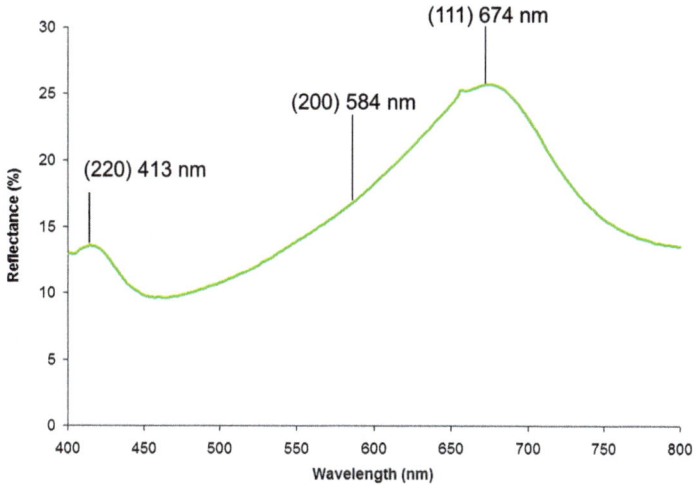

Figure 4.3 UV-vis spectrum of inverse opal silica with a (111) stop-band in the red region. A strong second reflection (220) can be seen at shorter wavelengths. A weak shoulder on the (111) corresponds to the (200) diffraction peak. The positions of the (200) and (220) peaks were calculated using the (111) position and an established fill fraction value for the solid component.

The stop-band position, as calculated using eqn (4.1), can be corrected by dynamical diffraction theory[27–29] using eqn (4.6),

$$\lambda_{\mathrm{D}} = \lambda_{\mathrm{B}}\left(1 - \frac{\psi_0'}{2\sin^2\theta}\right) \tag{4.6}$$

where ψ_0', the real portion of the crystal polarizability, is calculated with φ representing the high dielectric volume fraction and m being the ratio of the refractive index of both dielectric materials (eqn (4.7)).

$$\psi_0' = 3\varphi\frac{(m^2 - 1)}{(m^2 + 2)} \tag{4.7}$$

4.2.2   Stop-Band Shape and Intensity

While the stop-band position of a photonic crystal dictates the color expressed by the material, the shape of the stop-band, as seen in the UV-vis spectra of photonic crystals, governs the purity and intensity of that color. The shape of the peak can be broken down into two vital characteristics, the bandwidth and the reflection intensity.

The bandwidth of the stop-band in a photonic crystal affects the color expression of the crystal by including or excluding the wavelengths around the stop-band position. If the bandwidth is small, few wavelengths are included within the stop-band and the color expression will be limited to a small portion of the visible spectrum. The structural contribution to the perceived color of the material is limited to those few, grouped wavelengths, allowing the manifestation of specific colors found in the electromagnetic spectrum. If the stop-band is broad, however, the perceived color is more influenced by the wavelengths around the stop-band position, leading to color mixing as the stop-band extends out from the center. Kanai *et al.*[30] have proposed that increases in the width of the stop-band in a photonic crystal are likely the result of inhomogeneities in d-spacing or refractive index within the crystal layers, with the values for these properties fluctuating around a value that represents a perfectly homogeneous crystal. This fluctuation produces an effect that can be approximated as several overlapping stop-bands with small changes in n_{avg} and d_{hkl} giving rise to small shifts up or down in the spectrum, but always centered about the homogeneous crystal stop-band.

The shape of the stop-band is also dictated by the diffraction intensity,[28] which determines the relative magnitude of the stop-band. The ratio of diffraction intensity (I_H) and incident intensity (I_0) for a set of planes H can be expressed as

$$\frac{I_H}{I_0} = L - (L^2 - 1)^2 \tag{4.8}$$

where

$$L = y^2 + g^2 + \sqrt{(y^2 - g^2 - 1)^2 + 4g^2 y^2} \qquad (4.9)$$

with the components y and g described as

$$y = \frac{\psi_0' + 2\left(\frac{\lambda}{\lambda_B} - 1\right) \sin^2 \theta}{K_{\sigma,\pi} \psi_0' f} \qquad (4.10)$$

and

$$g = \frac{\psi_0''}{K_{\sigma\pi} \psi_0' f} \qquad (4.11)$$

In these expressions, λ is the wavelength of the incident light, ψ_0'' is a term that accounts for losses in intensity caused by scattering, and, for polarizations normal or parallel to the diffraction plane, $K_\sigma = 1$ and $K_\pi = \cos 2\theta$ can be used, respectively. The scattering factor (f) for spherical particles of radius a is provided by

$$f = \frac{3}{u^3} (\sin u - u \cos u) \qquad (4.12)$$

with

$$u = \frac{4\pi a}{\lambda} \sin \theta \qquad (4.13)$$

These equations show that the intensity of the stop-band is largely dependent on refractive-index contrast of the dielectric materials, with high contrast increasing the intensity of the reflection.

4.3 Dynamic Color Changes in Opaline Photonic Crystals

The structural color properties of opaline photonic crystals can be harnessed for applications that require optical switching or color shifting. The ability to dynamically change colors in response to mechanical stresses as well as chemical or environmental changes opens up numerous possibilities in regards to applications such as pigmentation and sensing.

The ability to alter the stop-band position of photonic crystals is dependent on two major properties of the material; the d-spacing between diffraction layers and the refractive-index contrast between those layers. The refractive-index contrast can be further broken down into the refractive index of the solid material portion and the void portion of the photonic crystal. Changes in any of these properties will lead to a measurable shift in the stop-band position according to eqn (4.1). This dynamic tunability will be discussed herein, emphasizing simple mechanochromic, thermochromic, chemochromic, and solvochromic responses of opaline materials and how they can be achieved, with a few relevant examples of nonclose-packed, body-centered cubic (bcc) structures as well.

4.3.1   Chromic Changes in Opal Structures by Modification of Lattice Parameters

In order to dynamically and reversibly change the lattice parameters of a photonic crystal, the material must be elastic and deformable. The necessary stretching or swelling of the material would break more inflexible materials, resulting in a degradation of the periodicity and stability of the photonic crystal and limiting its reuse. Indeed, opal structures themselves are relatively weak due to the fragile contact points between spheres within the structure. To remedy this, many deformable photonic crystals are made from colloidal crystal arrays encapsulated within a hydrogel or polymer matrix that fills the void space surrounding the colloidal crystal. The matrix serves to both lock in the colloidal particles and act as a medium for altering the lattice parameters of the crystal. The infilling matrix also allows for nonclose-packed lattices when the medium surrounding the electrostatically separated colloidal particles is polymerized *in situ*. The colloidal crystal encapsulated within such a matrix exhibits a stop-band that depends on the lattice spacing of the crystal. When the lattice is deformed, the stop-band position is changed to reflect the change in *d*-spacing within the material. This color change in response to an imposed stress is referred to as mechanochromism.

Asher *et al.*[31] first demonstrated a technique for encapsulating a photonic crystal inside a hydrogel by combining acrylamide, N,N'-methyl-enebis(acrylamide), and N-vinylpyrolidone with a polystyrene colloidal suspension and photopolymerizing the mixture, trapping the colloidal crystal particles in the hydrogel. The authors then demonstrated a decrease in diffraction wavelength as the film was uniaxially stretched. However, hydrogel-encapsulated colloidal crystal arrays are composed of ~30% water and, as such, are relatively fragile and possess varied stop-band positions with varying water content.

More robust colloidal crystals embedded within a thermoplastic polymer matrix have also been synthesized and exhibit mechanochromic properties. Silica spheres grafted with 3-(trimethoxysilyl)propyl methacrylate were embedded inside a poly(methyl acrylate) matrix. The resulting film exhibited a blueshift of ~50 nm as the film was stretched by 35%. However, the film only returned to its original length after a period of 2–4 h and could not withstand repeated stretching without tearing.[32]

A fully reversible mechanochromic effect was demonstrated by photopolymerization of poly(ethylene glycol) methacrylate and polystyrene spheres, forming a hydrogel with a nonclose-packed fcc array of polystyrene spheres.[33] The hydrogel was then fully dehydrated and reinfiltrated with 2-methoxyethyl acrylate along with a small amount of ethylene glycol dimethacrylate and a photoinitiator. A final photopolymerization produced a hydrogel photonic crystal that exhibited stop-band shifts of up to 93 nm under mechanical compression with an immediate return to its original stop-band position after the release of compression.[34] Repeated straining of the photonic crystal composite caused no structural damage due to the water-free highly crosslinked polymer matrix. A later study expanded this range of stop-band positions to 172 nm using tension and compression imparted by a piezoelectric actuator.[35] Another fully reversible mechanochromic effect was realized by embedding a colloidal crystal inside a poly(dimethylsiloxane) elastomer. The vivid color change associated with the imposed mechanical strain can be seen in Figure 4.4.[36]

By using low particle and gel concentrations in a hydrogel-embedded colloidal crystal, it is possible to synthesize a structure with a mechanochromically tunable stop-band that can be adjusted over most of the visible spectrum by compression of the material. Low particle concentration, along with careful control of any ionic species, enables the particles to experience electrostatic repulsions at longer distances in solution. The low gel concentration present during polymerization enables greater compression, and as a result, the stop-band of the photonic crystal matrix could be shifted by ~350 nm in a bcc crystal (Figure 4.5).[37]

The mechanical stresses that cause a mechanochromic response in photonic crystals are typically uniaxial or biaxial, induced by a single compression or tension on one or more axes. Three-dimensional changes in lattice parameters in an opaline photonic crystal can be accomplished by modifying the hydrogel or polymer filling matrices to respond to temperature or chemical environments by swelling or deswelling of the photonic crystal, changing lattice parameters without applied mechanical stress. Thermochromic, chemochromic, and solvochromic materials have been synthesized that take advantage of this response mechanism.

To create a thermochromic responsive photonic crystal, the photonic crystal array is embedded in a matrix material whose volume depends on temperature. Using a nonclose-packed bcc colloidal crystal array inside a poly(N-isopropylacrylamide) matrix, Asher and coworkers[38] fabricated a

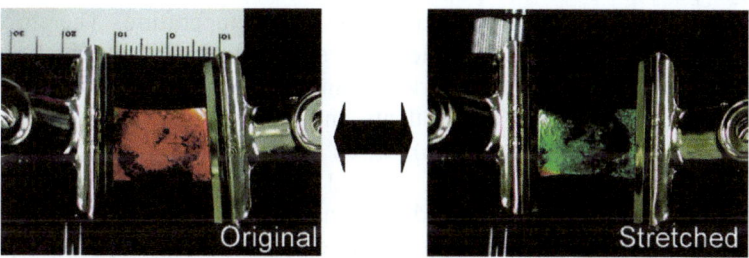

Figure 4.4 A colloidal crystal array of polymer spheres embedded in a poly(dimethylsiloxane) deformable matrix is stretched unixialy, resulting in a blueshifted stop-band position and a color shift from red to green. Reproduced with permission from Ref. 36, copyright 2005, the American Chemical Society.

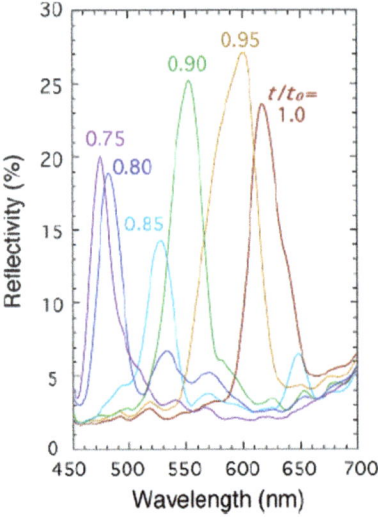

Figure 4.5   Reflectance spectrum of an encapsulated array of silica spheres being
deformed by mechanical stress. The terms t and t_0 represent the thickness
of the gel after and before compression.
Reproduced with permission from Ref. 37, copyright 2003, the American
Chemical Society.

responsive photonic crystal that alternated between a hydrated and dehydrated
state within a 25 °C temperature shift. During the temperature-induced dehy-
dration of the polymer, water was expelled from the matrix, shrinking the
embedding material and blueshifting the stop-band position as the lattice
parameters were shortened. This effect was reversible, and the color of the
photonic crystal could be effectively tuned throughout most of the visible
spectrum.

Thermoinitiated volume changes have also been realized in assembled
hydrogel colloidal particles without a functional encapsulating matrix.[39] The
hydrogel particles undergo a phase transition to a cloudy colorless liquid upon
heating to a specific volume phase transition temperature. Therefore, the
chromic response seen in these photonic crystals does not consist of a major
shift in stop-band position but rather an order–disorder transition as the
shrinking particles lose their periodicity and the material becomes more fluid
after water is expelled (Figure 4.6). Upon cooling, the crystal spontaneously
reorders and regains its original colored appearance. This order–disorder
transition can potentially be used for on-off optical switching at different
transition temperatures, depending on the composition of the hydrogel spheres.

Swelling and shrinking can also be induced within a colloidal crystal by a
chemical reaction. The mechanism for this swelling can be specifically tailored
to occur only in the presence of certain compounds or elements, thus enabling
colloidal photonic crystals to be used as optical sensors. This requires the use of
chemical recognition agents that undergo a chemical change in response to the

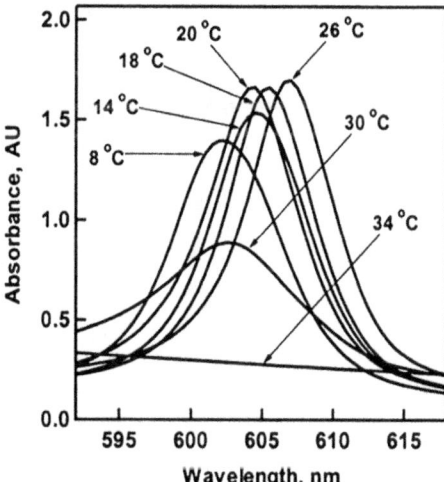

Figure 4.6 The stop-band peaks of an opal hydrogel at different temperatures leading up to the lower critical solution temperature of $\sim 32\,°C$. At higher temperatures, the crystal becomes disordered and the stop-band disappears. Reproduced with permission from Ref. 39, copyright 2000, the American Chemical Society.

presence of a specific analyte. For instance, Asher and coworkers[40–41] created a polymerized colloidal crystal array that was copolymerized with the crown ether, 4-acryloylaminobenzo-18-crown-6. The crown ether selectively bound Pb^{2+}, Ba^{2+}, and K^+, which, in turn, caused the hydrogel to swell due to the osmotic pressures within the gel. The magnitude of the swelling, and the subsequent change in d-spacing of the material, increased with increasing concentration of the analyte. A maximum lattice size was achieved when the crown ethers were saturated, with the lattice beginning to shrink with analyte supersaturation due to a decrease in osmotic pressure at high ionic strengths. This swelling behavior was reversible after the ions were extracted using deionized water. The behavior of these sensors at higher ionic strengths was later improved by incorporating nitrophenol groups that dissociated to form soluble phenolates, further swelling the hydrogel, and offsetting the shrinkage seen at high ionic strengths.[42] Sensors for pH[43] and the detection of specific compounds, such as glucose,[44] have been synthesized, as well as other sensors that employ novel molecularly imprinted hydrogels with high specificity for chosen analytes,[45–48] such as bisphenol A (Figure 4.7).[49]

Solvochromic changes in lattice parameters can also be achieved by swelling an immobilizing gel or polymer with solvent. Solvochromic responses such as these can be used in reversible photonic papers,[50,51] where a localized fluid infiltration into a poly(dimethylsiloxane) matrix embedded with a colloidal crystal changes the lattice constant of the material *via* swelling. A more permanent way to selectively alter lattice constants inside a photonic crystal gel has also been realized by creating a gel-immobilized colloidal crystal by

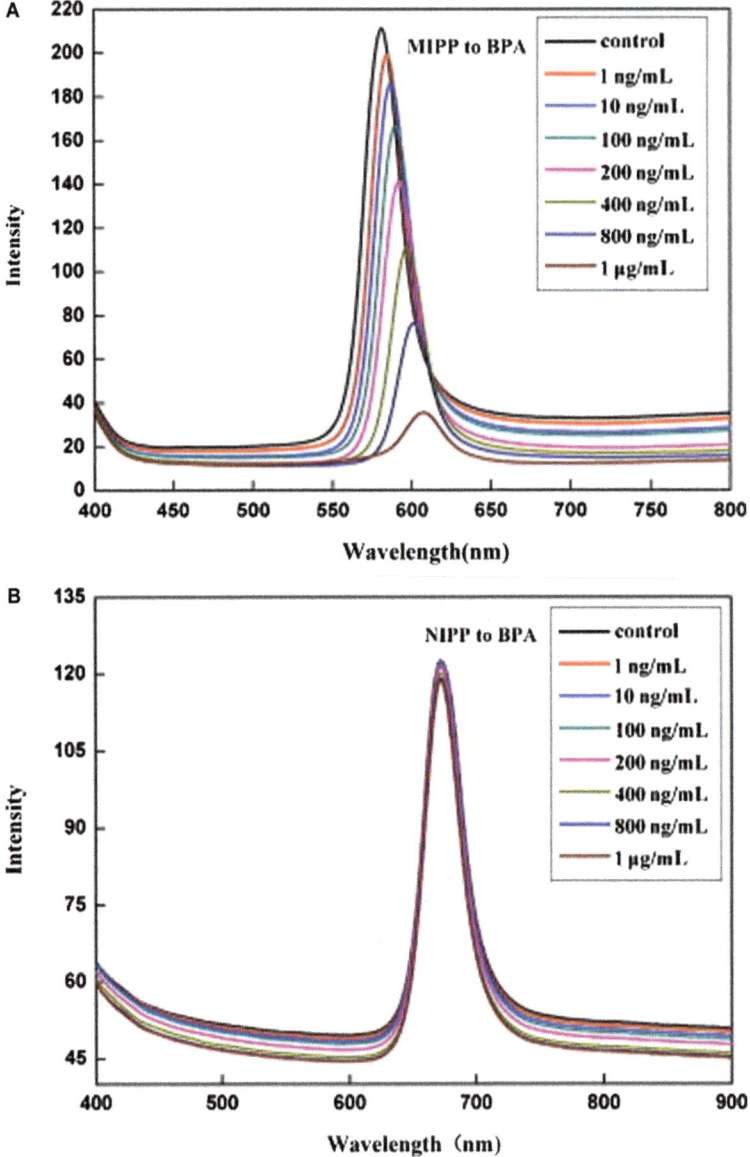

Figure 4.7   Reflectance spectra of a molecularly imprinted photonic crystal (MIPP) that is selective for bisphenol A and a nonimprinted photonic crystal (NIPP) when exposed to varying levels of bisphenol A.
Reproduced with permission from Ref. 49, copyright 2012, Elsevier.

photopolymerization. Subsequent swelling and photopolymerization using both positive and negative photomasks on specific locations of the film create regions with different color properties based on exposure to photopolymerization.[52]

Ionic liquids have also been proposed as swelling agents in gel-immobilized colloidal crystals as a nonvolatile alternative to other polar solvents. Ionic liquids not only eliminate swelling irregularities due to evaporation but, in some cases, also produce stop-band shifts greater than those seen after swelling with water or other polar solvents.[53]

4.3.2 Chromic Changes in Opal Structures by Modification of Refractive Indices

While the swelling events described above are responsible for the reversible color changes, they also modulate the bulk dimensions of a material. In applications where bulk changes cannot be tolerated, it is necessary to employ materials that have chromic responses due solely to refractive-index changes. The simplest approach to changing the refractive index within an opaline structure is to infiltrate a dry opal with various solvents.[54,55] This infiltration differs from solvent-induced swelling due to the lack of lattice changes upon infiltration. The replacement of air in the crystal void space with a higher refractive index solvent provides a simple redshift of the stop-band with no swelling or shrinking of the periodic structure. This allows for fully reversible color-changing properties as long as all infiltrated solvent can be evaporated or otherwise removed from the structure after infiltration. Due to this chromic effect, it is possible to determine the composition of mixed liquids with different refractive indices (Figure 4.8).[56]

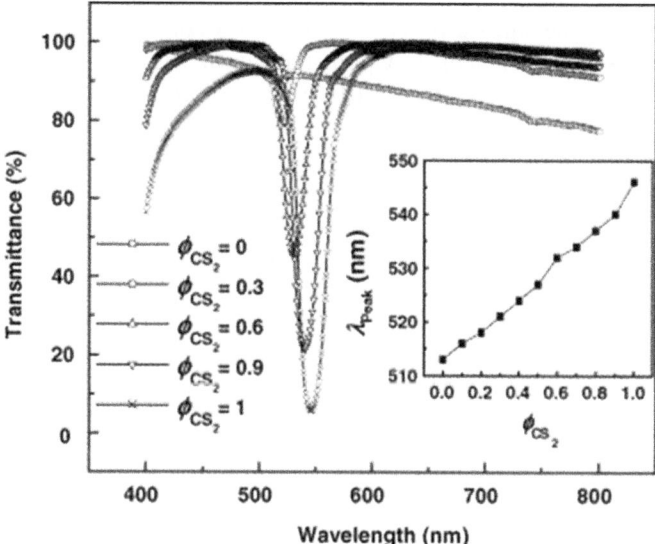

Figure 4.8 Transmission spectra of a silica opal consisting of 232 nm spheres infiltrated with a mixed liquid of ethanol and carbon disulfide. The stop-band is shifted to longer wavelengths with increasing carbon disulfide concentration.
Reproduced with permission from Ref. 56, copyright 2006, Elsevier.

The porous structure of opaline photonic crystals also serves as a convenient platform for the condensation of vapors. Vapors that condense within the void spaces of the crystal change the refractive index as condensation proceeds and can therefore be detected by observing the red-shift of the stop-band.[57] Mesoporous spheres can also be used in the colloidal crystal array to encourage condensation within the mesopores rather than in the void space between spheres.[58]

Phase changes in liquids infiltrated into opaline photonic crystals can be detected by observing a shift in the stop-band position upon a liquid–solid phase transition. For a silica opal with 292-nm spheres infiltrated with *para*-xylene, a stop-band shift of 24 nm was observed after a phase change from liquid to solid, providing a simple, temperature-based switching mechanism. Similar shifts of lesser magnitude were also seen using *tert*-butyl alcohol, water, and diphenyl ether as infiltrating solvents.[59]

The dynamic modification of the refractive index of the particles that make up the opaline photonic crystal also serves to alter the average refractive index and produce tunable stop-band shifts. For instance, Se/Ag_2Se core–shell particles were ordered into an fcc array and heated from room temperature to 150 °C. During this heating, different phase transitions for both the core material and the shell material occurred, changing their respective refractive indices and producing stop-band shifts at each transition event, in this case in the near-infrared region (Figure 4.9).[60]

The surface of the constituent opal spheres can also be modified to increase their refractive index and sensing capabilities. Gold nanoparticles were immobilized on the surface of silica spheres in a colloidal crystal in order to form a refractive index sensor. Upon infiltration with various solvents, the stop-band peak and the localized surface plasmon resonance (LSPR) peak in

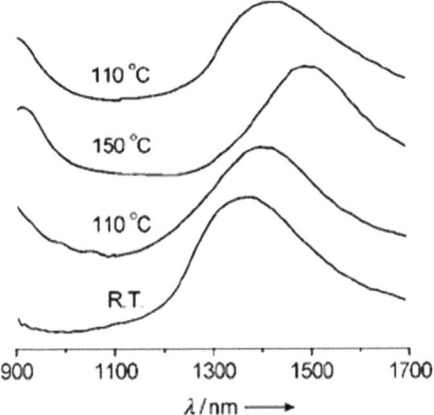

Figure 4.9   Reflectance spectra of a Se/Ag_2Se core–shell colloidal crystal at different temperatures with a stop-band in the near IR.
Reproduced with permission from Ref. 60, copyright 2005, Wiley-VCH.

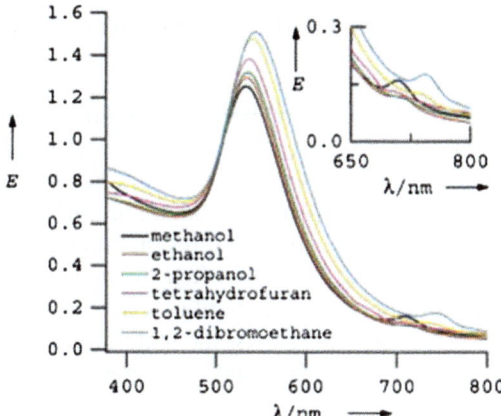

Figure 4.10   Spectra of gold-coated polymer opal film with peak shifts in both the stop-band and LSPR peak due to infiltration with various solvents. The inset highlights the area of the stop-band.
Reproduced with permission from Ref. 61, copyright 2002, Wiley-VCH.

the UV-vis spectrum were shifted (Figure 4.10). The stop-band peak proved to be more sensitive to changes in the surrounding index of refraction. However, as the index of refraction of the infilling solvent reached a value close to that of silica, the stop-band peak was greatly diminished. The presence of the LSPR peak enabled the sensor to continue to function, albeit at diminished sensitivity, even after the loss of the stop-band peak.[61]

Optical switching of infilling material is also possible using conducting polymers that respond to variations in applied voltage. As a voltage increase of 1.4 V was applied to a synthetic opal infiltrated with the conducting polymer, poly(3-alkylthiophene), the refractive index of the filling material decreased from 1.74 to 1.68. This change in refractive index produced a stop-band shift of <5 nm. A slightly greater stop-band shift was seen when a separate polymer, poly(3-octadecylthiophene), known to have a refractive-index change as a function of temperature, was infiltrated into a silica opal colloidal crystal. With a temperature increase of 125 °C, the refractive index of the polymer was lowered from 1.67 to 1.58, producing a stop-band shift of ~8 nm.[62]

Although opaline photonic crystals have proven to show dynamic and responsive color changes with liquid infiltration along with the subsequent alteration in the magnitude of the average refractive index, the size of the stop-band shift due to this infiltration remains low because of the small amount of void space that makes up the infiltrated portion of the crystal. The presence of only ~26% void space in a crystal with perfect packing efficiency means that the potential chromic change is severely limited. A greater response is realized by increasing the amount of void space available for infiltration. This can be accomplished by using an inverse opal structure.

4.4 Dynamic Color Changes in Inverse Opal Photonic Crystals

Inverse opals are photonic crystal materials that are created using opaline photonic crystals as a template. Typically, a filling material is infiltrated into the void spaces around the colloidal array of spheres and the sphere template is removed. The infilling material is processed before or during the removal of the sphere template in order to form wall structures that retain the periodic spacing of the template colloidal crystal. Because often the materials prepared by this process are not true replicates of the interstitial space in the opal template but may have thinner, more strut-like shapes, the materials are also referred to by the more general term of 3-dimensionally ordered macroporous (3DOM) materials.

While 3DOM materials can be used in many of the same ways as opaline photonic crystals for temperature and infiltration-induced lattice changes,[63–71] using hydrogel and polymer compositions, 3DOM materials have the greatest advantage over opaline materials in their response to refractive-index changes. To illustrate this, eqn (4.1) can be used to estimate the difference in stop-band shifts for a dry silica opal and a dry silica inverse opal after infiltration with ethanol (refractive index of silica $= 1.455$, ethanol $= 1.333$). Using the values for perfect packing and a d_{hkl} of 250 nm for both structures, the stop-band for the opal shifts by 43 nm, while the inverse opal experiences a shift of 123 nm to longer wavelengths.

In addition to the morphological advantage of having at least three times the void space as photonic crystal opals, 3DOM materials also offer much greater variety in potential compositions. While opaline photonic crystals are restricted to a limited number of materials capable of forming monodisperse colloidal spheres of sufficient size, 3DOM materials can be synthesized in the form of various polymers,[72–75] hydrogels,[76,77] inorganic oxides,[78–86] metals,[87–93] and other compositions. The ability to choose from a larger pool of potential materials enables researchers to increase the refractive-index contrast in a photonic crystal by employing materials with a higher value of n. This allows for much higher stop-band intensities and increased coloration, opening up opportunities for color-changing pigment applications.

The use of higher-n materials beyond the polymers and hydrogels used for opaline photonic crystals enables syntheses of high refractive index photonic crystals that do not undergo lattice changes, allowing for purely refractive-index-induced solvochromic changes of a much greater magnitude than those seen in opaline photonic crystals. Also, the greater thermal stability and, often, good chemical stability of metal- or metal-oxide-based inverse opals allow for dynamic color changes in a wider array of conditions and environments.

The first example of a solvent-induced color change in 3DOM materials was put forth by Stein and coworkers.[94–95] Colored 3DOM powders of various metal oxides were synthesized with stop-bands in the visible spectrum. The powders were then infiltrated with various solvents while the stop-band

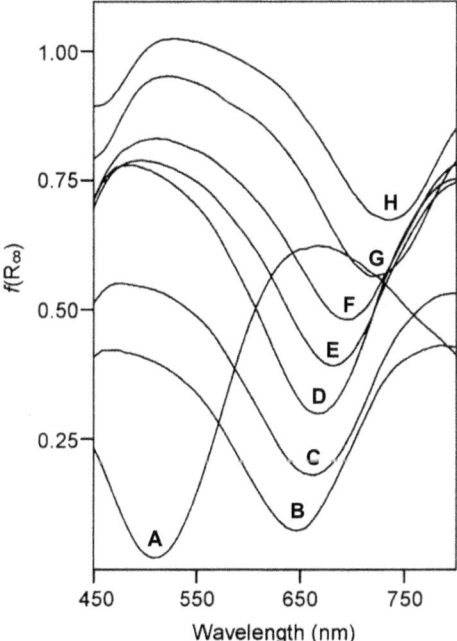

Figure 4.11 Diffuse reflectance UV-vis spectra of 3DOM zirconia with different infilling material. (A) Air $n = 1$, (B) methanol $n = 1.329$, (C) ethanol $n = 1.360$, (D) 2-propanol $n = 1.377$, (E) tetrahydrofuran $n = 1.407$, (F) DMF $n = 1.431$, (G) toluene $n = 1.496$, and (H) 1,2-bibromoethane $n = 1.538$.
Reproduced with permission from Ref. 95, copyright 2002, the American Chemical Society.

position was monitored by UV-vis spectroscopy (Figure 4.11). The 3DOM metal oxides exhibited a much larger shift in position than similar studies by Bogolomov *et al.* on opals of similar composition.[54,55] Such a large shift in color is a product of having a larger void space capable of accommodating solvent. With >74% of the structure being composed of air with a refractive index of 1, replacing that fraction of space with solvents of various refractive indices all larger than 1 raises the n_{avg} of the photonic crystal beyond what would be possible with an opal structure, increasing the potential optical signal shift.

The response of a colored 3DOM material to infiltration can be further tuned, as in the case of silica inverse opals modified to create regions of different wettability within the crystal (Figure 4.12).[96,97] Upon infiltration with a solvent, only regions that are wettable above a certain threshold allow for the infiltration of the solvent. The wetted areas are revealed by their stop-band shift from the refractive-index change in the pores of the material. Using this technique, it is possible to create patterned photonic crystals with regioselective optical switching that can function as selectively exposable pigmentation.

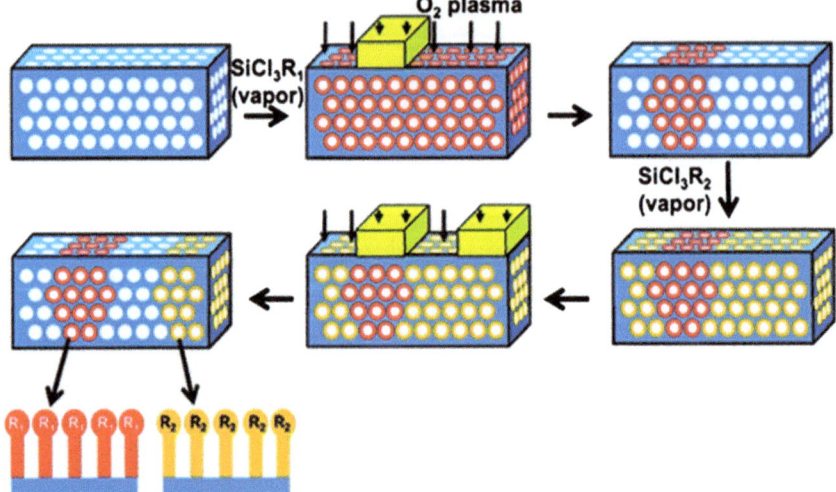

Figure 4.12 Schematic of the procedure for introducing variations in wettability
within a photonic crystal. Repeated exposure to alkylchlorosilane vapors
followed by oxygen plasma etching while using poly(dimethylsiloxane)
masks allows for the creation of different areas of wettability.
Reproduced with permission from Ref. 97, copyright 2011, the American
Chemical Society.

This sort of discriminatory stop-band shifting has numerous applications for
sensing of liquids with a specific surface tension.

This selective, solvochromic tunability of 3DOM materials over wide
wavelength ranges provides an advantage in optical sensing over opaline
photonic crystals. Sensors that can detect analytes in liquids or gases, or even
phase-detection sensors have been realized by harnessing the sensitivity and
functionalizability of these inverse opal structures. Numerous examples of these
sensors that detect shifts in stop-band position with changing refractive index
have been published.[98–100]

More sophisticated materials that tailor these optical sensors for detection of
specific analyte compounds or selection of specific solvents from a mixture have
also been realized. A 3DOM optical sensor based on a phenolic resin exhibited
both superoleophilic and superhydrophobic properties, allowing it to act as a
hydrocarbon sensor that does not suffer from the interference of water or other
polar solvents. The sensor exhibited a linear response in stop-band position
with increases in refractive index when exposed to several different petroleum
components such as octane, tetradecane, hexadecane, diesel oil, wax oil, and
isopropylbenzene. The uptake of solvents into the 3DOM structure was
reversible by washing with octane and drying.[101]

Inverse opal optical sensors can be further specialized for certain analytes by
attaching specific functional groups to the surface of the material. For example,
by attaching specific probe antibodies, it is possible to create functionalized
inverse opal photonic beads that exhibit average refractive-index changes upon

the binding of analyte proteins. The large void space present in the inverse opal increases the throughput of potential analyte throughout the structure and improves the response times of such devices.[102]

The large void space present in 3DOM structures is also ideal for infiltration with liquids whose own refractive index can be dynamically tuned. Optically birefringent nematic liquid crystals infiltrated into an inverse opal can provide instantaneous dynamic tunability through both electro-optical switching and a thermally driven phase transition of the liquid crystal.[103] An electro-optic shutter effect was realized by switching the orientation of nematic molecules inside an inverse opal structure using an external electric field.

Rapid photoinduced switching has been realized using liquid-crystal infiltration into an inverse opal lattice. Photochromic liquid crystal azo dyes were incorporated into a silica inverse opal and irradiatiated with UV light. A *cis–trans* isomerism induced by the irradiation caused a nematic-to-isotropic phase transition in the liquid crystal. The reverse was accomplished upon irradiation with visible light. In the nematic phase, the dielectric constant of the liquid crystal was not uniform throughout each inverse opal pore due to random orientations of the liquid crystal. This interrupted the periodicity of the Bragg diffraction planes, and a strong stop-band was not observed. Upon irradiation and a phase change to the isotropic phase, anisotropy was lost, the dielectric constant became uniform throughout the material, and a stop-band was produced. This same experiment was attempted with a silica opal structure, however, the small void space limited the effects that could be seen with the photoinduced switching.[104]

4.5 Color Optimization in Opal and Inverse Opal Materials

In order to increase the magnitude of the optical signal in photonic sensors or increase the coloration of photonic pigments, structural color optimization in 3DOM and opal photonic crystals is an important step toward achieving the full potential of these materials. A simple strategy towards improving the optical signal from a material standpoint is to increase the refractive-index contrast between the high and low dielectric portions of the crystal. This serves to increase the intensity of the stop-band (as seen in eqn (4.8)), improving response intensity. In sensor applications, little can be done about the refractive indices of the chosen analytes. Therefore, higher-*n* materials must be used in order to increase the magnitude of contrast between diffraction planes. By increasing the reflective response of these photonic crystals, it is also possible to retain a higher stop-band intensity after the decrease in reflectance associated with the infiltration of liquids. These infiltrations reduce the refractive-index contrast between the high and low dielectric portions of the crystal and decrease the intensity of the photonic reflection. However, strong optical absorption by the material itself in the stop-band can affect the intended color properties.

Most synthetic opals are constructed from either silica or polymer, as these two materials can be easily synthesized in large amounts with very low dispersity in the diameters of the particles. However, a few types of high-n spheres with diameters on the same order of the visible regime of wavelengths have also been synthesized. Materials such as titania,[105] selenium,[106] cadmium sulfide,[107] zinc selenide,[108] zinc sulfide,[109] and several different metals[110] have been synthesized in the form of monodisperse spheres that can be used to form colloidal crystals with high refractive-index contrast. These colloidal crystals can exhibit a larger stop-band than silica or polymer opals[111] and maintain a high refractive-index contrast even after infiltration with most solvents.

Confined growth of colloidal particles has also recently been demonstrated for the creation of new opaline materials. By growing crystalline materials inside a spherical template and then removing that template, monodisperse size ranges of these materials could be realized, adding a prospective synthesis tool for monodisperse particles of new compositions. Several examples of double inverse opals have been synthesized, creating either nonclose-packed opaline structures or spherical particles of monodisperse size.[112–116] Even templated zeolite spheres grown within a carbon inverse opal have been created with diameters approaching 350 nm.[117]

Hollow or filled core–shell sphere structures have also been investigated, which add additional control over the photonic stop-band. By varying shell thickness and the materials used, the stop-band in an opaline colloidal crystal can be greatly increased.[118] Colloidal crystal core–shell structures of varying compositions, mostly metal chalcogenides and polymers, have been synthesized, all with enhanced stop-band properties.[119–122]

In the case of inverse opals, the enhancement of the stop-band is similarly achieved by increasing the refractive-index contrast of the wall material. With a highly periodic structure, complete bandgaps are possible for 3DOM materials with an index of refraction contrast of 2.8 or greater.[123] Towards this end, high refractive index inverse opals have been prepared using materials such as tantalum nitride,[124] silicon,[125] and germanium[126] in order to achieve a complete bandgap.

Structural modification can also be used to improve the optical properties of inverse opals. Most high-n inverse opals have very limited penetration of light into the structure due to visible-light absorbance from the structure walls and limited window sizes between pores. By enlarging the windows and opening up the structure for penetration of photons, truly 3-dimensional optical effects start to appear rather than just surface effects commonly seen in structures with small window sizes (Figure 4.13).[127]

For applications such as pigmentation, where intense and vibrant colors are desired, increasing the reflectance of the photonic crystal through refractive-index contrast is not the only strategy to use in order to increase the coloration of the material. Some potential materials for opal or inverse opal photonic crystals may not be suitable for pigmentation due to their low or high absorbance, with low absorbers appearing mostly white and high absorbers appearing mostly black. Both still manifest opalescent color, but the effect is

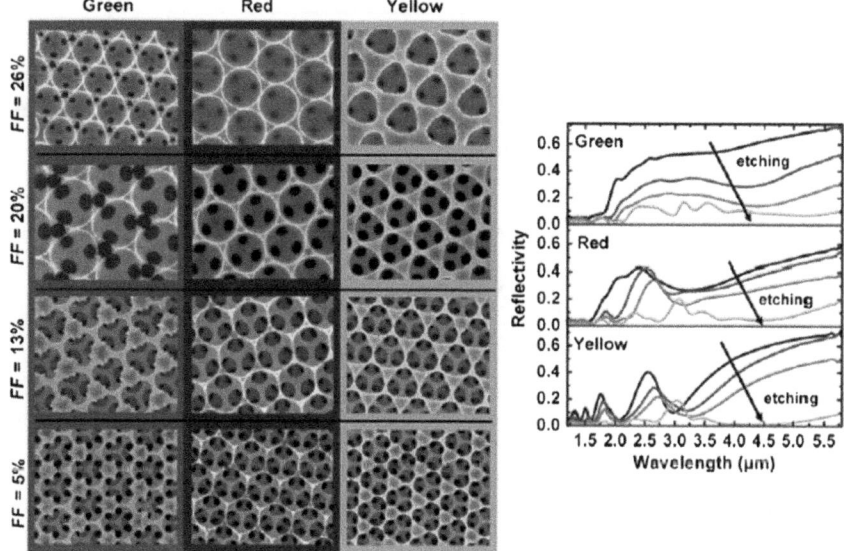

Figure 4.13 Left: SEM images of Ni inverse opals with varying fill fraction (*FF*). Right: Reflectivity as Ni *FF* reduces for each color presented. Spectra with black lines represent 26% *FF* with lighter lines representing decreasing *FF*.
Reproduced with permission from Ref. 127, copyright 2007, Wiley-VCH.

largely trumped by the materials' absorbance properties. It has been shown that predominantly white, opalescent, polymer colloidal crystals can become more intensely colored by the controlled incorporation of an absorbing material with the structure.[128] The absorbing material acts to decrease the number of photons that are scattered out of the crystal and are not part of a diffraction event, lowering the reflectance spectrum of the material at all wavelengths, but allowing the stop-band wavelengths to dominate. Several functional materials employing this concept have been recently synthesized. They exhibit a dramatic enhancement of color intensities as a result of the addition of an absorbing substance, usually carbon black or carbon nanoparticles.[129–134]

The effects of the amount of carbon black incorporated into a polymer opal structure were also studied, showing an increase in maximum stop-band reflectance above the background reflectance with the addition of a small amount (0.05 wt%) of carbon black. A maximum in stop-band peak height was reached at 0.15 wt%, with subsequent additions of carbon black reducing the intensity of the reflectance at the stop-band by absorbing photons even at the stop-band (Figure 4.14). The reduction in scattering due to the added absorber can be seen clearly in Figure 4.14 by the reduction in background reflectance relative to the stop-band peak.[134] This effect can also be seen in colloidal sphere arrays using magnetite particles that provide a dark background without the addition of an absorber.[135]

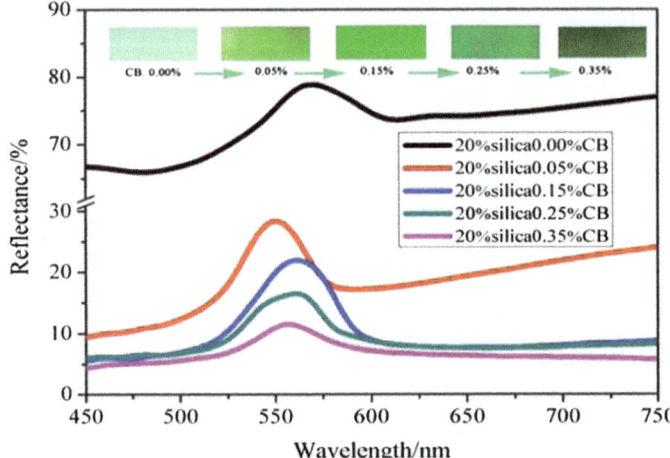

Figure 4.14   Reflectance spectra of colloidal crystal films consisting of 241-nm
polymer spheres mixed with 20 wt% colloidal silica and with differing
wt% composition of carbon black.
Reproduced with permission from Ref. 134, copyright 2012, the Royal
Society of Chemistry.

4.6   Conclusions

Both opal and inverse opal photonic crystals are capable of exhibiting brilliant
colors based on their structural characteristics through Bragg diffraction of
light. These properties make them interesting new pigmenting and decorative
materials. Pigments with structural color offer the advantage of providing long-
lasting colors without relying on potentially toxic organic molecules or metal
complexes. In particular, for chemically inert, oxide-based photonic crystals,
the colors do not degrade over time or with exposure to sunlight, as is often a
problem for molecular dyes. Colors can be vivid, and particles greater than
100 µm typically exhibit an opalescent shine. Their intensities depend on
refractive-index contrast and can be further enhanced by doping with light-
absorbing additives. Both opals and inverse opals can also exhibit rapid
chromic changes in response to different conditions, including temperature,
infiltration with various filling substances, and irradiation. Such changes come
about through the alteration of the refractive index of one or more components
or a change in the lattice spacing of the dielectric materials that make up the
photonic crystal. These reversible and stable color-change characteristics are
ideal for displays that do not require rapid switching and for optical sensors.

Each of the two classes of structurally colored pigments has its advantages
and disadvantages. Opaline photonic crystals, being denser, tend to be mechanically stronger than inverse opals. However, with the smaller available void
space for infiltration in opaline photonic crystals, the potential increase in
average refractive index after fluid infiltration is more restricted. Therefore,
shifts in color by this mechanism are limited in magnitude. By utilizing a

hydrogel or polymer-based encapsulating matrix as a platform for the dynamic tuning of the stop-band position, the color response is greatly enhanced due to changes in the lattice parameters of the material that are often reversible. As a result, it is possible for a tunable opal photonic crystal to span the entire visible range of wavelengths. In addition, polymer-based matrices provide structural flexibility to produce photonic crystal materials that are deformable.

On the other hand, inverse opal materials with their much larger void volumes can provide a platform for large stop-band shifts based on refractive-index changes alone. The more open, interconnected pore space also facilitates faster throughput of solvents and other analytes into and out of the structure, improving the response time of the induced color shifts. The synthesis of inverse opals is more general than that of opals by assembly of monodisperse spheres, allowing for a greater selection of material compositions. This permits the use of materials with higher refractive index to create a more complete photonic bandgap. However, double-templating techniques, *i.e.* using inverse opals as templates for synthetic opals, have begun to extend the compositional range for opals as well.

Further advances in both types of structures are expected in the future, with an emphasis on the optimization of the intensity, purity, and angle dependence of colors in these structurally colored materials, sensitivity, specificity, and rate of color response, as well as the commercial scale-up of optical sensors, pigments, and numerous other devices that take advantage of the dynamic tunability of opal and inverse opal photonic crystals.

References

1. E. Yablonovitch, *Phys. Rev. Lett.*, 1987, **58**, 2059.
2. S. John, *Phys. Rev. Lett.*, 1987, **58**, 2486.
3. K. Nassau, *The Physics and Chemistry of Color: The Fifteen Causes of Color, 2nd edn*, John Wiley & Sons Inc., New York, 2001.
4. S. Kinoshita, S. Yoshioka and K. Kawagoe, *Proc. R. Soc. London, Ser. B*, 2002, **269**, 1417.
5. A. R. Parker, R. C. McPhedran, D. R. McKenzie, L. C. Botten and N. A. Nicorovici, *Nature*, 2001, **409**, 36.
6. D. W. Lee, *Nature*, 1991, **349**, 260.
7. H. C. Bolton, L. A. Bursill, A. C. McLaren and R. G. Turner, *Phys. Status Solidi B*, 1966, **18**, 221.
8. J. Sanders, *Acta Crystallogr., Sect. A: Found. Crystallogr.*, 1968, **24**, 427.
9. B. J. Ackerson, S. E. Paulin, B. Johnson, W. van Megen and S. Underwood, *Phys. Rev. E: Stat. Nonlinear Soft Matter Phys.*, 1999, **59**, 6903.
10. P. N. Pusey, W. van Megen, P. Bartlett, B. J. Ackerson, J. G. Rarity and S. M. Underwood, *Phys. Rev. Lett.*, 1989, **63**, 2753.
11. P. G. Bolhuis, D. Frenkel, S.-C. Mau and D. A. Huse, *Nature*, 1997, **388**, 235.
12. L. V. Woodcock, *Nature*, 1997, **385**, 141.

13. S.-C. Mau and D. A. Huse, *Phys. Rev. E: Stat. Nonlinear Soft Matter Phys.*, 1999, **59**, 4396.
14. P. Jiang, J. F. Bertone, K. S. Hwang and V. L. Colvin, *Chem. Mater.*, 1999, **11**, 2132.
15. Y. K. Koh, C. H. Yip, Y.-M. Chiang and C. C. Wong, *Langmuir*, 2008, **24**, 5245.
16. B. J. Ackerson, *J. Rheol.*, 1990, **34**, 553.
17. M. D. Haw, W. C. K. Poon and P. N. Pusey, *Phys. Rev. E: Stat. Nonlinear Soft Matter Phys.*, 1998, **57**, 6859.
18. N. Koumakis, A. B. Schofield and G. Petekidis, *Soft Mater.*, 2008, **4**, 2008.
19. M. Holgado, F. García-Santamaría, A. Blanco, M. Ibisate, A. Cintas, H. Míguez, C. J. Serna, C. Molpeceres, J. Requena, A. Mifsud, F. Meseguer and C. López, *Langmuir*, 1999, **15**, 4701.
20. M. E. Leunissen, H. R. Vutukuri and A. van Blaaderen, *Adv. Mater.*, 2009, **21**, 3116.
21. J. Yamanaka, M. Murai, Y. Iwayama, M. Yonese, K. Ito and T. Sawada, *J. Am. Chem. Soc.*, 2004, **126**, 7156.
22. S. H. Im, Y. T. Lim, D. J. Suh and O. O. Park, *Adv. Mater.*, 2002, **14**, 1367.
23. A. van Blaaderen, R. Ruel and P. Wiltzius, *Nature*, 1997, **385**, 321.
24. Y. Yin, Z.-Y. Li and Y. Xia, *Langmuir*, 2003, **19**, 622.
25. Z. Cai, J. Teng, D. Xia and X. S. Zhao, *J. Phys. Chem. C*, 2011, **115**, 9970.
26. İ. İ. Tarhan and G. H. Watson, *Phys. Rev. Lett.*, 1996, **76**, 315.
27. P. A. Rundquist, P. Photinos, S. Jagannathan and S. A. Asher, *J. Chem. Phys.*, 1989, **91**, 4932.
28. Y. Monovoukas and A. P. Gast, *Langmuir*, 1991, **7**, 460.
29. L. Liu, P. Li and S. A. Asher, *J. Am. Chem. Soc.*, 1997, **119**, 2729.
30. T. Kanai, T. Sawada, A. Toyotama, J. Yamanaka and K. Kitamura, *Langmuir*, 2007, **23**, 3503.
31. S. A. Asher, J. Holtz, L. Liu and Z. Wu, *J. Am. Chem. Soc.*, 1994, **116**, 4997.
32. J. M. Jethmalani and W. T. Ford, *Chem. Mater.*, 1996, **8**, 2138.
33. S. H. Foulger, P. Jiang, A. C. Lattam, D. W. Smith and J. Ballato, *Langmuir*, 2001, **17**, 6023.
34. S. H. Foulger, P. Jiang, Y. Ying, A. C. Lattam, D. W. Smith and J. Ballato, *Adv. Mater.*, 2001, **13**, 1898.
35. S. H. Foulger, P. Jiang, A. Lattam, D. W. Smith, J. Ballato, D. E. Dausch, S. Grego and B. R. Stoner, *Adv. Mater.*, 2003, **15**, 685.
36. H. Fudouzi and T. Sawada, *Langmuir*, 2005, **22**, 1365.
37. Y. Iwayama, J. Yamanaka, Y. Takiguchi, M. Takasaka, K. Ito, T. Shinohara, T. Sawada and M. Yonese, *Langmuir*, 2003, **19**, 977.
38. J. M. Weissman, H. B. Sunkara, A. S. Tse and S. A. Asher, *Science*, 1996, **274**, 959.
39. J. D. Debord and L. A. Lyon, *J. Phys. Chem. B*, 2000, **104**, 6327.
40. J. H. Holtz and S. A. Asher, *Nature*, 1997, **389**, 829.

41. J. H. Holtz, J. S. W. Holtz, C. H. Munro and S. A. Asher, *Anal. Chem.*, 1998, **70**, 780.
42. A. C. Sharma, T. Jana, R. Kesavamoorthy, L. Shi, M. A. Virji, D. N. Finegold and S. A. Asher, *J. Am. Chem. Soc.*, 2004, **126**, 2971.
43. K. Lee and S. A. Asher, *J. Am. Chem. Soc.*, 2000, **122**, 9534.
44. D. Nakayama, Y. Takeoka, M. Watanabe and K. Kataoka, *Angew. Chem. Int. Ed.*, 2003, **42**, 4197.
45. X. Hu, G. Li, J. Huang, D. Zhang and Y. Qiu, *Adv. Mater.*, 2007, **19**, 4327.
46. X. Hu, G. Li, M. Li, J. Huang, Y. Li, Y. Gao and Y. Zhang, *Adv. Funct. Mater.*, 2008, **18**, 575.
47. Z. Wu, X. Hu, C.-a. Tao, Y. Li, J. Liu, C. Yang, D. Shen and G. Li, *J. Mater. Chem.*, 2008, **18**, 5452.
48. Z. Wu, C.-a. Tao, C. Lin, D. Shen and G. Li, *Chem. Eur. J.*, 2008, **14**, 11358.
49. C. Guo, C. Zhou, N. Sai, B. Ning, M. Liu, H. Chen and Z. Gao, *Sens. Actuators B Chem.*, 2012, **166–167**, 17.
50. H. Fudouzi and Y. Xia, *Adv. Mater.*, 2003, **15**, 892.
51. H. Fudouzi and Y. Xia, *Langmuir*, 2003, **19**, 9653.
52. P. Jiang, D. W. Smith, J. M. Ballato and S. H. Foulger, *Adv. Mater.*, 2005, **17**, 179.
53. T. Kanai, S. Yamamoto and T. Sawada, *Macromolecules*, 2011, **44**, 5865.
54. V. N. Bogomolov, S. V. Gaponenko, A. M. Kapitonov, A. V. Prokofiev, A. N. Ponyavina, N. I. Silvanovich and S. M. Samoilovich, *Appl. Phys. A-Mater.*, 1996, **63**, 613.
55. V. N. Bogomolov, S. V. Gaponenko, I. N. Germanenko, A. M. Kapitonov, E. P. Petrov, N. V. Gaponenko, A. V. Prokofiev, A. N. Ponyavina, N. I. Silvanovich and S. M. Samoilovich, *Phys. Rev. E: Stat. Nonlinear Soft Matter Phys.*, 1997, **55**, 7619.
56. J. Li, W. Huang and Y. Han, *Colloids Surf., A*, 2006, **279**, 213.
57. H. Yang, P. Jiang and B. Jiang, *J. Colloid Interf. Sci.*, 2012, **370**, 11.
58. Y. Yamada, T. Nakamura and K. Yano, *Langmuir*, 2008, **24**, 2779.
59. J. Li, W. Huang, Z. Wang and Y. Han, *Colloids Surf., A*, 2007, **293**, 130.
60. U. Jeong and Y. Xia, *Angew. Chem. Int. Ed.*, 2005, **44**, 3099.
61. Z.-Z. Gu, R. Horie, S. Kubo, Y. Yamada, A. Fujishima and O. Sato, *Angew. Chem. Int. Ed.*, 2002, **41**, 1153.
62. S. Satoh, H. Kajii, Y. Kawagishi, T. Tamura, A. Fujii, M. Ozaki, R. D. McCullough and K. Yoshino, *Synth. Met.*, 2001, **121**, 1503.
63. Y.-J. Lee, S. A. Pruzinsky and P. V. Braun, *Langmuir*, 2004, **20**, 3096.
64. J. Li, Y. Wu, J. Fu, Y. Cong, J. Peng and Y. Han, *Chem. Phys. Lett.*, 2004, **390**, 285.
65. A. C. Arsenault, T. J. Clark, G. von Freymann, L. Cademartiri, R. Sapienza, J. Bertolotti, E. Vekris, S. Wong, V. Kitaev, I. Manners, R. Z. Wang, S. John, D. Wiersma and G. A. Ozin, *Nature Mater.*, 2006, **5**, 179.

66. J. Y. Wang, Y. Cao, Y. Feng, F. Yin and J. P. Gao, *Adv. Mater.*, 2007, **19**, 3865.

67. M. Honda, K. Kataoka, T. Seki and Y. Takeoka, *Langmuir*, 2009, **25**, 8349.

68. J. Wang and Y. Han, *J. Colloid Interf. Sci.*, 2011, **357**, 139.

69. J. Wang and Y. Han, *J. Colloid Interf. Sci.*, 2011, **353**, 498.

70. L.-Q. Wang, F.-Y. Lin and L.-P. Yu, *Analyst*, 2012, **137**, 3502.

71. Y. Yuan, Z. Li, Y. Liu, J. Gao, Z. Pan and Y. Liu, *Chem. Eur. J.*, 2012, **18**, 303.

72. S. H. Park and Y. Xia, *Adv. Mater.*, 1998, **10**, 1045.

73. S. H. Park and Y. Xia, *Chem. Mater.*, 1998, **10**, 1745.

74. P. Jiang, K. S. Hwang, D. M. Mittleman, J. F. Bertone and V. L. Colvin, *J. Am. Chem. Soc.*, 1999, **121**, 11630.

75. S. A. Johnson, P. J. Ollivier and T. E. Mallouk, *Science*, 1999, **283**, 963.

76. Y. Takeoka and M. Watanabe, *Langmuir*, 2002, **18**, 5977.

77. Y. J. Lee and P. V. Braun, *Adv. Mater.*, 2003, **15**, 563.

78. O. D. Velev, T. A. Jede, R. F. Lobo and A. M. Lenhoff, *Nature*, 1997, **389**, 447.

79. B. T. Holland, C. F. Blanford and A. Stein, *Science*, 1998, **281**, 538.

80. O. D. Velev, T. A. Jede, R. F. Lobo and A. M. Lenhoff, *Chem. Mater.*, 1998, **10**, 3597.

81. J. E. G. J. Wijnhoven and W. L. Vos, *Science*, 1998, **281**, 802.

82. P. Yang, T. Deng, D. Zhao, P. Feng, D. Pine, B. F. Chmelka, G. M. Whitesides and G. D. Stucky, *Science*, 1998, **282**, 2244.

83. B. T. Holland, L. Abrams and A. Stein, *J. Am. Chem. Soc.*, 1999, **121**, 4308.

84. B. T. Holland, C. F. Blanford, T. Do and A. Stein, *Chem. Mater.*, 1999, **11**, 795.

85. H. Yan, C. F. Blanford, B. T. Holland, W. H. Smyrl and A. Stein, *Chem. Mater.*, 2000, **12**, 1134.

86. G. I. N. Waterhouse, J. B. Metson, H. Idriss and D. Sun-Waterhouse, *Chem. Mater.*, 2008, **20**, 1183.

87. P. Jiang, J. Cizeron, J. F. Bertone and V. L. Colvin, *J. Am. Chem. Soc.*, 1999, **121**, 7957.

88. O. D. Velev, P. M. Tessier, A. M. Lenhoff and E. W. Kaler, *Nature*, 1999, **401**, 548.

89. H. Yan, C. F. Blanford, B. T. Holland, M. Parent, W. H. Smyrl and A. Stein, *Adv. Mater.*, 1999, **11**, 1003.

90. J. E. G. J. Wijnhoven, S. J. M. Zevenhuizen, M. A. Hendriks, D. Vanmaekelbergh, J. J. Kelly and W. L. Vos, *Adv. Mater.*, 2000, **12**, 888.

91. H. Yan, C. F. Blanford, W. H. Smyrl and A. Stein, *Chem. Commun.*, 2000, 1477.

92. G. von Freymann, S. John, M. Schulz-Dobrick, E. Vekris, N. Tetreault, S. Wong, V. Kitaev and G. A. Ozin, *Appl. Phys. Lett.*, 2004, **84**, 224.

93. N. R. H. Denny, S. Turgeon, R. T. Lytle, J. C. Norris and D. J. Stein, *Proc. SPIE*, 2005, **6005**, 60050501.

94. C. F. Blanford, R. C. Schroden, M. Al-Daous and A. Stein, *Adv. Mater.*, 2001, **13**, 26.
95. R. C. Schroden, M. Al-Daous, C. F. Blanford and A. Stein, *Chem. Mater.*, 2002, **14**, 3305.
96. I. B. Burgess, N. Koay, K. P. Raymond, M. Kolle, M. Lončar and J. Aizenberg, *ACS Nano*, 2011, **6**, 1427.
97. I. B. Burgess, L. Mishchenko, B. D. Hatton, M. Kolle, M. Lončar and J. Aizenberg, *J. Am. Chem. Soc.*, 2011, **133**, 12430.
98. U. Kamp, V. Kitaev, G. von Freymann, G. A. Ozin and S. A. Mabury, *Adv. Mater.*, 2005, **17**, 438.
99. Y. Nishijima, K. Ueno, S. Juodkazis, V. Mizeikis, H. Misawa, T. Tanimura and K. Maeda, *Opt. Exp.*, 2007, **15**, 12979.
100. H. Li, L. Chang, J. Wang, L. Yang and Y. Song, *J. Mater. Chem.*, 2008, **18**, 5098.
101. H. Li, J. Wang, L. Yang and Y. Song, *Adv. Funct. Mater.*, 2008, **18**, 3258.
102. Y. Zhao, X. Zhao, J. Hu, M. Xu, W. Zhao, L. Sun, C. Zhu, H. Xu and Z. Gu, *Adv. Mater.*, 2009, **21**, 569.
103. K. Busch and S. John, *Phys. Rev. Lett.*, 1999, **83**, 967.
104. S. Kubo, Z.-Z. Gu, K. Takahashi, Y. Ohko, O. Sato and A. Fujishima, *J. Am. Chem. Soc.*, 2002, **124**, 10950.
105. X. Jiang, T. Herricks and Y. Xia, *Adv. Mater.*, 2003, **15**, 1205.
106. U. Jeong and Y. Xia, *Adv. Mater.*, 2005, **17**, 102.
107. E. Matijević and W. Deborah Murphy, *J. Colloid Interf. Sci.*, 1982, **86**, 476.
108. X. Wang, J. Zhu, Y. g. Zhang, J. Jiang and S. Wei, *Appl. Phys. A-Mater.*, 2010, **99**, 651.
109. D. M. Wilhelmy and E. Matijevic, *J. Chem. Soc., Faraday Trans. 1 F*, 1984, **80**, 563.
110. Y. Wang and Y. Xia, *Nano Lett.*, 2004, **4**, 2047.
111. M. Alexander and S. Charles, *J. Phys. Condens. Matter*, 1999, **11**, 997.
112. S. M. Yang, N. Coombs and G. A. Ozin, *Adv. Mater.*, 2000, **12**, 1940.
113. P. Jiang, J. F. Bertone and V. L. Colvin, *Science*, 2001, **291**, 453.
114. L. Xu, L. D. Tung, L. Spinu, A. A. Zakhidov, R. H. Baughman and J. B. Wiley, *Adv. Mater.*, 2003, **15**, 1562.
115. L. Wang, Q. Yan and X. S. Zhao, *J. Mater. Chem.*, 2006, **16**, 4598.
116. G. Guan, K. Kusakabe, H. Ozono, M. Taneda, M. Uehara and H. Maeda, *J. Mater. Sci.*, 2007, **42**, 10196.
117. W. C. Yoo, S. Kumar, R. L. Penn, M. Tsapatsis and A. Stein, *J. Am. Chem. Soc.*, 2009, **131**, 12377.
118. G.-R. Yi and S.-M. Yang, *J. Opt. Soc. Am. B*, 2001, **18**, 1156.
119. M. L. Breen, A. D. Dinsmore, R. H. Pink, S. B. Qadri and B. R. Ratna, *Langmuir*, 2001, **17**, 903.
120. K. P. Velikov and A. van Blaaderen, *Langmuir*, 2001, **17**, 4779.
121. K. P. Velikov, A. Moroz and A. van Blaaderen, *Appl. Phys. Lett.*, 2002, **80**, 49.
122. I. D. Hosein and C. M. Liddell, *Langmuir*, 2007, **23**, 2892.

123. K. Busch and S. John, *Phys. Rev. E Stat., Nonlinear Soft Matter Phys.*, 1998, **58**, 3896.
124. A. Rugge, J.-S. Park, R. G. Gordon and S. H. Tolbert, *J. Phys. Chem. B*, 2004, **109**, 3764.
125. Y. A. Vlasov, X.-Z. Bo, J. C. Sturm and D. J. Norris, *Nature*, 2001, **414**, 289.
126. F. Meseguer, A. Blanco, H. Míguez, F. García-Santamaría, M. Ibisate and C. López, *Colloids Surf., A*, 2002, **202**, 281.
127. X. Yu, Y. J. Lee, R. Furstenberg, J. O. White and P. V. Braun, *Adv. Mater.*, 2007, **19**, 1689.
128. C. I. Aguirre, E. Reguera and A. Stein, *ACS Appl. Mater. Interf.*, 2010, **2**, 3257.
129. O. L. Pursiainen, J. J. Baumberg, H. Winkler, B. Viel, P. Spahn and T. Ruhl, *Opt. Exp.*, 2007, **15**, 9553.
130. O. L. J. Pursiainen, J. J. Baumberg, H. Winkler, B. Viel, P. Spahn and T. Ruhl, *Adv. Mater.*, 2008, **20**, 1484.
131. J. J. Baumberg, O. L. Pursiainen and P. Spahn, *Phys. Rev. B: Condens. Matter*, 2009, **80**, 201103.
132. D. R. E. Snoswell, A. Kontogeorgos, J. J. Baumberg, T. D. Lord, M. R. Mackley, P. Spahn and G. P. Hellmann, *Phys. Rev. E: Stat. Nonlinear Soft Matter Phys.*, 2010, **81**, 020401.
133. C. E. Finlayson, P. Spahn, D. R. E. Snoswell, G. Yates, A. Kontogeorgos, A. I. Haines, G. P. Hellmann and J. J. Baumberg, *Adv. Mater.*, 2011, **23**, 1540.
134. Z. Shen, L. Shi, B. You, L. Wu and D. Zhao, *J. Mater. Chem.*, 2012, **22**, 8069.
135. L. He, M. Wang, J. Ge and Y. Yin, *Acc. Chem. Res.*, 2012, **45**, 1431.

CHAPTER 5

Optical Properties of Tunable Photonic Crystals Using Liquid Crystals

RYOTARO OZAKI,*[a] MASANORI OZAKI[b] AND
KATSUMI YOSHINO[c]

[a] Graduate School of Engineering and Science, Ehime University,
Matsuyama, Ehime 790-8577, Japan; [b] Division of Electrical, Electronic and
Information Engineering, Graduate School of Engineering, Osaka
University, Suita, Osaka 565-0871, Japan; [c] Shimane Institute for Industrial
Technology, Matsue, Shimane 690-0816, Japan
*Email: ozaki.ryotaro.mx@ehime-u.ac.jp

5.1 Introduction

Liquid crystals are mesophases between crystalline solids and isotropic liquids. They may flow like viscous fluids and also possess features that are characteristic of solid crystals. The constituents are rod-like or disk-like organic molecules that normally have self-assembled characteristics. These molecules exhibit different physical properties between the long and short molecular axes. Owing to their molecular shape, liquid crystals have various anisotropic properties, such as permittivity, refractive index, and viscosity. There are many types of liquid-crystal states that are classified by the long-range orientational order of the molecules.[1,2] In a nematic phase, rod-like molecules are aligned to a certain direction but they do not have a positional order. The nematic phase is the most commonly used in flat-panel displays.[3,4] This is because nematic liquid

RSC Smart Materials No. 5
Responsive Photonic Nanostructures: Smart Nanoscale Optical Materials
Edited by Yadong Yin
© The Royal Society of Chemistry 2013
Published by the Royal Society of Chemistry, www.rsc.org

crystals have high fluidity and large dielectric anisotropy that are key properties in the liquid-crystal devices. Since the fluidity and dielectric anisotropy allow liquid-crystal molecules to align parallel to or perpendicular to an electric field, the molecular director can be controlled by application of an electric field. On the basis of this field-induced molecular reorientation, almost all liquid-crystal devices act as optical shutters.

Liquid-crystal infiltrated photonic crystals have attracted considerable attention from both fundamental and practical points of view, because they can be regarded as tunable photonic crystals.[5–11] Photonic crystals are one-dimensional (1D), two-dimensional (2D) or three-dimensional (3D) ordered structures with a periodicity comparable to an optical wavelength. They are composed of two or more different dielectrics and their periodicity opens up a photonic bandgap in which the existence of photons is forbidden.[12–15] This can be explained by an analogy to electrons in a solid-state crystal. In a photonic bandgap, electromagnetic fields cannot propagate due to destructive interference between waves scattered from the periodically modulated refractive-index structure. This is similar to destructive interference of electron waves from the periodic potential of an atomic lattice at a certain frequency.

When there is a defect that disturbs the periodicity of a photonic crystal structure, localized photonic states appear in the photonic bandgap. The states are so-called defect modes at which photons are confined in the defect. In particular, 3D photonic bandgap materials with a defect allow us to achieve a 3D photon confinement.[12,13] Such an optical confinement is the most important feature of photonic crystals because the electric field in a defect can be strongly enhanced at a defect mode resonance frequency. An appropriate line defect in a 2D or 3D photonic crystal serves as a waveguide that can guide light in a desired direction by photonic bandgap confinement.[16,17] Such a defect acting as a microcavity or a waveguide is very important in certain applications, for example, low-threshold lasers, microwaveguides, optical circuits, and so on.[16–22] Furthermore, tunable photonic crystals having variable photonic bandgaps by controlling optical periodicities are also utilized for functional applications, for example, tunable lasers[8,9,23–27] and tunable optical wave-guides.[28] In particular, liquid-crystal infiltrated photonic crystals have been studied for the realization of tunable photonic crystals. When a liquid crystal is used as a material of a periodic structure, photonic bandgaps can be controlled because liquid crystals have a large optical anisotropy and are sensitive to external conditions such as temperature, electric fields, and magnetic fields. The field-induced molecular reorientation gives rise to a change in the optical length of a photonic crystal. In the cases of opals or inverse opals filled with a nematic liquid crystal, a tunable photonic bandgap has been demonstrated by applying an electric field or by controlling temperature.[5,6]

Tunable defect modes are also an attractive subject. However, introduction of a defect in a 3D photonic crystal using nanofabrication remains a major technical challenge.[29–32] Although 1D photonic crystals do not have a complete photonic bandgap, there are plenty of applications using extraordinary wave-length dispersion and a localized photonic state in the defect layer. So far, intensive studies on 1D photonic crystal applications have been reported: air-bridge microcavities,[33,34] photonic band-edge lasers,[20] nonlinear optical diodes,[35] and enhancements of optical nonlinearity.[36,37] In this chapter, we mainly focus on 1D tunable photonic crystals using a liquid crystal.

This chapter is divided into three parts. In the first part, fundamentals of photonic crystals are described to explain photonic band structure, photonic bandgap, and defect mode. In the second part, we focus on a 1D photonic crystal with a liquid-crystal defect layer and study its applications as a tunable laser, an electro-optics switch, and a controllable polarizer. In the third part, we focus on the liquid-crystal-infiltrated photonic crystal without a defect and study a tunable band-edge laser and tunable negative refraction.

5.2　Fundamentals of 1D Photonic Crystals

Photonic crystals have photonic bandgaps that forbid electromagnetic wave propagation in a certain direction. To understand the optical characteristics of a photonic crystal, photonic band analysis is particularly important. The photonic band diagram of a photonic crystal, which allows the survey of all the dispersion characteristics of the photonic crystal, can be obtained by solving Maxwell's equations directly. There are several numerical methods for photonic band analysis, such as the plane wave expansion (PWE) method[38,39] and the finite-difference time-domain (FDTD) method.[40,41] These methods allow us to calculate not only the photonic band diagram but also electric-field profiles in the photonic crystal. Here, we discuss some of the basic properties of a 1D photonic crystal using calculation results.

Figure 5.1(a) shows the schematic view of a 1D photonic crystal. The darker and lighter gray layers represent the high- and low-index regions, respectively.

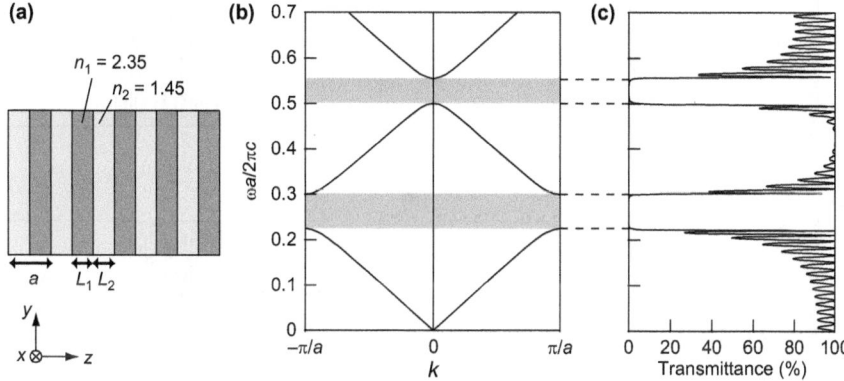

Figure 5.1　(a) Schematic view of a 1D photonic crystal. (b) Photonic band structure of the 1D photonic crystal with calculated by the PWE method. (c) Transmission spectrum of the 1D photonic crystal calculated by the transfer-matrix method. The parameters are as follows: refractive indices $n_1 = 2.35$, $n_2 = 1.45$; layer thickness $L_1 = L_2 = 70$ nm.

Figure 5.1(b) shows the photonic band structure of the 1D photonic crystal calculated by the PWE method with the following parameters: refractive indices $n_1 = 2.35$, $n_2 = 1.45$; layer thickness $L_1 = L_2 = 70$ nm. In Figure 5.1(b), the vertical axis represents normalized frequency $\omega a/2\pi c$ and the horizontal axis represents wave vector k, where a is a lattice constant and c is the speed of light. In general, normalized frequency is used as a vertical axis because of the existence of a simple scaling law concerned with the lattice constant and wavelength. In the diagram, there are some gaps in which the wave vector does not exist over a certain frequency range. The gaps are the so-called photonic bandgaps. To improve understanding of the diagram, Figure 5.1(c) shows the transmission spectrum of the 1D photonic crystal calculated by the transfer-matrix method.[15] It is clear that the dips in transmittance appear over the frequency ranges of the photonic bandgaps. The diagram also shows small group velocities near the band edges, which is determined by the slope of the curve, *i.e.* $v_g = d\omega/dk$.

Electric-field profiles in the 1D photonic crystal for various frequencies are shown in Figure 5.2. The electric fields are calculated by the transfer-matrix method with the same parameters as the transmission spectrum. Figure 5.2(a) shows the electric field for $\omega a/2\pi c = 0.15$, where the darker and lighter gray layers represent $n_1 = 2.35$ and $n_2 = 1.45$, respectively. The incident light enters from the left, and then propagates through the 1D photonic crystal without disturbance. This is because the light propagation is hardly affected by the dielectric periodic structure at the photonic band frequency. The electric field at a bandgap frequency $\omega a/2\pi c = 0.22$ is shown in Figure 5.2(b). The incident light exponentially decreases inside the 1D photonic crystal. At this frequency the

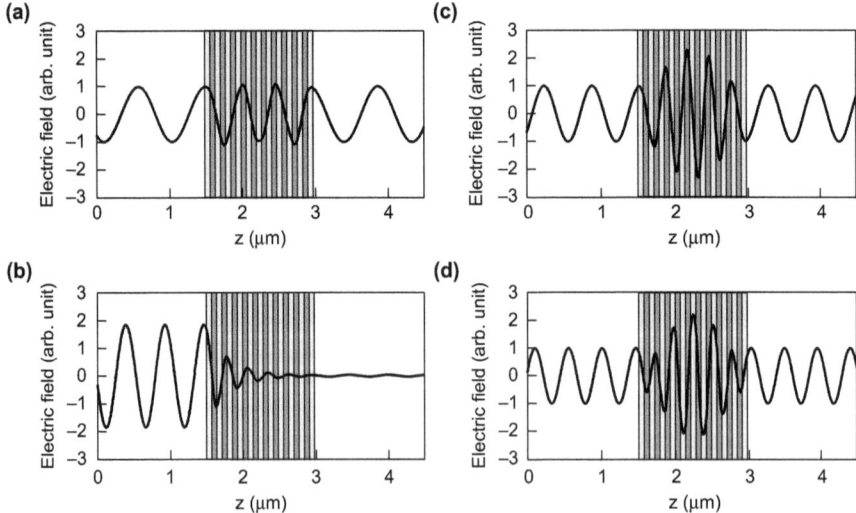

Figure 5.2 Electric-field profiles in the photonic crystal for various situations: (a) photonic band, (b) photonic bandgap, (c) lower band edge, and (d) upper band edge.

phases of reflected light from boundaries are matched, and then the inphase reflections give rise to strong light reflection. That is, the internal interference forbids the existence of light at the photonic bandgap frequency.

When the light with a photonic band-edge frequency enters the 1D photonic crystal, the electric field is enhanced in the 1D photonic crystal. Figures 5.2(c) and 5.2(d) show the electric-field profiles at the lower and upper band edge, respectively. Since their group velocities approach zero at the band edges, the electric and magnetic fields form a standing wave in the 1D photonic crystal. At both band edges, the electric field is enhanced because the photonic crystal acts as a resonant cavity. In Figure 5.2(c), the peaks of the inside electric field are located on the darker layers. This means that the photons are concentrated in the higher-index regions; $n_1 = 2.35$. In contrast, the peaks of the inside electric field in Figure 5.2(d) are located on the lighter layers. The refractive index n_2 of the lighter gray layer is 1.45. Since the energy of a photon becomes lower in a higher-index medium, the difference in energy between the lower and upper band edges can be explained by where photons are concentrated. In general, such an electric-field enhancement at the band edge is used for a distributed feedback (DFB) laser.

One of the most attractive features of photonic crystals is a defect mode in a photonic bandgap. When a defect of periodicity is introduced into the lattice, photons are strongly localized in the defect. Figure 5.3(a) shows the transmission spectrum of a 1D photonic crystal with an air gap. The following parameters are used: $n_1 = 2.35$, $n_2 = 1.45$, $n_d = 1.0$, $L_1 = 64\,\mathrm{nm}$, $L_2 = 103\,\mathrm{nm}$, and $L_d = 75\,\mathrm{nm}$, where n_d is the refractive index of the air defect and L_d is a defect length. The photonic bandgap is formed from 510 nm to 780 nm in which the sharp peak appears at 640 nm due to the introduction of the air defect. The narrow peak corresponds to a defect state that is a localized mode in the photonic bandgap. Figure 5.3(b) shows the electric-field profile at the defect mode frequency. It is clear that the electric fields are strongly enhanced in the defect. Note that the peak intensity for the defect mode is larger than those for the band-edge modes shown in Figures 5.2(c) and 5.2(d). The reason for this is that photons are confined in the narrow defect at the defect mode, whereas photons are confined in the whole photonic crystal at the band-edge modes. The electric-field enhancement allows strong interactions between photons and

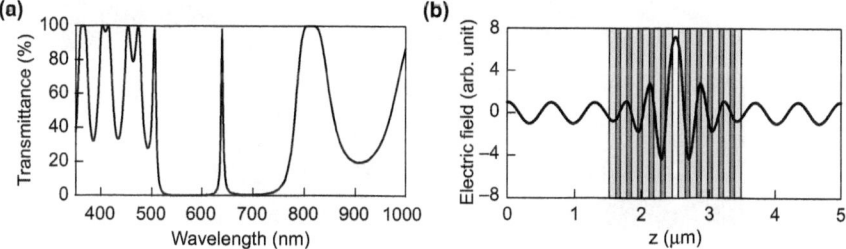

Figure 5.3   (a) Transmission spectrum of a 1D photonic crystal with an air defect. (b) Electric-field profile at the defect-mode wavelength of 640 nm.

materials. Such photonic crystals with a defect are expected to enhance an optical gain, a nonlinear-optical effect, or a magneto-optical effect.

5.3 Tunable Defect Mode in Photonic Crystals with a Liquid-Crystal Defect

Optical properties of a photonic crystal, such as photonic bandgap and defect mode, depend on the structural periodicity and the refractive indices of the materials. If we can tune the structural periodicity or refractive index in the photonic crystal, the optical properties can be controlled. There are several ways to change the optical periodicity. For example, an external stress will induce a deformation in the periodic structure,[25] and temperature control is also effective for changing the refractive index of a material.[5,6] However, from the standpoint of application to optical devices, the precision and speed of tuning by stress or temperature are inferior compared to an electrical control. Liquid crystals have a large birefringence and their refractive indices can be electrically controlled by molecular reorientation. Therefore, a liquid crystal is one of the most suitable materials for tunable photonic crystals.

5.3.1 Tunable Defect Mode

Let us consider a nematic liquid crystal introduced into a 1D photonic crystal, as shown in Figure 5.4. The photonic crystal is composed of dielectric multilayers with the nematic liquid-crystal defect layer located at the center. We assume that the initial orientation of the liquid-crystal molecules is set along the y-axis and that the molecules are reoriented parallel to the z-axis by applying an electric field. In general, there exists a threshold electric field E_{th} above which the molecules can be redirected from the initial orientation. It is well known that a nematic liquid crystal has such a threshold caused by the Frederiks transition.[1,2] The threshold is given by

$$E_{th} = \frac{\pi}{d} \sqrt{\frac{K_{11}}{\varepsilon_0 \Delta \varepsilon}} \qquad (5.1)$$

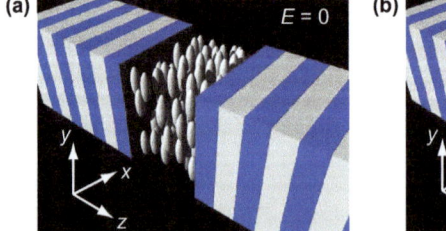
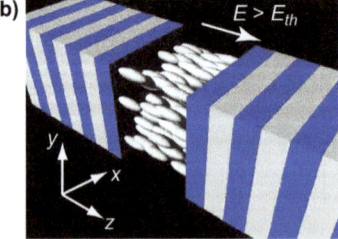

Figure 5.4   Schematic views of a 1D photonic crystal with a nematic liquid crystal for (a) $E = 0$ and (b) $E > E_{th}$. The liquid-crystal director can be controlled using electric-field-induced molecular reorientation.

where d is thickness of the liquid-crystal layer, K_{11} is the splay elastic constant, ε_0 is the vacuum permittivity, $\Delta\varepsilon$ is dielectric anisotropy of the liquid crystal. The threshold voltage V_{th} is easily obtained from simple calculation and is approximately 1 V for a typical nematic liquid crystal.

When light linearly polarized along the y-axis enters the photonic crystal, the incident light encounters the extraordinary refractive index n_e in the liquid-crystal layer in the absence of an applied electric field. Above E_{th}, the molecules tilt towards the z-axis with an angle θ that is defined from the y-axis. Using the angle, the effective refractive index $n'(\theta)$ for the y-polarization is determined by

$$n'(\theta) = \frac{n_o n_e}{\sqrt{n_e^2 \sin^2 \theta + n_o^2 \cos^2 \theta}} \tag{5.2}$$

When the molecules are aligned parallel to the z-axis by a sufficient voltage, the effective refractive index becomes equal to the ordinary index n_o; $n'(90°) = n_o$. Since nematic liquid crystals have a large anisotropy in refractive index, the molecular reorientaion in the defect layer gives rise to a change in the optical length. This indicates that the defect-mode wavelength can be controlled by applying voltage to the liquid-crystal defect layer.

We here examine how the liquid-crystal layer is related to tuning of defect modes. Figure 5.5 shows the transmission spectra of the 1D photonic crystal for different defect thicknesses. These calculations were carried out using the 4×4 matrix method.[42] This method is a numerical analysis based on the Maxwell equations that can be used to quantitatively calculate the light propagation in an anisotropic medium with refractive index varying along one direction. The calculation parameters are summarized as follows: $n_1 = 2.35$, $n_2 = 1.45$, $n_e = 1.7$, $n_o = 1.5$, $L_1 = 64$ nm, and $L_2 = 103$ nm. The defect layer thicknesses in Figs. 5(a), 5(b), and 5(c) are 0.2 µm, 0.6 µm, and 1.2 µm, respectively. Here, the solid lines represent the transmission spectra for $n_e = 1.7$, while the broken lines represent those for $n_o = 1.5$. The two refractive indices correspond to $n'(0°)$ and $n'(90°)$. As is evident from the results, the number of defect modes in a photonic

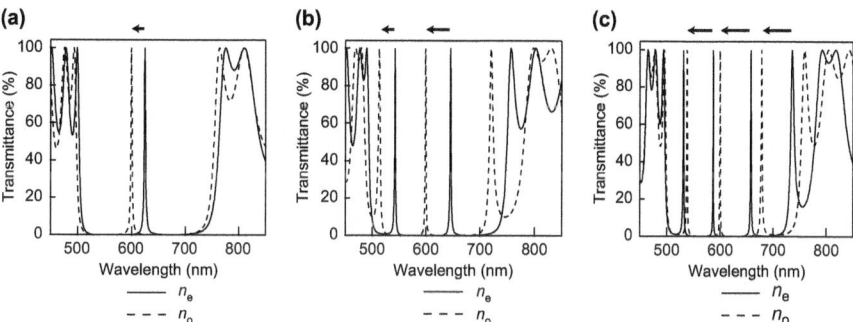

Figure 5.5 Transmission spectra of the 1D photonic crystals for different defect thicknesses: (a) 0.2 µm, (b) 0.6 µm, and (c) 1.2 µm. Peak shifts of defect modes clearly depend on defect thickness.

bandgap increases with length of the liquid-crystal defect layer. When the refractive index in the liquid-crystal layer changes from n_e to n_o, the defect-mode peaks shift to shorter wavelengths. The arrows on the transmission spectra indicate the direction of the peak shift from $\theta = 0°$ to $\theta = 90°$. In these calculations, the defect thicknesses are chosen so that one of the defect-mode wavelengths for n_o is 600 nm. The shift amounts for thicknesses of 0.2, 0.6, and 1.2 μm are 25, 46, and 58 nm, respectively. That is, a thicker defect layer provides a wider peak shift.

Figure 5.6(a) shows the experimentally measured transmission spectra of the 1D photonic crystal with a nematic liquid-crystal defect layer. A dielectric multilayer consisting of an alternating stack of SiO_2 and TiO_2 layers deposited on an In-Sn oxide (ITO)-coated glass substrate was used as a 1D photonic crystal. The refractive indices of SiO_2 and TiO_2 are 1.45 and 2.35, respectively. The center wavelength of the photonic band was adjusted to be 600 nm by setting the optical thickness of both SiO_2 and TiO_2 to be one-quarter of 600 nm; the thicknesses of the SiO_2 and TiO_2 layers were 103 nm and 64 nm, respectively. The number of SiO_2–TiO_2 pairs on each substrate was five. Polyimide was spin coated on the top surface of the multilayer, and then it was unidirectonally rubbed along the y-axis. The polyimide would then align liquid-crystal molecules parallel to the y-direction. To introduce the defect layer, a nematic liquid crystal was sandwiched between two multilayers using 1-μm spacers. The ordinary and extraordinary refractive indices of the liquid crystal are approximately $n_o = 1.5$ and $n_e = 1.7$, respectively. The transmission spectra in Figure 5.6(a) were measured with y-polarized light that corresponds to the initial orientation of the liquid-crystal molecules in the defect layer. The solid and broken lines represent the transmission spectra at 0 V and 8 V, respectively. An application of a voltage to ITO electrodes changes the molecular alignment of the liquid-crystal defect layer. The defect-mode peaks shift to shorter wavelengths upon applying voltage. The peak shifts agree with the calculation results in Figure 5.6(b). That is, the incident light encounters n_e at the initial

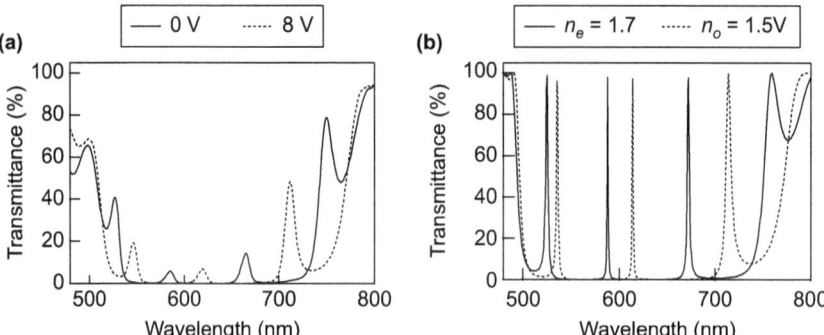

Figure 5.6   (a) Experimentally measured transmission spectra of the 1D photonic crystal with a nematic liquid-crystal defect layer. (b) Calculated transmission spectra of the photonic crystal by the 4×4 matrix method.

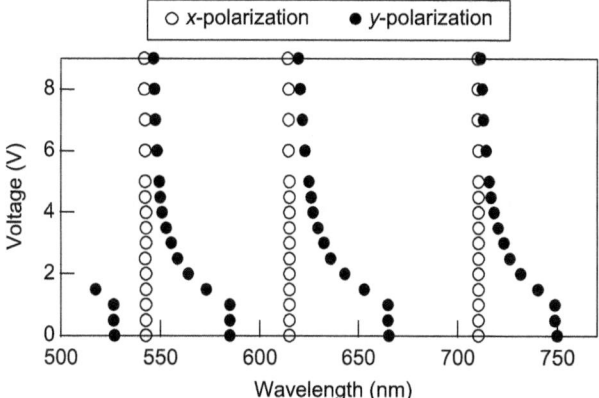

Figure 5.7 Voltage dependences of defect-mode peaks of the 1D photonic crystal with a nematic liquid crystal. The open and filled circles represent the peak wavelengths for *x*- and *y*-polarizations.

orientation and then encounters n_o at 8 V. The measured peak widths are broader than the calculated widths due to nonuniformity of the physical thickness of the defect layer.

The voltage dependences of the defect-mode wavelengths for light polarized along the *x*- and the *y*-axes are summarized in Figure 5.7. The open and filled circles represent the peak wavelengths for *x*- and *y*-polarizations, respectively. For *y*-polarized light, the defect-mode peaks begin to shift to shorter wavelengths with increasing voltage above 1 V, which is associated with the Frederiks transition mentioned above. All peaks shift in the same manner by about 40 nm, even upon applying low voltage. The peak located around 530 nm at 0 V shifts to a shorter wavelength and disappears above 1.8 V, which means that the peak is out of the bandgap at higher voltages and is no longer a defect mode. In contrast, the defect modes for *x*-polarized light remain unchanged upon applying voltage. This indicates that the *x*-polarized light is always affected by n_o, regardless of the applied voltage. This is because the *x* component of the refractive index is not changed by the reorientation shown in Figure 5.4. Therefore, for *x*-polarized light, the optical length of the defect layer is independent of the liquid-crystal molecular orientation.

5.3.2 Defect-Mode Laser

Although 1D photonic crystals do not have a complete photonic bandgap, there are many applications using the slow group velocity at a band edge and the localized photonic state in a defect layer. In particular, the study of defect mode lasing is one of the most attractive subjects because low-threshold lasing at a defect-mode resonance is expected. Since defect-mode wavelengths can be controlled upon applying electric field using a liquid-crystal defect, lasing wavelength can also be tuned. Electrically wavelength-tunable defect mode lasing has been demonstrated using this system.[8,43] In the experiments, an active medium was introduced inside the 1D photonic crystal with a liquid-crystal defect to achieve laser action. As an active laser medium, a laser dye[8] or a conducting polymer[43] was used. To excite the active medium, the second-harmonic light of a Q-switched Nd:YAG laser irradiated the sample. Figure 5.8(a) shows the pump energy dependence of emission intensity at the defect-mode wavelength. In this case, [2-[2-4-(dimethylamino)phenyl]-ethenyl]-6-methyl-4H-pyran-4-ylidene propanedinitrile was used as a laser dye dopant in the nematic liquid crystal. Above the threshold at a pump-pulse energy of about 5 μJ/pulse, the emission intensity significantly increases. This indicates that there exists a lasing threshold above which a light amplification accrues. In order to control the emission wavelength, the orientation of the liquid-crystal molecules in the defect layer was changed upon applying a square-wave voltage at 1 kHz. Figure 5.8(b) shows the emission spectra of the 1D photonic crystal with the dye-doped liquid-crystal defect as a function of applied voltage. It should be noted that a sharp lasing peak shifts toward shorter wavelengths with increasing voltage in the same manner as the defect modes shift (see Figure 5.7). The wavelength shift of the lasing peak is about 25 nm, even upon applying a low voltage. The lasing in a 1D photonic crystal with a liquid-crystal defect layer can be tuned over a wide range upon applying an electric field.

The density of states (DOS), which is the reciprocal of the group velocity of light, plays an important role in photonic crystal lasers.[44,45] The photonic DOS $\rho(\omega)$ is defined as the inverse slope of the dispersion relation,

$$\rho(\omega) = \frac{1}{v_g(\omega)} = \frac{dk(\omega)}{d\omega} \tag{5.3}$$

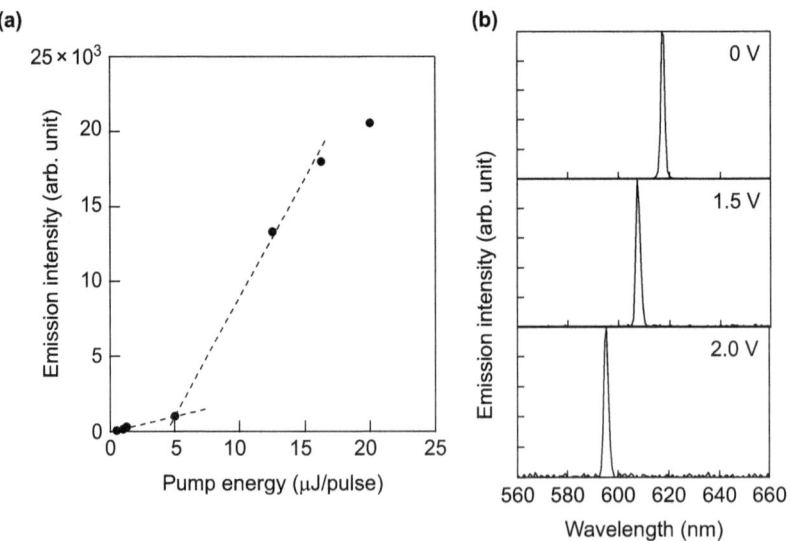

Figure 5.8   (a) Pump-energy dependence of emission intensity at the defect-mode wavelength. (b) Emission spectra of the 1D photonic crystal with dye-doped liquid-crystal defect as a function of applied voltage.

As is evident from the equation, a slow group velocity in a photonic crystal provides a large DOS, which is the same meaning as strong confinement. The dispersion relation of a photonic crystal can be calculated by the PWE method or other numerical methods. However, these methods can obtain a photonic band diagram for an infinite period but not for a finite period. Here, the DOS in a finite periodic structure is calculated with the complex transmission coefficient t written as

$$t = t_1 + it_2 \left(= \frac{E_t}{E_i} \right) \tag{5.4}$$

where t_1 and t_2 are the real and imaginary parts of the transmission coefficient, and E_t and E_i are the amplitudes of the transmitted and incident electric fields. Using the complex transmission coefficient, the DOS is given by

$$\rho(\omega) = \frac{dk(\omega)}{d\omega} = \frac{1}{D} \frac{\left(\dfrac{dt_2}{d\omega}\right) t_1 - \left(\dfrac{dt_1}{d\omega}\right) t_2}{t_1^2 + t_2^2} \tag{5.5}$$

where D is the thickness of the sample. The DOS for an isotropic medium ρ_{iso} is

$$\rho_{iso} = \frac{dk}{d\omega} = \frac{n}{c} \tag{5.6}$$

Figure 5.9(a) shows the transmission spectrum and the normalized DOS of a 10-layer SiO_2–TiO_2 multilayer without a defect. Here, t_1 and t_2 are calculated using the 4×4 matrix method. The DOS is increased at the band edges because of low group velocities. We can estimate a lasing wavelength in the DFB cavity from the peak of the DOS. Figure 5.9(b) shows the transmission spectrum and the normalized DOS of a 10-layer SiO_2–TiO_2 stack with a defect. The defect is sandwiched between two 5-layer stacks, and its refractive index and length are 1.7 and 190 nm, respectively. As is evident from Figure 5.9(b), the DOS at the defect-mode wavelength is significantly increased. That is, the group velocity at the defect-mode wavelength is much slower than that at the band edges. This result is consistent with the electric-field calculations shown in Figures 5.2 and

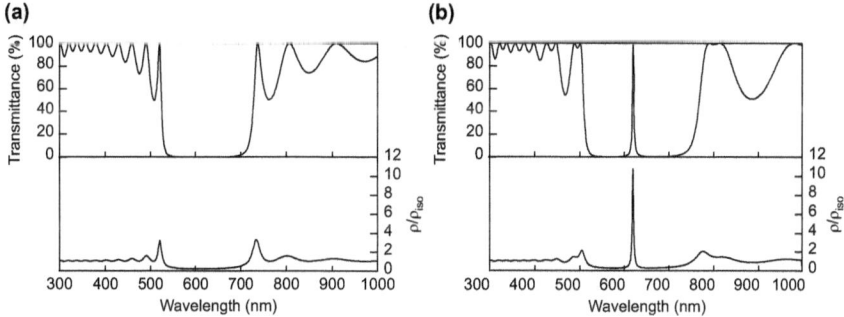

Figure 5.9 Transmission spectra and normalized DOSs of a 10-layer SiO_2–TiO_2 multilayer: (a) without a defect and (b) with a defect.

5.3. Such a slow group velocity and an electric-field enhancement are major advantages of a defect mode because an electric field can interact more strongly with a functional material in the defect. Thus, the analysis of the DOS is one of the most useful tools to investigate optical confinement in a photonic crystal.

The FDTD method also plays an important role in computer-aided analysis for photonic crystal lasers, in which the Maxwell equations are solved on the basis of the Yee algorithm in discrete time and lattices.[40] FDTD is a dependable method to simulate a wide range of problems for electromagnetics, photonics, acoustics, and so on.[41] Using this method, the dynamics of light propagation and photon localization can be numerically simulated. To investigate laser dynamics, an auxiliary differential equation FDTD (ADE-FDTD) method has also been developed, in which the FDTD method is coupled with the classical electron oscillator model and the rate equation in a four-level energy system.[46–50] The ADE-FDTD method can simulate light emission from a photonic crystal, such as a defect-mode laser or band-edge laser.

We here consider light propagation in an active gain medium that induces a polarization. The Maxwell equations are written as

$$\nabla \times \mathbf{E} = -\mu_0 \frac{\partial \mathbf{H}}{\partial t} \tag{5.7}$$

$$\nabla \times \mathbf{H} = -\varepsilon_0 \varepsilon \frac{\partial \mathbf{E}}{\partial t} + \frac{\partial \mathbf{P}}{\partial t} \tag{5.8}$$

where $\mathbf{P}$ is a macroscopic polarization. On the basis of a classical electron oscillator model, the polarization $\mathbf{P}$ for an isotropic medium can be described by

$$\frac{\partial^2 P}{\partial t^2} + \Delta\omega_a \frac{\partial P}{\partial t} + \omega_a^2 P = \frac{\gamma_r}{\gamma_c} \frac{e^2}{m} \Delta N E \tag{5.9}$$

where $\Delta\omega_a = 1/\tau_{21} + 2/T_2$ is the full width at half-maximum linewidth of atomic transition. T_2 is the mean time between dephasing events which is taken to be 5.0×10^{-15} s. $\Delta N(z,t) = N_1(z, t) - N_2(z, t)$ and $\gamma_r = 1/\tau_{21}$ is the real decay rate of the second level, and γ_c is the classical rate.

In a four-level atomic system, the rate equations are expressed as follows:

$$\frac{dN_3(z, t)}{dt} = P_r N_0(z, t) - \frac{N_3(z, t)}{\tau_{32}} \tag{5.10}$$

$$\frac{dN_2(z, t)}{dt} = \frac{N_3(z, t)}{\tau_{32}} + \frac{E(z, t)}{\hbar\omega_a} \frac{dP(z, t)}{dt} - \frac{N_2(z, t)}{\tau_{21}} \tag{5.11}$$

$$\frac{dN_1(z, t)}{dt} = \frac{N_2(z, t)}{\tau_{21}} - \frac{E(z, t)}{\hbar\omega_a} \frac{dP(z, t)}{dt} - \frac{N_1(z, t)}{\tau_{10}} \tag{5.12}$$

$$\frac{dN_0(z, t)}{dt} = \frac{N_1(z, t)}{\tau_{20}} - P_r N_0(z, t) \tag{5.13}$$

where N_0, N_1, N_2, and N_3 are the electron numbers at each energy level. The lifetimes at each energy level, τ_{32}, τ_{21}, and τ_{10}, are chosen to be 1×10^{-13}, 1×10^{-9}, and 1×10^{-11} s, respectively, and are similar to those of laser dyes such as coumarine or rhodamine. The total electron density $N_0^0 = N_0 + N_1 + N_2 + N_3$ is $5.5 \times 6.02 \times 10^{23}$, and initially, all of them are at the ground state ($N_0 = N_0^0$). Pumping rate P_r is a controlled variable that should be tuned by the pumping intensity in the real experiment.

In Figure 5.10(a), the solid line is the calculated emission spectrum of a dye in a bulk at a low pump rate by the ADE-FDTD method. The center wavelength of the dye spectrum is 570 nm and the half-width at half-maximum is 65 nm. The broken line is the transmission spectrum of a 1D photonic crystal without a defect. The physical properties of the 1D photonic crystal are as follows: $n_1 = 1.5$, $n_2 = 1.65$; $L_1 = 100$ nm, $L_2 = 100$ nm; and the number of layers is 20. Figure 5.10(b) shows the calculated dye emission spectra from the 1D photonic crystal without a defect as a function of pumping rate, where the spatial distribution of the dye is assumed to be uniform in the 1D photonic crystal. At a low pumping rate of 1×10^{10}, a broad spectrum is observed in which light emissions are suppressed by the photonic band. With increasing pumping rate, the spectral peak becomes higher and narrower at the band-edge wavelength. Above a threshold, the single narrow peak appears at the band-edge wavelength. The pumping-rate dependence of the emission intensity at the band edge is plotted in Figure 5.10(c). It is clear that there is a threshold at a pumping rate of approximately 1.5×10^{11}. Thus, the ADE-FDTD simulation has been a powerful tool because it can reproduce emission spectra and provide a laser threshold.

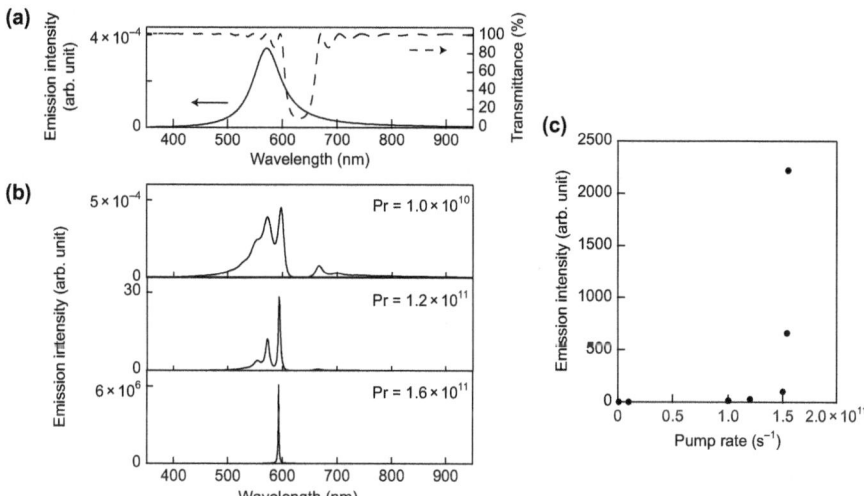

Figure 5.10 ADE-FDTD simulation results: (a) Emission spectrum of a dye in bulk and transmission spectrum of a 1D photonic crystal, (b) Emission spectra of a dye-doped 1D photonic crystal as a function of pumping rate, and (c) Pumping-rate dependence of emission intensity at the lower band-edge wavelength.

5.3.3   Electro-Optic Switch Using a Tunable Defect Mode

A tunable defect mode can be utilized to develop and functionalize optical devices. An electro-optic switch using a tunable defect mode has also been demonstrated, in which the response time is much faster than that of a conventional liquid-crystal shutter.[51] The response time of a twisted nematic liquid-crystal device is generally of the order of ten milliseconds, whereas that of the defect mode switch is submillisecond. Figure 5.11(a) shows transmission spectra of a 1D photonic crystal with a 1-μm nematic liquid-crystal defect layer. The solid and broken lines correspond to the transmission spectra at 0 V and 12 V, respectively. The defect-mode peaks shift to shorter wavelengths with increasing voltage. For example, the 690 nm peak at 0 V shifts to 633 nm at 12 V. As mentioned above, the peak shifts originate from the decrease in the optical length of the defect layer caused by the field-induced liquid-crystal molecular reorientation. Here we focus on the transmission intensity of the defect-mode wavelength of 633 nm. In the absence of a voltage, a light beam of 633 nm cannot propagate through the 1D photonic crystal because the wavelength is in the photonic bandgap. However, when a voltage of 12 V is applied to the liquid-crystal defect layer, the defect-mode peak at 633 nm allows light to propagate through the photonic crystal. In experiments, a 632.8-nm wavelength He-Ne laser beam was used as a light source to measure the electro-optic response based on defect mode shift. The He-Ne laser beam entered into the photonic crystal with normal incident angle and was linearly polarized along the y-direction, which corresponded to the initial liquid-crystal molecular orientation. Figure 5.11(b) shows the electro-optic responses of the defect-mode switch to voltage pulses. High-speed submillisecond response time is observed even at relatively low voltage. Here, the response time is defined as the time required for the change from 10% to 90% of the total transmission change. The rise time is proportional to V^{-2} and decreases to the order of microseconds with increasing voltage. On the other hand, the fall time has little dependence on voltage. These results indicate that the driving force of the

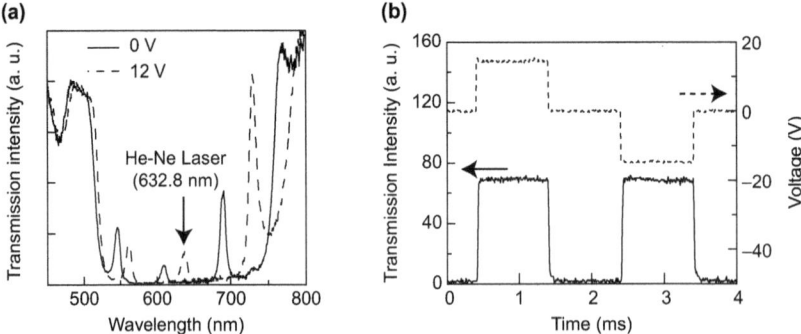

Figure 5.11   (a) Transmission spectra of a 1D photonic crystal with a 1-μm nematic liquid-crystal defect layer. (b) Defect-mode switch's 632.8 nm electro-optic responses to voltage pulses.

electro-optic switching originates from the dielectric anisotropy of a nematic liquid crystal. Generally, the fall time of a nematic liquid crystal released from a director deformation is of the order of ten milliseconds due to a slow molecular relaxation. However, in the case of defect-mode switching, the fall time is also of the order of several tens of microseconds that is notably short for a nematic liquid-crystal device. These submillisecond responses can be achieved by the use of a narrow spectral line width defect mode.

To analyze the electro-optic switch using a tunable defect mode, it is necessary to investigate the field-induced director dynamics in the nematic liquid-crystal defect layer.[52] The director configuration in the nematic liquid crystal can be calculated on the basis of the Franck continuum theory in which liquid crystals are regarded as an elastic continuum.[1,2,4,53] An orientational deformation in a liquid crystal can be described in terms of three basic types of deformation: splay, twist, and bend. There are three independent elastic constants K_{11}, K_{22}, and K_{33} for the splay, the twist, and the bend deformations. Using elastic constants K_{11} and K_{33}, director vector $\mathbf{n}$, and electric field $\mathbf{E}$, the free-energy density g of a splay-bend deformation is given by

$$g = \frac{1}{2} K_{11} (\nabla \cdot \mathbf{n})^2 + \frac{1}{2} K_{33} [\mathbf{n} \times (\nabla \times \mathbf{n})]^2 - \frac{1}{2} \varepsilon_0 \varepsilon_a (\mathbf{E} \cdot \mathbf{n})^2 \qquad (5.14)$$

where ε_0 is the permittivity of vacuum and ε_a is the dielectric anisotropy between the ordinary and extraordinary axes of the nematic liquid crystal. Using the director angle θ, the free energy density g is written as

$$g = \frac{1}{2} (K_{11} \cos^2 \theta + K_{33} \sin^2 \theta) \left(\frac{d\theta}{dz}\right)^2 - \frac{1}{2} \varepsilon_0 \varepsilon_a E \sin^2 \theta \qquad (5.15)$$

The dynamic characteristics of the director are determined by solving the following torque balance equation:

$$\gamma \frac{\partial \theta}{\partial t} = -\left(\frac{\partial g}{\partial \theta} - \frac{d}{dz} \frac{\partial g}{\partial (d\theta/dz)}\right) \qquad (5.16)$$

where γ is the rotational viscosity of the nematic liquid crystal. To solve eqn (5.16) numerically, it is approximated in time and space domains using a difference method.

Figure 5.12 shows the director configuration of the nematic liquid-crystal molecules in the defect layer calculated on the basis of the Franck continuum theory as a function of time. In the calculations, the physical parameters of 5CB are used, which are listed as follows: $\varepsilon_a = 11$; $K_1 = 6.37$ pN, $K_3 = 8.6$ pN; and $\gamma = 0.06$ Pa s. The field-induced director deformations with a voltage of 15 V are shown in Figure 5.12(a). The initial orientation is along the y-axis with a pretilt angle of 0.1°. By applying a voltage of 15 V, the director angle θ increases with time, but the director angles at both interfaces are fixed due to a strong anchoring. After applying the voltage of 15 V for 300 μs, almost all of the liquid-crystal molecules reorient parallel to the z-axis ($\theta = 90°$), and the molecular directors do not change any longer. Figure 5.12(b) shows the

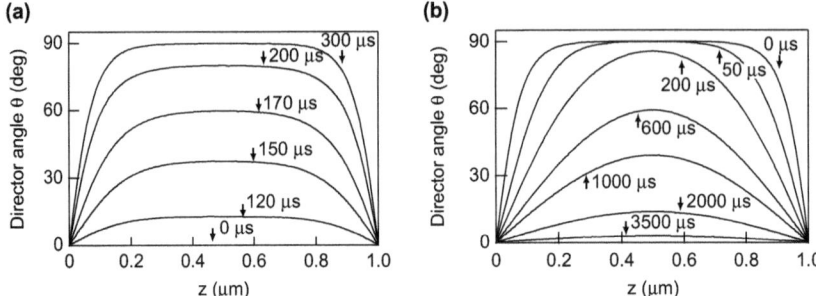

Figure 5.12   Director configuration of the nematic liquid-crystal molecules in the defect layer calculated on the basis of the Franck continuum theory as a function of time: (a) electric-field-induced molecular reorientation and (b) director relaxation in the absence of electric field.

relaxations of the nematic liquid-crystal molecules after removing the applied voltage of 15 V. The relaxations start at both ends of the liquid-crystal layer, and the director angles gradually return to 0°. Since the relaxation is caused by an elastic distortion, the relaxation speed gradually decreases with time. Therefore, the reorientation time in the relaxation process is much longer than that with the applied voltage. After 5000 μs, almost all of the nematic liquid-crystal molecules are oriented along the y-axis ($\theta = 0°$).

Now, we know the dynamic response of the liquid-crystal molecules in the defect. The time response of the transmittance can also be calculated using the change in the refractive-index spatial distribution with time. Figure 5.13 shows the transmission spectra of the 1D photonic crystal with the nematic liquid-crystal defect layer for the applied voltage of 15 V as a function of time. The spectra are calculated using the y-polarized light that exhibits a large anisotropy in refractive index when the liquid-crystal molecules are reoriented from the y- to z-axes by applying a voltage. The defect-mode peaks shift to shorter wavelengths with elapsed time. This can be explained by the fact that the refractive index is reduced from the extraordinary to ordinary refractive indices by applying a voltage, and the change in refractive index shortens the optical length of the defect layer. In this case, light with a wavelength of 610 nm cannot propagate through the photonic crystal in the absence of the voltage, but applying a voltage allows light propagation through the photonic crystal at 610 nm.

Figure 5.13(b) shows the calculated 610 nm wavelength electro-optic responses obtained by applying a square-wave voltage of 15 V with a frequency of 100 Hz. In this case, the defect-mode peak shifts from 657 to 610 nm with the application of 15 V, and the peak returns to 657 nm after removing the applied voltage. As is evident from Figure 5.13(b), both the rise and fall response times are less than 1 ms. This is caused by two reasons: one is that the whole process of molecular reorientation is not required because it is based on a narrow peak shift of a defect mode, the other is that the switching uses a faster director change in the relaxation process after removing the voltage. As shown in

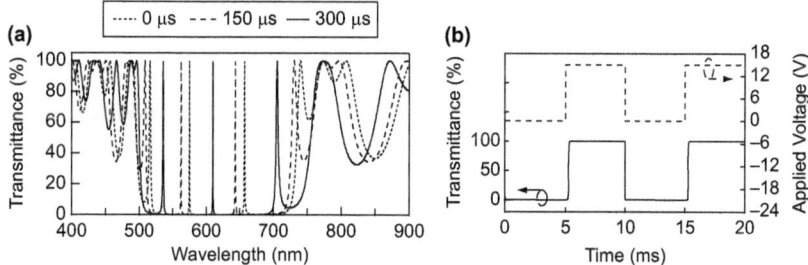

Figure 5.13 (a) Calculated transmission spectra of a 1D photonic crystal with a 1-μm liquid-crystal defect layer for the applied voltage of 15 V as a function of time. (b) Calculated electro-optic responses of the defect mode switch to voltage pulses at 610 nm.

Figure 5.12(b), the relaxation speed gradually decreases with time. Therefore, microsecond responses can be achieved by using the narrow peak shift and the fast part of the relaxation.

5.3.4 Polarization Device Using a Liquid-Crystal Defect

Optical characteristics of the 1D photonic crystal with a nematic liquid crystal depend on polarization as mentioned above. In Figure 5.7, two groups of defect modes appear in transmission spectra. One group of defect mode shifts to shorter wavelengths with an increase of voltage, whereas the other group does not shift. Their polarizations are linear and are perpendicular to each other. They are associated with the extraordinary and ordinary refractive indices of the nematic liquid crystal in the defect. Therefore, if the liquid-crystal director rotates in plane, the polarization direction at a defect-mode wavelength also rotates with the director rotation.[54]

Figure 5.14(a) shows a 1D photonic crystal with six electrodes to control the azimuthal angle ϕ of the liquid-crystal director in the defect. A homeotropic alignment is used here to eliminate an anisotropy of the azimuthal anchoring on the surface. In the defect layer, the molecular director aligns perpendicular to the surface in the absence of an electric field, while the molecules reorient parallel to the surface under the field. To measure the polarization of transmitted light at a defect-mode wavelength, circularly polarized light is used as incident light and an analyzer is placed in front of a detector. Figure 5.14(b) shows the transmission spectra at $\phi = 0°$ as a function of the analyzer angle Φ. The peak maxima at 582 and 596 nm are $\Phi = \pm 90°$ and $\Phi = 0°$, respectively. Besides, the two peaks are orthogonal to each other. Figure 5.14(c) shows the transmission spectra at $\phi = 60°$. In this case, the peak maxima at 582 and 596 nm are $\Phi = 30°$ and $\Phi = 60°$. These results indicate that the polarization of transmitted light at a defect mode can be electrically controlled by inplane switching.

In tunable lasing and electro-optic switching, the defect-mode wavelengths are continuously changed by applying a voltage. In both cases, the

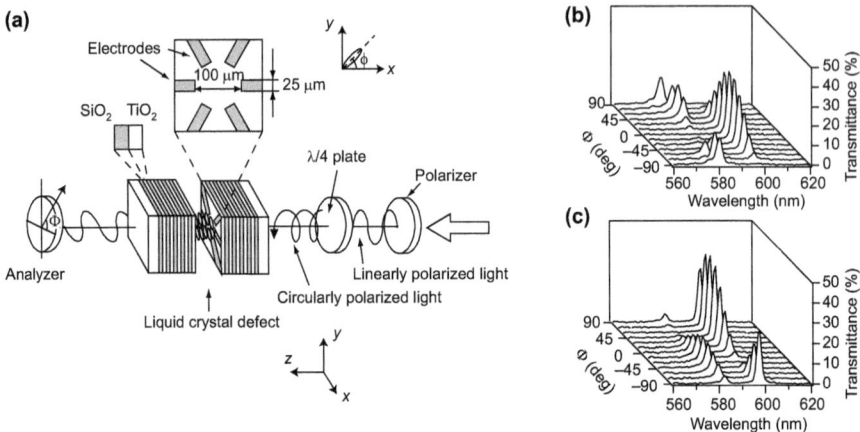

Figure 5.14   (a) 1D photonic crystal with six electrodes to control the azimuthal angle ϕ of the liquid-crystal director in the defect. (b) Transmission spectra at $\phi = 0°$ as a function of the analyzer angle Φ. The peak maxima at 582 and 596 nm are $\Phi = \pm 90°$ and $\Phi = 0°$, respectively. (c) Transmission spectra at $\phi = 60°$. The peak maxima at 582 and 596 nm are $\Phi = -30°$ and $\Phi = 60°$, respectively.

liquid-crystal molecules are originally parallel to the substrates and tilted toward the z-axis and there is no inplane component in the molecular realignment. On the other hand, in this geometry, the molecular long axis rotates in the x–y plane and the optical length of the defect layer does not vary with the reorientation. Therefore, the defect-mode wavelengths for the inplane switching do not shift.

Several applications of the 1D photonic crystal with a nematic liquid-crystal defect layer have been described. However, there are more studies on various types of defect modes using cholesterics,[55–58] ferroelectrics,[59,60] and radially-oriented[61] or twisted nematic liquid crystals.[62] They have unique features as well. For further details, please refer to the original articles.

5.4   Defect-Free Photonic Crystal Infiltrated with a Liquid Crystal

The important features of a defect-free photonic crystal are a photonic bandgap,[14,15] a slow group velocity at a photonic band edge,[20] and unusual light propagation at a high-order photonic band.[63–71] In this section, we study a tunable band-edge laser and tunable negative refraction in a defect-free photonic crystal infiltrated with a liquid crystal.

5.4.1   Tunable Laser in Liquid-Crystal/Polymer Grating

Nanoimprint lithography is well suited for the fabrication of photonic nano-structures. Nanoimprint lithography, which is a method of nanopattern transfer, has been of considerable practical interest for low-cost technology and mass production.[70–74] The technique is expected to be used for development of nanofabricated organic optical devices such as a flexible DFB laser.[75,76] Micropatterns fabricated by nanoimprint lithography have been used not only as optical devices but also for the study of liquid-crystal alignment, because the gratings play an important role in liquid-crystal molecular alignment.[77,78] Here, we study an electrically tunable band-edge laser using a nanoimprinted 1D photonic crystal infiltrated with a liquid crystal.

Figure 5.15(a) shows the scanning electron microscopy images of a replica fabricated by UV-nanoimprint lithography. The quartz nanoimprint mold has 100-nm line and 100-nm space patterns, and its pattern depth is also 100 nm. In the photopolymerized resin, a laser dye is doped to obtain laser emission. When the nanoimprinted grating was optically pumped by second-harmonic light of a Q-switched Nd:YAG laser, laser emissions were observed, as shown in Figure 15.5(b). Here, the pump laser beam was focused at the center of the grating by a spherical lens. Figure 15(c) shows the emission spectrum from the dye-doped 1D grating. At 591 nm, a narrow peak appears on the broad dye spectrum. The emission peak is caused by the DFB cavity.

To infiltrate a nematic liquid crystal into the trenches of the grating, a sandwich structure was prepared using two ITO-coated glasses with and without the imprinted grating, as shown in Figure 5.16(a). The surface of the nonpatterned glass was coated with a polyimide film to align liquid-crystal molecules and rubbed unidirectionally. In contrast, the glass with the grating was not coated with a polyimide film and was not rubbed. The molecules on the nonpatterned glass and the polymer grating were aligned by the rubbed polyimide film and the surface shape of the grating, respectively.

The 532-nm Nd:YAG laser was focused by a cylindrical lens and the sample was irradiated with the beam. In this case, a longer-wavelength dye was used because the resonance wavelength in the DFB cavity shifted to more red by filling the liquid crystal into the trenches. A broad emission from the sample was observed at a low pump energy. The emission spectrum having a 660-nm

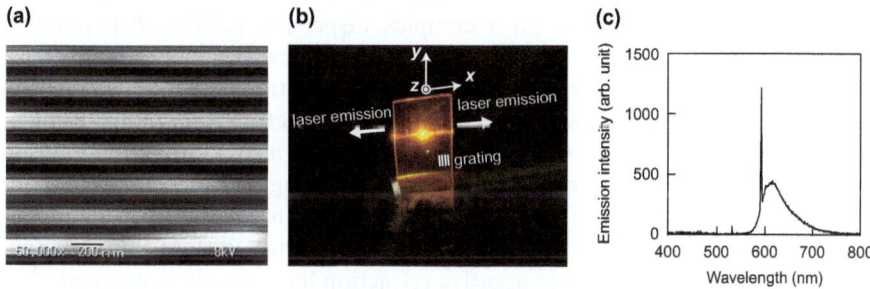

Figure 5.15   (a) Scanning electron microscopy images of a nanoimprinted grating. The pattern is a 100-nm line and a 100-nm space, and its depth is also 100 nm. (b) Photograph of laser action from the dye-doped polymer grating. (c) Emission spectrum from the dye-doped polymer grating without a liquid crystal.

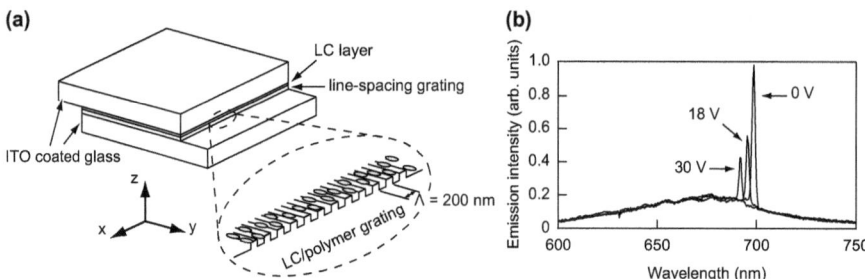

Figure 5.16   (a) Schematic illustration of a liquid-crystal/polymer grating fabricated by nanoimprint lithography. (b) Lasing spectra from the liquid-crystal/ polymer grating for various voltages.

peak conformed to the spontaneous emission of the doped dye. By increasing the pump energy, a sharp peak appeared at 698 nm above a certain threshold energy. To control the refractive index of the LC layer, a square-wave voltage with a frequency of 1 kHz was applied to the ITO electrodes. The sharp peak shifted to shorter wavelengths with increasing voltage, whereas the broad spontaneous emissions of the dye were unchanged. Figure 5.16(b) shows the emission spectra from the grating infiltrated with the liquid crystal for various voltages. The peak shift is due to the molecular reorientation in the liquid crystal by voltage application. Before voltage application, the liquid-crystal molecules in the trench of the polymer grating align in the trench direction, which corresponds to the y-axis of Figure 5.16(a). After voltage application, the molecules are reoriented from the y-axis to the z-axis.

5.4.2   Tunable Negative Refractive Index

In the liquid-crystal-infiltrated nanoimprinted 1D photonic crystal, the band-edge wavelength could be controlled by field-induced reorientation of the liquid-crystal molecules. On the other hand, 2D or 3D photonic crystals have more unique features based on a complex dispersion relation between the frequency and wave vector, for example, suitably designed photonic crystals exhibit a negative refractive index.[65] Before the invention of photonic crystals, unusual light propagation in a dielectric periodic structure was studied using a diffraction grating.[79] Not only a photonic crystal but also a diffraction grating provides complex light propagation due to folding wave vectors. Liquid crystals also show extraordinary beam propagation in which the propagation direction is determined by an ellipsoidal wave-vector surface. By using a nematic liquid crystal, tunable negative refraction has been demonstrated.[80,81] In both cases, an equifrequency surface (EFS), which represents the locus in the reciprocal space of propagating wave vectors at a fixed frequency, is key for the direction of the light propagation.

To study tunable negative refraction, we here deal with a GaAs 2D triangular lattice photonic crystal infiltrated with a nematic liquid crystal. The reason that

GaAs ($n = 3.6$) is chosen is that a large index contrast is necessary to give rise to an effective refractive index in a photonic crystal. Although GaAs is not transparent for visible light, GaAs is one of the highest index materials in the infrared region. Here we assume the GaAs photonic crystal is infiltrated with a dual-frequency nematic liquid crystal having a large birefringence ($\Delta n > 0.3$).[82] Dual-frequency liquid crystals exhibit a positive dielectric anisotropy at low frequencies and a negative dielectric anisotropy at high frequencies. are the illustrations of liquid-crystal molecular alignment in the photonic crystal. In this configuration, when a low-frequency electric field is applied along the z-direction, the liquid-crystal molecules align parallel to the z-direction as shown in Figure 5.17(a). The liquid-crystal molecules should be reoriented perpendicular to the z-direction when a high-frequency electric field is applied along the z-direction.

Let us consider TM mode light propagation consisting of E_z, H_x, and H_y. The electric field E_z is affected by only the extraordinary index when the molecular directors align parallel to the z-direction. In contrast, the E_z is affected by only the ordinary index when the molecular directors are perpendicular to the z-direction. In this configuration, the propagating light is not affected by both the ordinary and extraordinary indices simultaneously. Thus, we could avoid the complicated propagation owing to the optical anisotropy of the liquid crystal because the light is affected by only the ordinary or extraordinary index. This is the important key to obtain a circular-shaped EFS in k-space. Figure 5.18(a) shows the photonic band diagram of the structure calculated by the PWE method. The solid and broken lines are for the ordinary and extraordinary refractive indices, respectively. An effective refractive index of a photonic crystal can be defined when a photonic band forms a bell shape near the Γ point.[65] As shown in Figure 5.18(b) which is an enlarged view of Figure 5.18(a), the 5th and 7th bands are bell-shaped and inverse-bell-shaped profiles near the Γ point. The sign of the effective index is determined by the upward or downward direction of the bell-shaped band. An upward bell-shaped band means the photonic crystal could have positive refractive indices, while a downward bell-shaped band means that the photonic crystal could have negative refractive indices. The effective refractive index of the photonic crystal

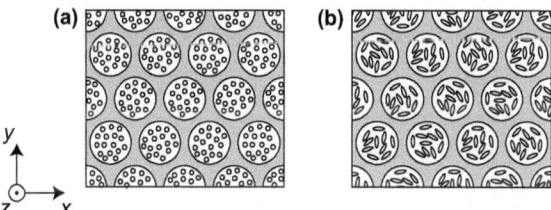

Figure 5.17   A GaAs 2D triangular lattice photonic crystal infiltrated with a dual-frequency nematic liquid crystal. (a) The liquid-crystal directors are parallel to the z-axis when a low-frequency electric field is applied along the z-axis. (b) The directors are perpendicular to the z-axis when a high-frequency electric field is applied along the z-axis.

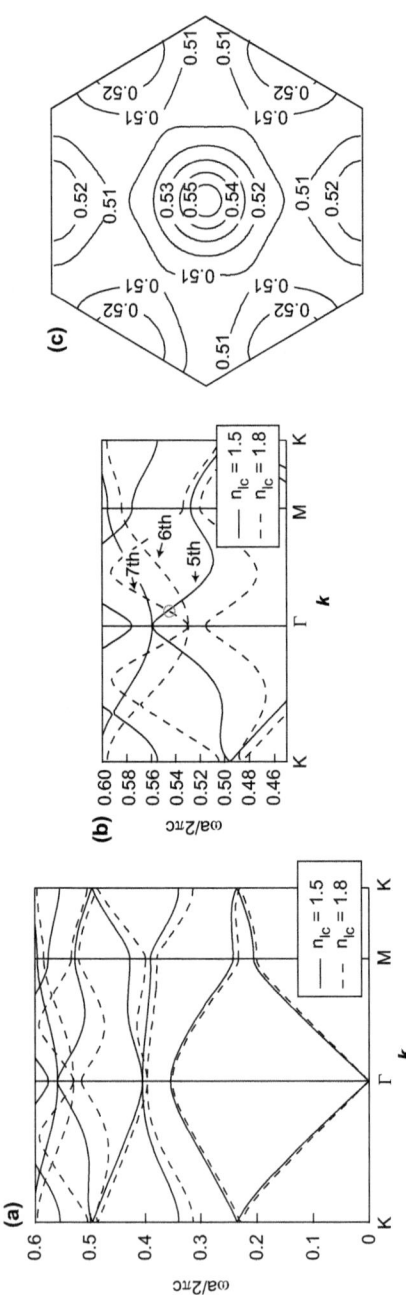

Figure 5.18   Photonic band diagram of the tunable photonic crystal calculated by a plane-wave method. (b) is the enlarged diagram of Figure 18(a). The R/a ratio of the photonic crystal is 0.42, where a is the lattice constant and R is the radius of the holes. (c) is EFS of the 5th band for $n_{lc} = 1.5$.

is determined by $|n_{eff}| = ck/\omega$ that is definable when an EFS is circular. Figure 5.18(c) shows the EFS of the 5th band for $n_{lc} = 1.5$. The EFS corresponds to a contour plot of the photonic band, and a hexagon represents the first Brillouin zone of the 2D triangular photonic crystal. It is clear that the contours become circular above the normalized frequency $\omega a/2\pi c$ of 0.52 at the center. These calculations indicate that the photonic crystal could have an effective refractive index between 0.52 and 0.55. Here, we again emphasize that the effective index is determined by $|n_{eff}| = ck/\omega$. Note that there are some intersection points in Figure 5.18(b). The marked intersections mean that both the bands for $n_{lc} = 1.5$ and $n_{lc} = 1.8$ have the same k and ω. That is, the photonic crystal with $n_{lc} = 1.5$ and $n_{lc} = 1.8$ have the same $|n_{eff}|$ under these conditions. In contrast, the sign of the effective index is not the same because the directions of the bell-shaped bands are opposite between $n_{lc} = 1.5$ and $n_{lc} = 1.8$. Therefore, this indicates that the tunable photonic crystal can change the sign of the effective refractive index by using the liquid crystal.

Figure 5.19 shows the FDTD simulated propagation from the liquid crystal into the tunable photonic crystal at $\omega a/2\pi c = 0.545$ with incident angle of 3°. Figure 5.19(a) is a schematic illustration of tunable diffraction in which the upper medium is liquid crystal, and the lower medium is the liquid-crystal-infiltrated photonic crystal. If the refractive index of the photonic crystal can be controlled from positive to negative, the light propagation direction should be changed as shown in Figure 5.19(a). FDTD results for $n_{lc} = 1.5$ and $n_{lc} = 1.8$ are shown in Figs 19(b) and 19(c), respectively. As is evident in the figures, the refraction direction in the photonic crystal depends on the refractive index of the liquid crystal. This indicates that we could control the refraction direction between positive and negative by using reorientation of the liquid crystal. The effective refractive indices of the photonic crystal with $n_{lc} = 1.5$ and $n_{lc} = 1.8$ are -0.132 and 0.132, respectively, which are determined by $|n_{eff}| = ck/\omega$. Although the value of the effective refractive index is less than the index of air $n_{air} = 1$, it is not wrong. This is because the effective refractive index is for the phase velocity, not for the group velocity. According to Snell's law, the refractive angles are $-36.5°$ and $45.5°$ for $n_{lc}/n_{eff} = -1.5/0.132$ and $n_{lc}/n_{eff} = 1.8/0.132$, respectively. The arrows in Figure 5.19 are written using the angles determined by Snell's law. It is clear that the arrow directions agree well with FDTD simulated propagation.

Light propagation through a combination of positive and negative index media is also examined. Several groups have studied the optical splitter using a combination of triangular-shaped photonic crystals.[83,84] We here consider an optical switch using two triangular shaped photonic crystals shown in Figure 5.20. When the two triangular-shaped photonic crystals have the same effective index, an incident light propagates straight. However, if we use two triangular-shaped photonic crystals having the same absolute value of effective index but opposite signs, an incident light could turn 90° at the interface due to negative refraction. Figure 5.20 shows the two types of light propagation in the system. In Figure 5.20(a), the light propagates straight because they have the same effective index. In contrast, when the two

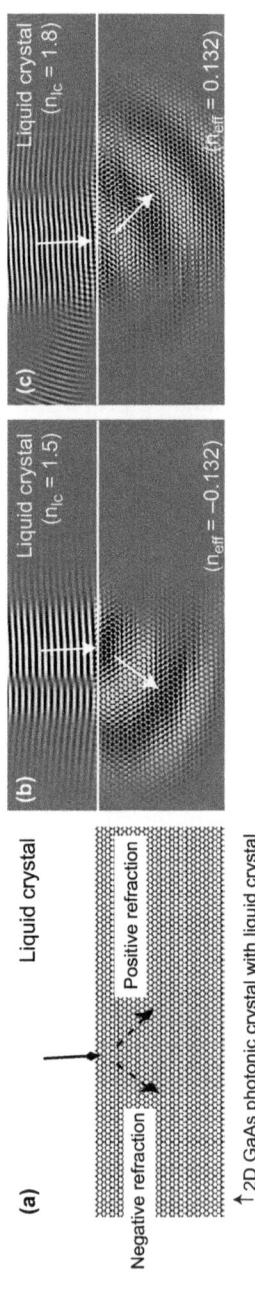

Figure 5.19 FDTD simulated propagation from the liquid crystal into the photonic crystal at $\omega a/2\pi c = 0.545$ with incident angle of $3°$: (a) Schematic illustration of tunable diffraction. (b) $n_{lc} = 1.5$ and (c) $n_{lc} = 1.8$. The tunable photonic crystal is the lower part.

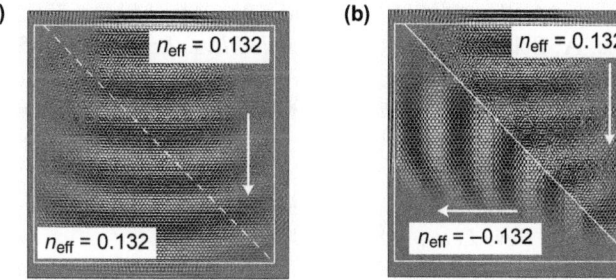

Figure 5.20 Optical switch using two triangular photonic crystals having positive or negative refractive index. FDTD simulated propagation in the combination of two triangular photonic crystals. (a) Photonic crystals with the same effective index. (b) Photonic crystals with opposite effective refractive index signs.

triangular-shaped photonic crystals have positive and negative indices, the light turns 90° to the left. Note that the light is deflected at right angle without a waveguide structure.

In summary, light propagation in the liquid-crystal-infiltrated 2D photonic crystal at the high-order photonic bands has been numerically investigated. The photonic band calculation indicated that the photonic crystal could have a positive or negative effective refractive index depending on liquid-crystal refractive index. The FDTD simulation showed that positive or negative refraction could be controlled by reorientation of the liquid crystal. Further, by combining two triangular photonic crystals having opposite index signs, incident light could be turned 90° without a waveguide structure.

References

1. P. G. de Gennes and J. Port, *The Physics of Liquid Crystals*, Oxford University Press, Oxford, 1995.
2. S. Chandrasekhar, *Liquid Crystals*, Cambridge University Press, Cambridge, 1992.
3. H. Kawamoto, *Proc. IEEE*, 2002, **90**, 460.
4. D. K. Yang and S. T. Wu, *Fundamentals of Liquid Crystal Devices*, Wiley, Hoboken, 2006.
5. K. Yoshino, Y. Shimoda, Y. Kawagishi, K. Nakayama and M. Ozaki, *Appl. Phys. Lett.*, 1999, **75**, 932.
6. K. Yoshino, S. Satoh, Y. Shimoda, Y. Kawagishi, K. Nakayama and M. Ozaki, *Jpn. J. Appl. Phys.*, 1999, **38**, L961.
7. S. W. Leonard, J. P. Mondia, H. M. van Driel, O. Toader, S. John, K. Busch, A. Birner, U. Gösele and V. Lehmann, *Phys. Rev. B*, 2000, **61**, R2389.
8. R. Ozaki, T. Matsui, M. Ozaki and K. Yoshino, *Appl. Phys. Lett.*, 2003, **82**, 3593.

9. B. Maune, M. Lončar, J. Witzens, M. Hochberg, T. Baehr-Jones, D. Psaltis, A. Scherer and Y. Qiu, *Appl. Phys. Lett.*, 2004, **85**, 360.

10. T. S. Perova, V. A. Tolmachev, E. V. Astrova, Y. A. Zharova and S. M. O'Neill, *Phys. Status Solidi C*, 2007, **4**, 1961.

11. R. Ozaki, T. Shinpo, K. Yoshino, M. Ozaki and H. Moritake, *Appl. Phys. Exp.*, 2008, **1**, 012003.

12. E. Yablonovitch, *Phys. Rev. Lett.*, 1987, **58**, 2059.

13. S. John, *Phys. Rev. Lett.*, 1987, **58**, 2486.

14. J. D. Joannopoulos, P. R. Villeneuve and S. Fan, *Solid State Commun.*, 1997, **102**, 165.

15. K. Sakoda, *Optical Properties of Photonic Crystals*, Springer, Berlin, 2005.

16. A. Mekis, J. C. Chen, I. Kurland, S. Fan, P. R. Villeneuve and J. D. Joannopoulos, *Phys. Rev. Lett.*, 1996, **77**, 3787.

17. J. D. Joannopoulos, P. R. Villeneuve and S. Fan, *Nature*, 1997, **386**, 143.

18. K. Srinivasan and O. Painter, *Opt. Exp.*, 2002, **10**, 670.

19. Y. Akahane, T. Asano, B.-S. Song and S. Noda, *Nature*, 2003, **425**, 944.

20. J. P. Dowling, M. Scalora, M. J. Bloemer and C. M. Bowden, *J. Appl. Phys.*, 1994, **75**, 1896.

21. O. Painter, R. K. Lee, A. Scherer, A. Yariv, J. D. O'Brien, P. D. Dapkus and I. Kim, *Science*, 1999, **284**, 1819.

22. K. Sakoda, K. Ohtaka and T. Ueta, *Opt. Exp.*, 1999, **4**, 481.

23. M. Ozaki, M. Kasano, D. Ganzke, W. Haase and K. Yoshino, *Adv. Mater.*, 2002, **14**, 306.

24. M. Ozaki, M. Kasano, T. Kitasho, D. Ganzke, W. Haase and K. Yoshino, *Adv. Mater.*, 2003, **15**, 974.

25. H. Finkelmann, S. T. Kim, A. Muñoz, P. Palffy-Muhoray and B. Taheri, *Adv. Mater.*, 2001, **13**, 1069.

26. A. Chanishvili, G. Chilaya, G. Petriashvili, R. Barberi, R. Bartolino, G. Cipparrone, A. Mazzulla and L. Oriol, *Adv. Mater.*, 2004, **16**, 791.

27. Y. Inoue, H. Yoshida, K. Inoue, Y. Shiozaki, H. Kubo, A. Fujii and M. Ozaki, *Adv. Mater.*, 2011, **23**, 5498.

28. H. Takeda and K. Yoshino, *Phys. Rev. B*, 2003, **67**, 073106.

29. S. Y. Lin, J. G. Fleming, D. L. Hetherington, B. K. Smith, R. Biswas, K. M. Ho, M. M. Sigalas, W. Zubrzycki, S. R. Kurtz and J. Bur, *Nature*, 1998, **394**, 251.

30. A. Blanco, E. Chomski, S. Grabtchak, M. Ibisate, S. John, S. W. Leonard, C. Lopez, F. Meseguer, H. Miguez, J. P. Mondia, G. A. Ozin, O. Toader and H. M. Driel, *Nature*, 2000, **405**, 437.

31. S. Noda, K. Tomoda, N. Yamamoto and A. Chutinan, *Science*, 2000, **289**, 604.

32. H. B. Sun, S. Matsuo and H. Misawa, *Appl. Phys. Lett.*, 1999, **74**, 786.

33. J. S. Foresi, P. R. Villeneuve, J. Ferrera, E. R. Thoen, G. Steinmeyer, S. Fan, J. D. Joannopoulos, L. C. Kimerling, H. I. Smith and E. P. Ippen, *Nature*, 1997, **390**, 143.

34. M. Notomi, E. Kuramochi and H. Taniyama, *Opt. Exp.*, 2008, **16**, 11095.

35. M. D. Tocci, M. J. Bloemer, M. Scalora, J. P. Dowling and C. M. Bowden, *Appl. Phys. Lett.*, 1995, **66**, 2324.
36. T. Hattori, N. Tsurumachi and H. Nakatsuka, *J. Opt. Soc. Am.*, 1997, **14**, 348.
37. Y. Dumeige, P. Vidakovic, S. Sauvage, I. Sgnes and J. A. Levenson, *Appl. Phys. Lett.*, 2001, **78**, 3021.
38. K. M. Ho, C. T. Chan and C. M. Soukoulis, *Phys. Rev. Lett.*, 1990, **65**, 3152.
39. M. Plihal and A. A. Maradudin, *Phys. Rev. B*, 1991, **44**, 8565.
40. K. S. Yee, *IEEE Trans. Antennas Propag.*, 1966, **45**, 354.
41. A. Taflove and S. C. Hagness, *Computational Electrodynamics: The Finite-Difference Time-Domain Method, 3rd edn*, Artech House, London, 2005.
42. D. W. Berreman, *J. Opt. Soc. Am.*, 1971, **62**, 502.
43. R. Ozaki, Y. Matsuhisa, M. Ozaki and K. Yoshino, *Appl. Phys. Lett.*, 2004, **84**, 1844.
44. J. M. Bendickson, J. P. Dowling and Michael Scalora, *Phys. Rev. E*, 1996, **53**, 4107.
45. J. Schmidtkea and W. Stille, *Eur. Phys. J. B*, 2003, **31**, 179.
46. A. S. Nagra and R. A. York, *IEEE Trans. Antennas Propag.*, 1998, **46**, 334.
47. X. Jiang and C. M. Soukoulis, *Phys. Rev. Lett.*, 2000, **85**, 70.
48. S.-H. Chang and A. Taflove, *Opt. Exp.*, 2004, **12**, 3827.
49. S. Shi and D. W. Prather, *Opt. Exp.*, 2007, **15**, 10294.
50. T. Matsui and M. Kitaguchi, *Appl. Phys. Exp.*, 2010, **3**, 061701.
51. R. Ozaki, M. Ozaki and K. Yoshino, *Jpn. J. Appl. Phys.*, 2003, **42**, L669.
52. R. Ozaki, H. Moritake, K. Yoshino and M. Ozaki, *J. Appl. Phys.*, 2007, **101**, 033503.
53. F. C. Frank, *Faraday Soc.*, 1958, **25**, 19.
54. R. Ozaki, M. Ozaki and K. Yoshino, *Jpn. J. Appl. Phys.*, 2004, **43**, L1477.
55. M. H. Song, B. Park, K. C. Shin, T. Ohta, Y. Tsumoda, H. Hoshi, Y. Takanishi, K. Ishikawa, J. Watanabe, S. Nishimura, T. Toyooka, Z. Zhu, T. M. Swager and H. Takezoe, *Adv. Mater.*, 2004, **16**, 779.
56. Y. Matsuhisa, R. Ozaki, M. Ozaki and K. Yoshino, *Jpn. J. Appl. Phys.*, 2005, **44**, L629.
57. J. Hwang, M. H. Song, B. Park, S. Nishimura, T. Toyooka, J. W. Wu, Y. Takanishi, K. Ishikawa and H. Takezoe, *Nature Mater.*, 2005, **4**, 383.
58. N. Y. Ha, Y. Ohtsuka, S. M. Jeong, S. Nishimura, G. Suzaki, Y. Takanishi, K. Ishikawa and H. Takezoe, *Nature Mater.*, 2008, **7**, 43.
59. Y. Matsuhisa, W. Hasse, A. Fujii and M. Ozaki, *Appl. Phys. Lett.*, 2006, **89**, 201112.
60. C. H. Chen, V. Y. Zyryanov and W. Lee, *Appl. Phys. Exp.*, 2012, **5**, 082003.
61. K. Tagashira, H. Yoshida, H. Kubo, A. Fujii and M. Ozaki, *Appl. Phys. Exp.*, 2010, **3**, 062002.
62. Y. T. Lin, W. Y. Chang, C. Y. Wu, V. Y. Zyryanov and W. Lee, *Opt. Exp.*, 2010, **18**, 26959.

63. H. Kosaka, T. Kawashima, A. Tomita, M. Notomi, T. Tamamura, T. Sato and S. Kawakami, *Phys. Rev. B*, 1998, **58**, R10096.
64. L. Wu, M. Mazilu, T. Karle and T. F. Krauss, *IEEE J. Quantum. Electron.*, 2002, **38**, 915.
65. M. Notomi, *Phys. Rev. B*, 2000, **62**, 10696.
66. P. V. Parimi, W. T. Lu, P. Vodo and S. Sridhar, *Nature*, 2003, **426**, 404.
67. E. Cubukcu, K. Aydin, E. Ozbay, S. Foteinopoulou and C. M. Soukoulis, *Nature*, 2003, **423**, 604.
68. C. Monzon, P. Loschialpo, D. Smith, F. Rachford, P. Moore and D. W. Frester, *Phys. Rev. Lett.*, 2006, **96**, 207402.
69. A. Berrier, M. Mulot, M. Swillo, M. Qiu, L. Thylén, A. Talneau and S. Anand, *Phys. Rev. Lett.*, 2004, **93**, 073902.
70. H. Takeda and K. Yoshino, *Phys. Rev. E*, 2003, **67**, 056607.
71. R. Ozaki, H. Moritake, K. Yoshino and A. A. Zakhidov, *Mol. Cryst. Liq. Cryst.*, 2011, **545**, 1291.
72. S. Y. Chou, P. R. Krauss and P. J. Renstrom, *Appl. Phys. Lett.*, 1995, **67**, 3114.
73. S. Y. Chou, P. R. Krauss and P. J. Renstrom, *J. Vac. Sci. Technol. B*, 1996, **14**, 4129.
74. L. J. Guo, *Adv. Mater.*, 2007, **19**, 495.
75. M. Berggren, A. Dodabalapur, R. E. Slusher, A. Timko and O. Nalamasu, *Appl. Phys. Lett.*, 1998, **72**, 410.
76. R. Ozaki, T. Shinpo, K. Yoshino, M. Ozaki and H. Moritake, *Appl. Phys. Exp.*, 2008, **1**, 012003.
77. Y.-T. Kim, S. Hwang, J.-H. Hong and S.-D. Lee, *Appl. Phys. Lett.*, 2006, **89**, 173506.
78. Y. Yi, M. Nakata, A. R. Martin and N. A. Clark, *Appl. Phys. Lett.*, 2007, **90**, 163510.
79. P. S. J. Russell, *Phys. Rev. A*, 1986, **33**, 3232.
80. Q. Zhao, L. Kang, B. Li, J. Zhou, H. Tang and B. Zhang, *Appl. Phys. Lett.*, 2006, **89**, 221918.
81. O. P. Pishnyak and O. D. Lavrentovich, *Appl. Phys. Lett.*, 2006, **89**, 251103.
82. J. Sun, H. Xianyu, S. Gauza and S. Y. Wu, *Liq. Cryst.*, 2009, **36**, 1401.
83. D. M. Pustai, S. Shi, C. Chen, A. Sharkawy and D. W. Prather, *Opt. Exp.*, 2004, **12**, 1823.
84. V. Zabelin, L. A. Dunbar, N. L. Thomas, R. Houdre, M. V. Kotlyar, L. O'Faolain and T. F. Krauss, *Opt. Lett.*, 2007, **32**, 530.

CHAPTER 6

Tunable Colloidal Crystals Immobilized in Soft Hydrogels

TOSHIMITSU KANAI

Faculty of Engineering, Yokohama National University, 79-5 Tokiwadai, Hodogaya, Yokohama, Kanagawa 240-8501, Japan
*Email: tkanai@ynu.ac.jp

6.1 Introduction

Three-dimensional periodic arrays of monodisperse colloidal particles are called colloidal crystals, by analogy with atomic crystals.[1–10] They have attracted considerable attention in the field of materials science and in technological applications because of their novel optical applications as photonic crystals.[11–15] Photonic crystals are periodic structures with a refractive index on a scale comparable to the wavelength of light. They form photonic band structures in which the motion of photons can be controlled, similar to the way in which the periodicity of a semiconductor crystal can control the motion of electrons. The periodic structure can be artificially fabricated by a top-down method, such as a lithographic technique, and various distinct optical phenomena have been verified.[16–19] Colloidal crystals have spatial periodicity of the refractive index between the particles and the surroundings, and hence they act as photonic crystals. Since they are prepared by a bottom-up self-assembly process in which the particles spontaneously assemble to form the crystalline structure, the periodic structure cannot be designed as well as through the lithographic technique. However, colloidal crystals have a great advantage in that large crystals can be prepared at low cost and are expected to be mass-producible photonic crystalline materials. In addition,

RSC Smart Materials No. 5
Responsive Photonic Nanostructures: Smart Nanoscale Optical Materials
Edited by Yadong Yin
© The Royal Society of Chemistry 2013
Published by the Royal Society of Chemistry, www.rsc.org

three-dimensional photonic crystals with optical stop-bands in the range of ultraviolet (UV) and visible light can be prepared simply using submicrometer particles, which cannot be achieved straightforwardly by the top-down method.

The optical stop-band of colloidal crystals is observed as the Bragg reflection. As atomic crystals exhibit X-ray diffraction, a similar diffraction occurs in colloidal crystals according to Bragg's law (Figure 6.1),

$$m\lambda = 2n_c d_{hkl} \sin\theta \tag{6.1}$$

where m is an integer, λ is the wavelength of light, n_c is the refractive index of the crystal, d_{hkl} is the interplanar spacing of the hkl lattice planes, and θ is the Bragg diffraction angle. In the case of X-ray diffraction from atomic crystals, n_c can be regarded as unity. As shown in eqn (6.1), the Bragg wavelength of the colloidal crystal at a specific incident light angle is determined from the value of the refractive index and lattice constant of the crystal. Since it is difficult to drastically change the refractive index of materials, adjusting the lattice constant is essential for setting the desired Bragg wavelength.

Colloidal crystals may be classified into two categories on the basis of their crystallization mechanism and packing (Figure 6.2). One comprises opal-type colloidal crystals,[1–5] in which monodisperse particles are aligned with the closest-packing in a dry state. By evaporating the liquid medium in the colloidal suspension on a substrate, the particles spontaneously assemble to form the closest-packed structure on the substrate. Their inverse structure, which is called the inverse opal,[20–22] can be obtained by infiltration of a different material with a high refractive index into the void space in the opal structure. Since a high contrast in refractivity between the particles and the surroundings is preferable in applications as photonic bandgap materials, this type of colloidal crystal has been extensively studied. However, although various methods for preparing opal structures, such as cell packing,[23] graphoepitaxy,[24] substrate dipping,[25,26] and centrifugation,[21] have been proposed, the preparation of large single crystals, which are desirable in many applications, is difficult because crack networks, ranging up to a few hundred micrometers in size, are generally formed during the drying process. In addition, since the

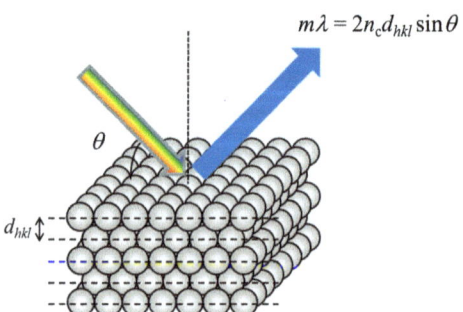

Figure 6.1   Schematic image of Bragg diffraction in colloidal crystals.

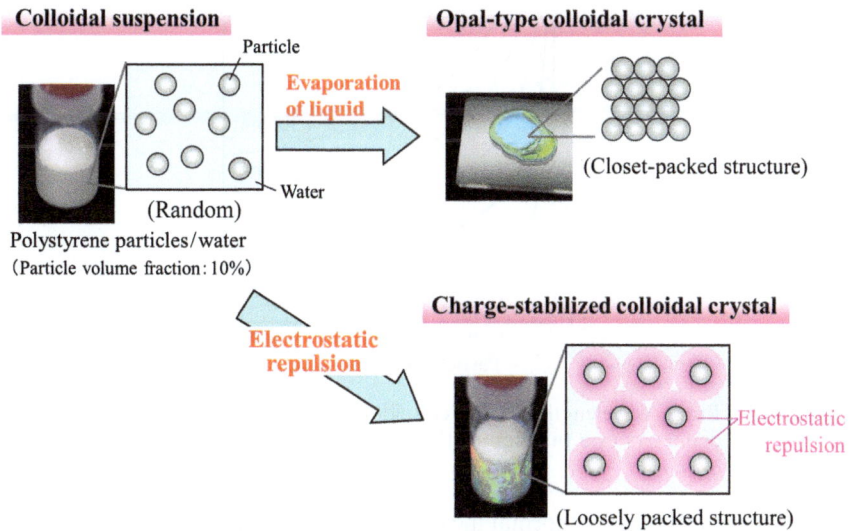

Figure 6.2   Schematic images of two types of colloidal crystals.

particles in the opal structure touch one another, the crystal lattice constant is uniquely determined by the particle diameter. Therefore, the preparation of colloidal crystals with significantly different Bragg wavelengths requires colloidal particles of different sizes.

The other type of colloidal crystal comprises charge-stabilized colloidal crystals, in which charged particles form crystalline structures, but with a low packing density in the liquid medium because of the repulsive electrostatic interactions between the particles.[6–9] Since the electrostatic interaction reaches over a wide range, the charge-stabilized system seems to have a high possibility of forming large single crystals. Furthermore, an advantage of the charge-stabilized colloidal crystals is that the lattice constant can be significantly altered by changing the particle volume fraction. Thus, the Bragg wavelength of the crystals can be tuned over a wide range. For example, the range of the Bragg wavelength for polystyrene colloidal crystals with face-centered-cubic (FCC) structure in water can be estimated as follows. The lattice constant a for FCC structure is derived from a geometrical consideration, using the particle volume fraction ϕ and the particle diameter d as

$$a=\left(\frac{2\pi}{3\phi}\right)^{1/3} d \tag{6.2}$$

The value of n_c can be approximated by a volume-weighted average of the refractive indices of the particles, n_p, and the medium, n_{md}, as follows:[27,28]

$$n_c = n_p\phi + n_{md}(1 - \phi) \tag{6.3}$$

where the refractive indices of the polystyrene particles and water are $n_p = 1.59$ and $n_{md} = 1.33$, respectively. Considering the Bragg diffraction from the FCC

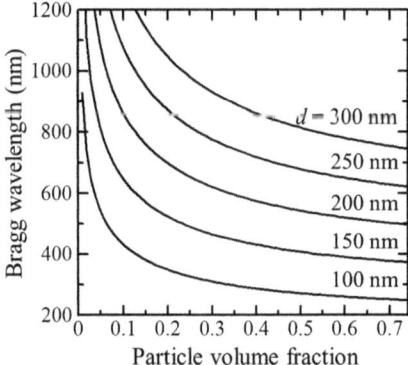

Figure 6.3 Bragg wavelengths of the colloidal crystals with representative particle diameters as a function of the particle volume fraction.

(111) planes in the case of normal incidence ($\theta = 90°$), the Bragg wavelengths of crystals with different particle sizes can be estimated as a function of the particle volume fraction using eqns (6.1)–(6.3) and $d_{111} = \frac{a}{\sqrt{3}}$, as shown in Figure 6.3. For particles with diameters of 100–300 nm, the Bragg wavelengths can be widely tuned in the range of UV, visible, and near-infrared light by changing the particle volume fraction.

A drawback of the charge-stabilized system, in which charged particles are suspended in a liquid medium, is its instability against influences such as mechanical shocks and chemical contamination. Therefore, a method to immobilize the unstable particle arrays in a hydrogel has been developed.[29,30] Hydrogels are covalently crosslinked three-dimensional polymer networks that contain a large amount of water filling the interstitial network spaces at the molecular level. By dissolving gelation reagents that include a monomer, a crosslinker, and a polymerization initiator in an aqueous suspension of charged colloidal crystals and polymerizing them, the crystalline particle arrays can be immobilized in a hydrogel network. More interestingly, the gel-immobilized colloidal crystals thus prepared exhibit an additional very attractive feature: the Bragg wavelength can be manipulated due to the looseness in packing and the stimulus-response properties of the hydrogel (Figure 6.4). Since hydrogels undergo volume changes in response to external stimuli such as temperature, pH, or mechanical stress, the lattice constant or Bragg wavelength of the crystals embedded in the gel can be tuned by adjusting the volume of the hydrogel through an external stimulus.[31–33] This is useful for applications such as tunable photonic crystals and biological and chemical sensors for monitoring the changes in the environment through the Bragg diffraction wavelength or diffraction color. In order to realize the practical uses of such attractive features, a reasonable method for preparing large single-crystalline charged colloidal crystals and immobilizing them in hydrogel networks, while preserving crystalline quality, is required.

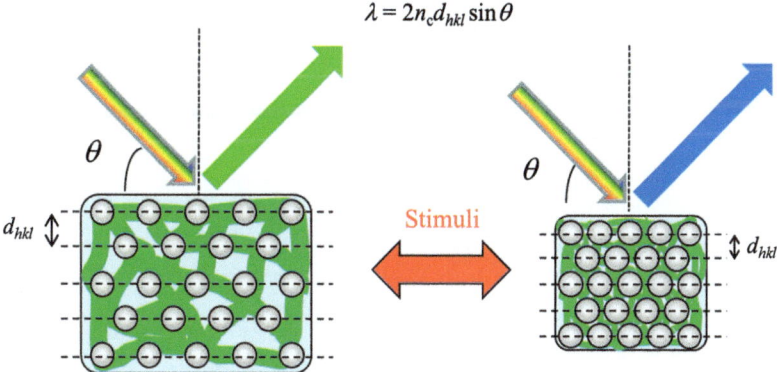

$$\lambda = 2n_c d_{hkl} \sin\theta$$

Figure 6.4 Schematic image of a tunable colloidal crystal immobilized in a gel network.

This chapter will provide an overview of our recent work on gel-immobilized colloidal crystals with high optical quality, including preparation methods and tuning properties. Two preparation methods for gel-immobilized colloidal crystals that combine microfluidics and a photopolymerization technique will be described. The first produces a centimeter-sized single-crystal-like colloidal crystal film through use of an air-pulse-drive system. We found that polycrystalline charged colloidal crystals could be converted into single-crystal-like colloidal crystals by running the suspension through a flat capillary cell, according to a flow-induced shear effect. The second, for spherical colloidal crystals such as colloidal crystal microspheres and shells, uses emulsion droplets as templates. Although colloidal crystals cannot form single crystals in the spherical shape, they exhibit a useful characteristic for applications, as will be described later. These crystals can be immobilized in a hydrogel network by a photopolymerization technique without deterioration of crystalline quality. Then, tuning by external stimuli, including temperature, mechanical compression, and solvent exchange using ionic liquids, will be described. In addition, the control of the effective bandwidth in gel-immobilized colloidal crystals will be shown. Finally, the conversion of the gel-immobilized colloidal crystal film into a dry film of densely packed colloidal crystals without cracks will be demonstrated.

6.2 Fabrication of Gel-Immobilized Colloidal Crystals

6.2.1 Single-Crystal-Like Colloidal Crystal Film

We developed an air-pulse-drive system to prepare large single-crystal-like colloidal crystal films (Figure 6.5).[34] This system is primarily composed of a flat capillary cell made of fused quartz (internal dimensions: 0.1 mm thick, 9 mm wide, 70 mm long), a digital pulse regulator (Musashi Engineering Inc., ML-606GX), and an air compressor. An aqueous suspension of charged colloids in a polycrystalline state is loaded into a plastic syringe, which is

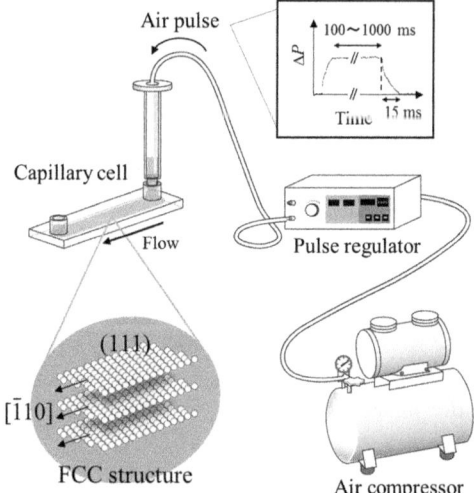

Figure 6.5   Schematic image of the air-pulse-drive system for the preparation of single-crystal-like colloidal crystals.
Reproduced with permission from *Adv. Funct. Mater.*, 2005, **15**, 25.[34]
Copyright 2005 Wiley-VCH Verlag GmbH & Co. KGaA.

connected to the pulse regulator attached to the air compressor. The syringe is connected to one side of the capillary cell, and then, the suspension is injected into the cell by a short air pulse generated by electronically controlled valve action in the regulator. The profile of the air pulse pressure (ΔP, difference from atmospheric pressure), with its stepwise shape and a sharply terminated tail (inset, Figure 6.5), is an important feature of this system. The strong flow can be abruptly stopped; otherwise, the slow tapering of the flow deteriorates the quality of the crystalline texture. Adjusting the pulse height in terms of pressure can quantitatively control the strength of the pulsed flow of the colloidal crystals. We used an aqueous suspension of charged polystyrene particles (particle diameter: 198 nm; particle volume fraction: 10%) in the poly-crystalline state.

Figures 6.6(a) and (b) show optical microscopy images and transmission spectra of the colloidal crystals prepared in the cells at various ΔP values, respectively. At a low ΔP of 1 kPa, an irregular polycrystalline texture was observed through the microscope. However, upon increasing ΔP, textural uniformity increased, and at $\Delta P = 15$ kPa, the prepared texture appeared almost uniform. In the spectra in Figure 6.6(b), this ΔP-dependent change was clearly observable in the region with wavelengths shorter than the dip at 845 nm, which was caused by the Bragg diffraction from the (111) lattice planes of the FCC structure parallel to the cell surface. The transmittance of the sample prepared at $\Delta P = 1$ kPa was as low as 30% in this region. However, upon increasing ΔP, the transmittance markedly increased and then saturated at a high value (80%) at $\Delta P > 10$ kPa.

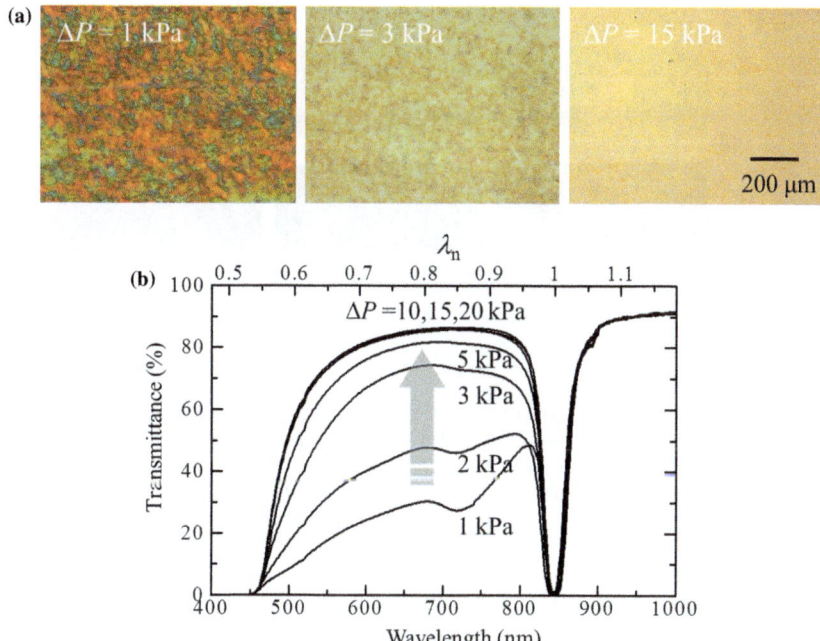

Figure 6.6 (a) Transmission optical microscopy images of colloidal crystals processed at the indicated ΔPs. (b) Transmission spectra for colloidal crystals processed at various ΔPs. λ_n represents a value normalized by the dip wavelength.
Reproduced with permission from *Adv. Funct. Mater.*, 2005, **15**, 25.[34]
Copyright 2005 Wiley-VCH Verlag GmbH & Co. KGaA.

Figure 6.7 shows imaging spectrographs of the same samples over a wide sample area, taken with an inplane image resolution of 100×25 micrometers, to spectrally confirm spatial uniformity. The images represent the spatial distribution of the transmittance as brightness, where the absence of uniformity in the optical properties is detected by contrasts in brightness.[35] For wavelengths above and at the dip ($\lambda_n = 1.1$ and 1.0), the transmittance of the entire area of the texture was very uniform regardless of ΔP. For wavelengths below the dip ($\lambda_n = 0.8$), however, a nonuniform texture with low brightness was observed at $\Delta P = 1$ kPa, whereas both uniformity and brightness increased with ΔP. These data indicate that the crystal prepared at $\Delta P > 10$ kPa had excellent uniformity from the spectroscopic point of view.

The observed ΔP dependence is simply explained as a consequence of the decrease in the fraction of irregular polycrystalline domains with increasing flow rate, according to the shear effect.[36–38] Considering the Bragg conditions of the FCC structure, the longest Bragg wavelength occurred when the incident light was normal to the (111) planes, which corresponds to the dip in Figure 6.6(b). Since the sample prepared at low ΔP included polycrystalline domains with random orientations, various Bragg reflections occurred at

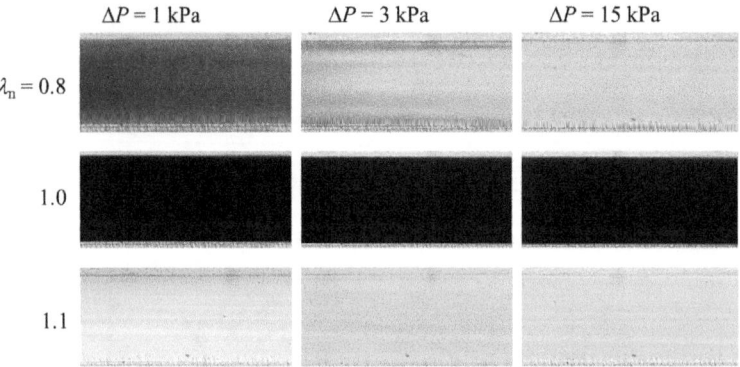

Figure 6.7 Single-wavelength images for representative wavelengths taken under transmission illumination for the colloidal crystals in the flat capillary processed at the indicated ΔPs.
Reproduced with permission from *Adv. Funct. Mater.*, 2005, **15**, 25.[34] Copyright 2005 Wiley-VCH Verlag GmbH & Co. KGaA.

wavelengths shorter than the dip. Therefore, a reduction in transmittance at low ΔP occurred in the shorter-wavelength region but not in the longer one. Upon increasing ΔP, the polycrystalline domains of the crystals were aligned by the shear flow generated in the cell, resulting in an increase in the transmittance. The saturation behavior above $\Delta P = 10$ kPa suggests that the polycrystalline domains were oriented highly enough to be regarded as a single-crystal-like structure.

The crystal thus prepared in the cell showed a distinct laser diffraction pattern, *i.e.* a Kossel pattern.[8,39] As shown in Figure 6.8, clear hexagonally symmetric patterns were observed. Quantitative analysis of the patterns indicated that almost the entire capillary space was filled with a single-domain crystal with a fixed crystallographic orientation determined by the cell geometry. The crystal had a cubic-close-packed structure of colloids with stacking faults, or a twinned FCC structure with a fixed orientation; the FCC (111) lattice plane was parallel to the cell face, and one of the closest-packed directions in the (111) plane was parallel to the cell axis. The crystal also showed angle-dependent uniform diffraction colors (Figure 6.9) and a transmission spectrum with angle-dependent sharp dips (Figure 6.10). These data demonstrated the high optical quality of the sample as a photonic crystal, *i.e.* the excellent impermeability at the stop-band (about 25 dB), independent of the tilting angle. This feature can be readily used as a tunable optical filter.

Although the single-crystal-like colloidal crystal thus formed in the cell was fragile, we succeeded in its immobilization in a hydrogel film by photo-polymerization, preserving its high crystalline quality.[40,41] Before injecting the colloidal suspension into the capillary cell, gelation reagents (*N*-methyl-olacrylamide (NMAM) as a monomer, *N*,*N'*-methylenebisacrylamide as a crosslinker, and camphorquinone as a polymerization photoinitiator) were dissolved in the suspension. In order to prevent deactivation of the radical

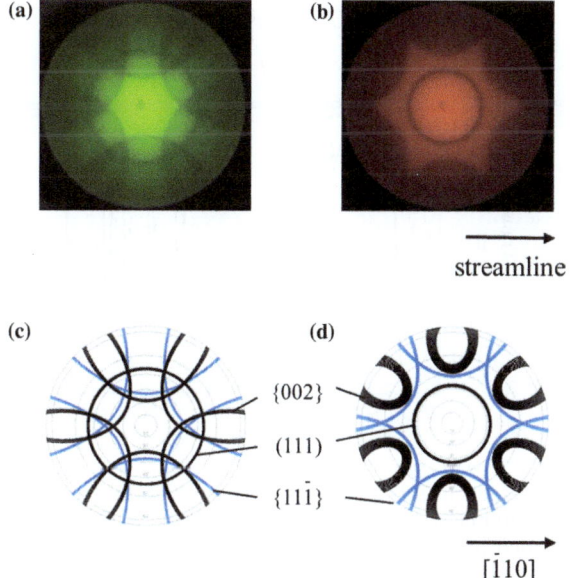

Figure 6.8 The experimentally obtained Kossel patterns for the colloidal crystal using (a) green (543.5 nm) and (b) red (632.8 nm) laser beams. The simulated Kossel patterns at (c) 543.5 nm and (d) 632.8 nm.
Reprinted with permission from *Jpn. J. Appl. Phys.*, 2003, **42**, L655.[39]
Copyright 2003 The Japan Society of Applied Physics.

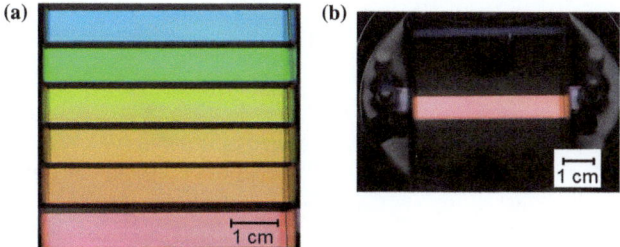

Figure 6.9 (a) Photographs of the colloidal crystal formed in the flat capillary cell taken from different angles for (111) Bragg diffraction. (b) To reduce light reflection at the cell surface, the cell was immersed in water with a prism on top.
Reprinted with permission from *Adv. Funct. Mater.*, 2005, **15**, 25.[34]
Copyright 2005 Wiley-VCH Verlag GmbH & Co. KGaA.

polymerization initiator, dissolved oxygen was removed from the suspension by Ar bubbling. After running the suspension in the cell, the single-domain crystal obtained by the shear effect was uniformly irradiated with two sets of high-brightness blue LED arrays from both sides of the cell surface through light diffusers to polymerize the dissolved gelation reagents (Figure 6.11).

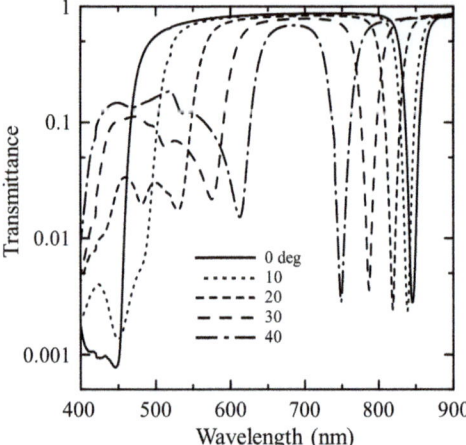

Figure 6.10 Angle-dependent transmission spectra of the colloidal crystal on the logarithmic scale. Here, the incident angle was swung from the direction normal to the flat face of the capillary to the direction of the elongated axis of the capillary, and the probe light was polarized perpendicularly to the plane of incidence, *i.e.* s-polarized light.
Reprinted with permission from *Adv. Funct. Mater.*, 2005, **15**, 25.[34] Copyright 2005 Wiley-VCH Verlag GmbH & Co. KGaA.

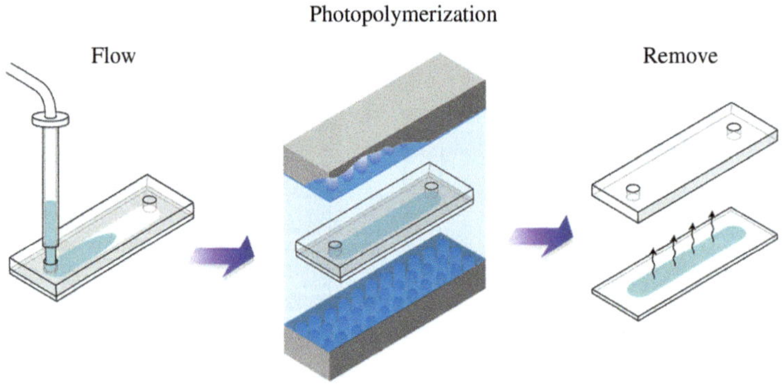

Figure 6.11 Schematic image of the procedures used for the preparation of gel-immobilized single-crystal-like colloidal crystal films using the flow method and photopolymerization.

Figure 6.12(a) shows the transmission spectra of the colloidal crystals before and after gelation. By properly adjusting the gelation conditions, including the oxygen-removal procedure and the irradiation intensity and time, an excellent spectral profile, *i.e.* a deep dip due to the stop-band and high transmittance at the passband wavelength, could be preserved after gelation. In addition, single-wavelength images indicated that high uniformity over a large area was preserved after gelation (Figure 6.12(b)). Figure 6.12(c) shows a photograph of the obtained gel film. The crystal could be removed from the cell as a

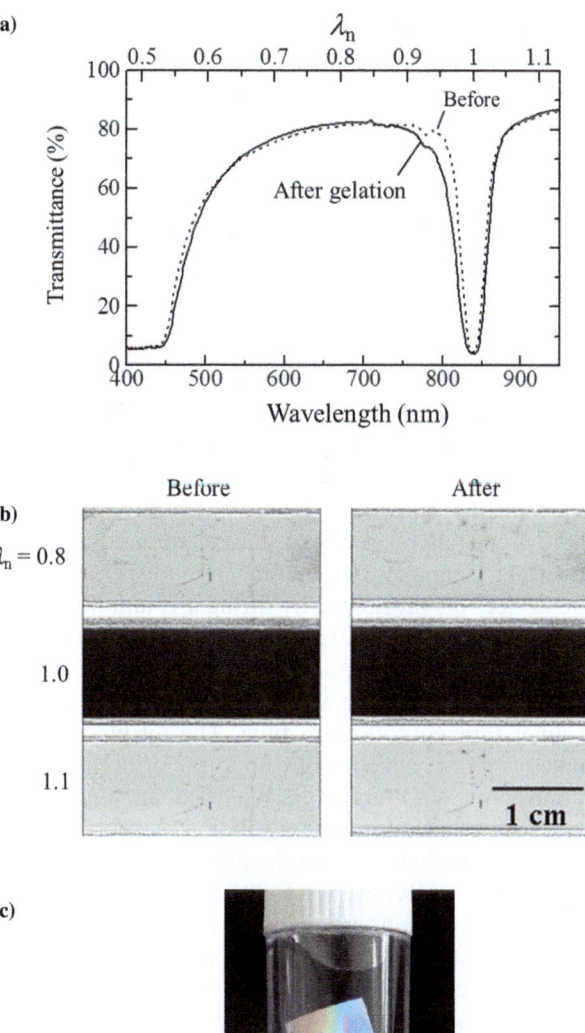

Figure 6.12 (a) Transmission spectra of the colloidal crystals before and after gelation. (b) Single-wavelength images taken under transmission illumination before and after gelation. The incident light was normal to the cell face. (c) Photograph of the gel-immobilized colloidal crystal film soaked in water.
Reprinted with permission from *J. Ceram. Soc. Jpn.*, 2012, **120**, 87.[9]
Copyright 2012 The Ceramic Society of Japan.

self-standing film with sufficient solidity to be handled with tweezers. Since the curved surface of the film was illuminated from various angles by room lighting, it showed multiple diffraction colors.

6.2.2 Colloidal Crystal Spheres

Although colloidal crystals cannot form single crystals that are spherically shaped, they have a unique characteristic: the Bragg diffraction wavelength is independent of the rotation under surface illumination at a fixed incident angle of the light.[42,43] Therefore, spherical colloidal crystals are potentially useful not only for conventional photonic crystal applications, but also as new types of spectrum-encoding carriers for biomolecular screening,[44,45] nonbleachable color pigments,[46] and refractive color displays.[47] We prepared spherical colloidal crystals immobilized in a gel network by the combination of microfluidics[48–50] and photopolymerization techniques.[51] The preparation process is schematically shown in Figure 6.13. In order to form charge-stabilized colloidal crystals in spherical shapes, we used a microfluidic device with a flow-focusing geometry to generate monodisperse water-in-oil (W/O) emulsions. The device was fabricated by fitting a round tapered glass capillary tube with an outer diameter of 1.0 mm into a square capillary with an inner diameter of 1.0 mm. The aqueous phase was pumped from one end of the outer square capillary while the oil phase was pumped from the opposite end into the orifice of the inner tapered collection tube. The oil phase hydrodynamically focused the aqueous phase, which broke up at the orifice of the collection tube to form monodisperse W/O droplets. We used an aqueous suspension of charged polystyrene particles containing gelation reagents (acrylamide (AAM) as a monomer, N,N'-methylenebisacrylamide as a crosslinker, and a polymerization photoinitiator (IRGACURE 2959®, Ciba)) as the aqueous phase. The oil phase was a polydimethylsiloxane oil (PDMS) containing a surfactant (Dow Corning 749) to stabilize the oil/water interfaces without destroying the crystalline state of the charged colloids. The solutions were filtered and then infused into the microfluidic device by positive displacement syringe pumps. By properly adjusting the conditions for fabrication, including the flow rates and the concentrations of the particles, surfactant, and the gelation reagents, stable

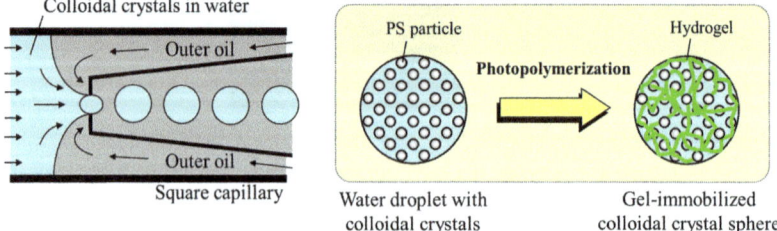

Figure 6.13 Schematic image of the procedures used for the preparation of gel-immobilized colloidal crystal spheres using capillary microfluidics and photopolymerization.

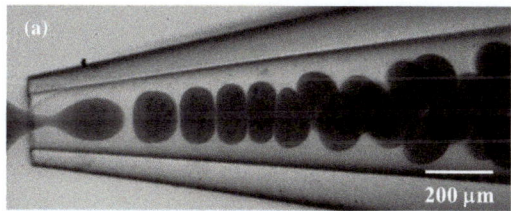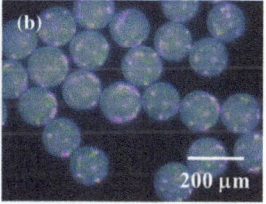

Figure 6.14 (a) Formation of water droplets with colloidal crystals in the microfluidic device. The flow rates of the inner aqueous and outer oil phases were 1000 and 2000 μL h^{-1}, respectively. (b) Optical microscopy image of the spherical colloidal crystals immobilized in PAAM hydrogels after UV light irradiation.
Reproduced with permission from *Small*, 2010, **6**, 807.[51] Copyright 2010 Wiley-VCH Verlag GmbH & Co. KGaA.

monodisperse aqueous droplets with colloidal crystals were formed periodically, as shown in Figure 6.14(a). The colloidal crystal droplets could be subsequently immobilized in polyacrylamide (PAAM) gel spheres by photo-irradiation, without losing the crystalline structure of the colloids; this was evident from the preservation of the diffraction colors shown in Figure 6.14(b).

The sizes of the gel spheres were determined by those of the pregel droplets, which could be controlled by adjusting the flow rates or the size of the capillary orifice. In addition, the packing density of the colloidal crystals was adjustable; the diffraction colors or Bragg wavelengths could be tuned by varying the concentration of particles in the droplets before gelation, as shown in Figure 6.15. Upon illumination by visible light, the gel-immobilized colloidal crystal spheres showed the characteristic colors of diffraction patterns;[42,52] a bright spot could be seen at the center of the sphere, while many spots or lines on a series of concentric rings could also be observed (Figures 6.15(a)–(c)). This was due to the spherical shape and relatively high contrast between the refractive indices of the colloidal particles and the surroundings.[42,47] In the reflection spectra of the spheres, reflection peaks were observed from the bright dots at the centers of the colloidal crystal spheres (Figure 6.15(d)), which were assigned to the Bragg diffractions from the (111) lattice planes of the FCC structure aligned normal to the incident light. The detection of only the 111 reflection peaks suggests that the colloidal particles in the sphere were well oriented parallel to the spherical interfaces as the FCC (111) lattice planes.

6.2.3 Colloidal Crystal Shells

In order to prepare gel-immobilized colloidal crystal shells, we used a microfluidic device that combined coflow and flow-focusing geometries to generate monodisperse oil-in-water-in-oil (O/W/O) double emulsions (Figure 6.16).[53] In comparison with the device for generating single emulsions, as shown in Figure 6.13, an injection tube for forming inner oil droplets was added in the device. The PDMS oil was pumped through the injection tube, while the

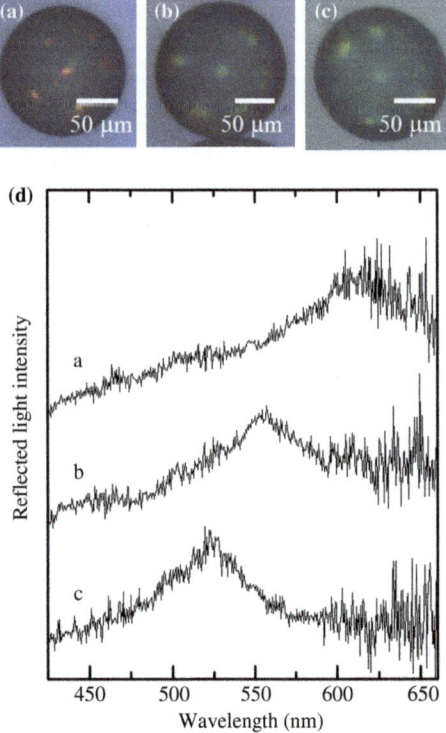

Figure 6.15   Photograph of the PAAM-immobilized colloidal crystal spheres with different particle concentrations ((a) $\phi = 0.31$, (b) $\phi = 0.46$, (c) $\phi = 0.59$) under light at normal incidence. (d) Normal reflection spectra of the PAAM-immobilized colloidal crystal spheres shown in (a)–(c).
Reproduced with permission from *Small*, 2010, **6**, 807.[51] Copyright 2010 Wiley-VCH Verlag GmbH & Co. KGaA.

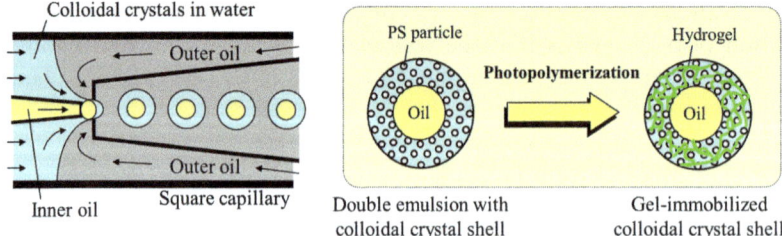

Figure 6.16   Schematic image of the procedures used for the preparation of microcapsules with gel-immobilized colloidal crystal shells using capillary microfluidics and photopolymerization.
Reproduced with permission from *Adv. Mater.*, 2010, **22**, 4998.[53] Copyright 2010 Wiley-VCH Verlag GmbH & Co. KGaA.

aqueous colloidal suspension in the crystalline state, containing gelation reagents and a surfactant (Tween 20), was pumped through the region between the injection tube and the outer square capillary in the same direction. The outer oil phase flowed through the outer capillary in the opposite direction, and hydrodynamically focused the inner and middle phases, which broke up at the orifice of the collection tube to form monodisperse double emulsion droplets with colloidal crystal shells.

When the fabrication conditions were not set appropriately, stable double emulsions with colloidal crystal shells could not be obtained. For example, an

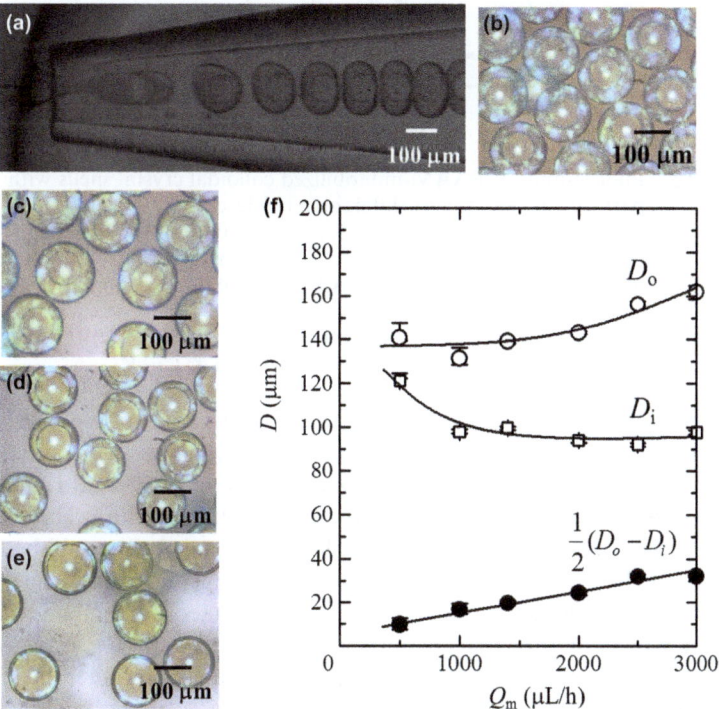

Figure 6.17 (a) Formation of double emulsions with colloidal crystal shells in the microfluidic device. The diameter of the polystyrene colloids was 120 nm. The flow rates of the inner oil, middle aqueous, and outer oil phases were 800, 2500, and 10 000 μL h^{-1}, respectively. (b) Optical microscopy image of the double emulsions with colloidal crystal shells after collection. (c)–(f) Dependence of the flow rate of the middle phase, Q_m, on the size of the double emulsions with colloidal crystal shells. The flow rate of the inner and outer oil phases was kept constant at 800 and 10 000 μL h^{-1}, respectively. Optical microscopy images of the double emulsions with colloidal crystal shells at a middle phase flow rate of (c) 3000, (d) 1400, and (e) 500 μL h^{-1}. (f) Plots of the diameter for the inner drop D_i, the outer drop D_o, and the thickness of the colloidal crystal shell $1/2(D_o - D_i)$ as a function of the flow rate of the middle phase.
Reprinted with permission from *Adv. Mater.*, 2010, **22**, 4998.[53]
Copyright 2010 Wiley-VCH Verlag GmbH & Co. KGaA.

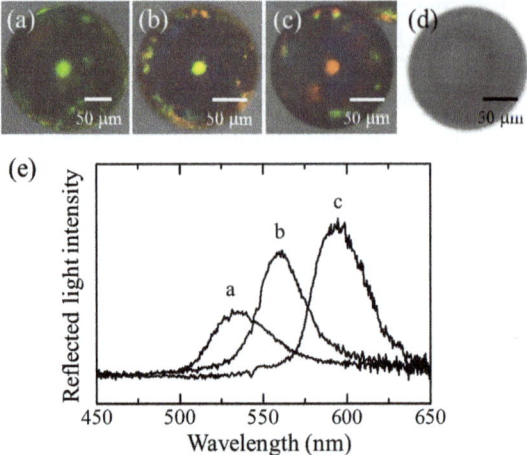

Figure 6.18   Photographs of PAAM-immobilized colloidal crystal shells with 198-nm particle diameters and different particle concentrations under light at normal incidence. ((a) $\phi = 0.55$, (b) $\phi = 0.45$, (c) $\phi = 0.35$) (d) Photograph of the sample (b) under transmission light. (e) Reflection spectrum at normal incidence of the PAAM-immobilized colloidal crystal shells shown in (a)–(c).
Reprinted with permission from *Adv. Mater.*, 2010, **22**, 4998.[53] Copyright 2010 Wiley-VCH Verlag GmbH & Co. KGaA.

insufficient amount of surfactant in the middle aqueous phase led to coalescence of the inner oil drops with the outer oil phase. In contrast, excess surfactant resulted in coalescence of the middle aqueous phases and destruction of the crystalline structure of the charged colloids. Under optimized conditions, however, stable double emulsions with colloidal crystal shells could be formed (Figure 6.17). The sizes of the inner oil and outer aqueous drops, and thus the thickness of the colloidal crystal shells, could be tuned by changing the flow rates of each phase and/or the size of the capillary orifices. For example, double emulsions with thinner colloidal crystal shells and larger inner drops were formed by gradually decreasing the flow rate of the middle phase (Figure 6.17(f)). Double emulsion droplets with colloidal crystal shells were then irradiated with UV light to photopolymerize the middle phases, thus immobilizing the colloidal crystals in the hydrogel shells (Figure 6.18). The gel-immobilized colloidal crystal shells showed color diffraction patterns and reflection spectral properties that were similar to those of the spheres.

6.3   Tuning Properties of the Gel-Immobilized Colloidal Crystals

6.3.1   Tuning by Temperature

The Bragg wavelength or optical stop-band wavelength of the gel-immobilized colloidal crystals can be altered on demand by using a stimuli-sensitive hydrogel. For example, the poly(*N*-isopropylacrylamide) (PNIPAM) gel is a thermosensitive polymer that undergoes a volume phase transition at around 32 °C.[54] If colloidal crystals were immobilized in the PNIPAM gel, the lattice constant or Bragg wavelength of the crystals embedded in the gel would be changed by changing the temperature. We prepared the PNIPAM-immobilized colloidal crystal films and spheres by the microfluidics and photopolymerization techniques using *N*-isopropylacrylamide (NIPAM) instead of NMAM and AAM as described in Section 6.2, and measured the temperature-dependent change in the size and the Bragg wavelength. As shown in Figure 6.19, during heating from 10 to 60 °C, the film began to shrink significantly at around 32 °C, resulting in a large shift in the Bragg diffraction wavelength. By contrast, colloidal crystals immobilized in the poly(*N*-methylolacrylamide) (PNMAM) gel film did not exhibit such thermosensitivity. The thermosensitivity of the gel-immobilized colloidal crystals can be changed by varying the mixing ratio of NMAM and NIPAM monomers in the gel network. As shown in Figure 6.20, upon increasing the ratio of the NMAM monomer x_{NMAM}, the change in Bragg wavelength with regard to temperature became gradual and shifted to higher temperature. Upon increasing x_{NMAM} to more than 0.5, the Bragg wavelength exhibited linear dependence on the temperature, although the change in the Bragg wavelength was not large.

Interestingly, the PNIPAM-immobilized colloidal crystal shells were found to exhibit greater sensitivity and a wider range of spectral changes compared to a bulk colloidal crystal or spherical gels of the same composition, as shown in Figure 6.21.[53] For the colloidal crystals immobilized in a conventional bulk gel, the crystal structure remained FCC during shrinkage because the lattice spacing decreased uniformly in all directions. Therefore, the ultimate Bragg

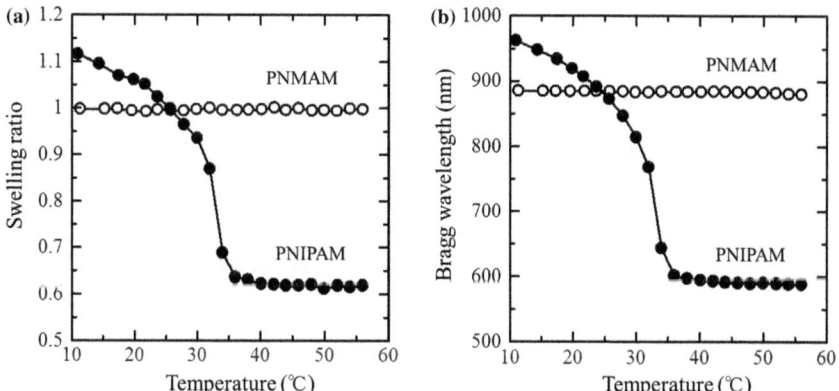

Figure 6.19 (a) A plot of the diameter of the gel-immobilized colloidal crystals as a function of temperature (●: PNIPAM-immobilized colloidal crystals; ○: PNMAM-immobilized colloidal crystals). (b) A plot of the wavelength of the Bragg reflection for the gel-immobilized colloidal crystals as a function of temperature (●: PNIPAM-immobilized colloidal crystals; ○: PNMAM-immobilized colloidal crystals). The curves in (a) and (b) are guides to the eye.

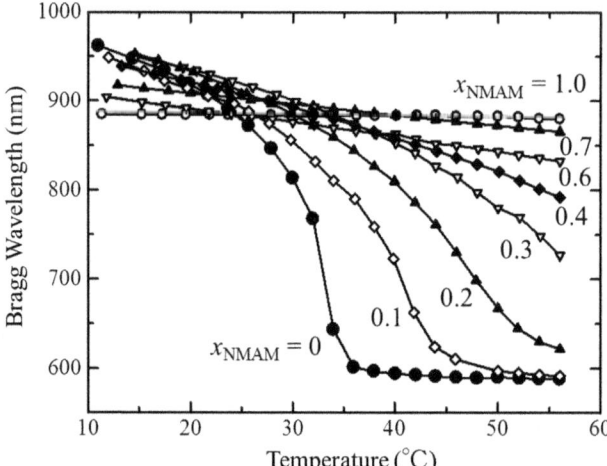

Figure 6.20 A plot of the wavelength of the Bragg reflection for colloidal crystals immobilized in the mixed gel of PNMAM and PNIPAM at various mixing ratios as a function of temperature. The parameter x_{NMAM} is the ratio of the NMAM monomer. The curves are guides to the eye.

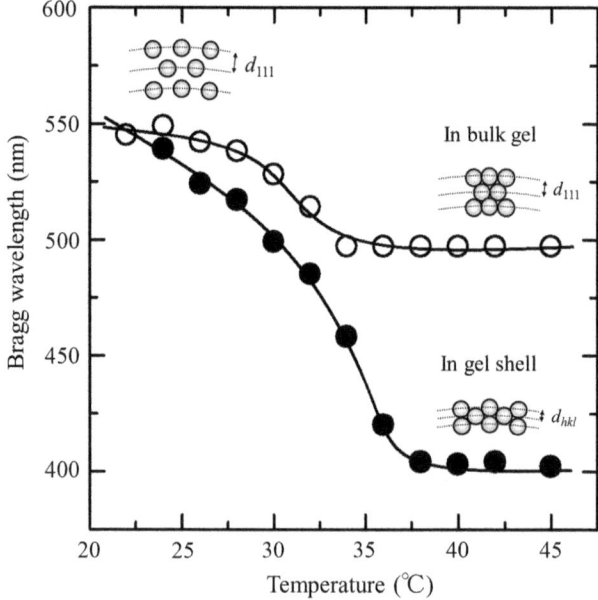

Figure 6.21 Plot of the wavelengths of the Bragg diffraction for the PNIPAM-immobilized colloidal crystal shell as a function of temperature (●: experimentally measured Bragg wavelength; ○: calculated Bragg wavelength assuming isotropic shrinkage of the gel). The curves are guides to the eye.
Reproduced with permission from *Adv. Mater.*, 2010, **22**, 4998.[53]
Copyright 2010 Wiley-VCH Verlag GmbH & Co. KGaA.

wavelength for the colloidal crystals in the conventional bulk gel was attained at an interplanar spacing of $d_{111} = \sqrt{\frac{2}{3}}\, d$, which is the closest packing condition for the FCC structure. However, in the case of the colloidal crystal shells, the shrinkage in the radial direction was greater than that in the direction along the interfaces of the shell, which led to a smaller interplanar spacing than that in the isotropically shrunken gel, as schematically shown in Figure 6.21. Thus, the Bragg wavelength of colloidal crystals immobilized in a gel shell could be manipulated with higher sensitivity and over a wider range of wavelengths. These materials are potentially useful as microcapsules that can monitor changes inside and outside the capsules through the Bragg diffraction wavelength or diffraction color.

6.3.2 Tuning by Mechanical Stress

The Bragg wavelength of gel-immobilized colloidal crystals can be tuned easily by applying mechanical stress. For example, the gel-immobilized colloidal crystal film was cut into a circular fragment with a diameter of 7 mm, and placed between two parallel glass plates (Figure 6.22(a)). Upon uniaxial compression, the lattice spacing normal to the film surface, which determines the observed Bragg wavelength, could be tuned. Figure 6.22(b) shows the transmission spectrum of the gel-immobilized colloidal crystal film under compression. The deep dip due to the optical stop-band could be tuned over a wavelength range of about 100 nm, preserving the spectral shape. In addition, the imaging spectrogram indicated that the spectral uniformity in the entire sample area was preserved under compression (Figure 6.22(c)). This excellent property of the film is potentially useful as a tunable photonic crystal and a stress sensor.

When the gel film was compressed at a tilt, it became a graded photonic crystal (Figure 6.23). Figure 6.23(c) shows the transmission spectra at points A to D on the inclined compressed film shown in Figure 6.23(b). While moving from A to D, the Bragg wavelength continuously shifted depending on the degree of compression. The imaging spectrogram also indicated that the impermeability area owing to the Bragg reflection moved with increases in the wavelength of the incident light (Figure 6.23(d)).

6.3.3 Tuning by Ionic Liquids

Gel-immobilized colloidal crystals that contain a swelling solvent work as tunable photonic crystals, as described above. However, if the solvent in the gel evaporates, the crystalline structure of the colloids is destabilized, which is an issue for applications. Ionic liquids, which are composed solely of anions and cations, have recently received increasing attention as new solvents because of their many attractive features, such as nonvolatility, noncombustibility, high ionic conductivity, high thermal stability, and favorable solvation properties.[55] Therefore, they have the potential for use in fuel cells (as nonvolatile

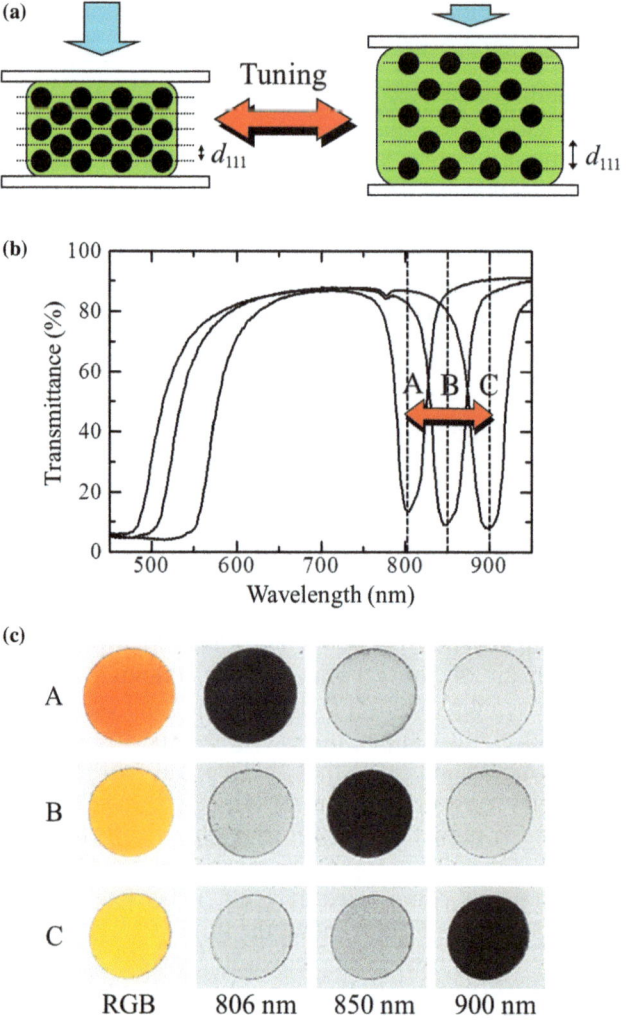

Figure 6.22   (a) Schematic image of the tuning of the lattice spacing in the gel-immobilized colloidal crystals by mechanical compression. (b) Transmission spectrum of the gel-immobilized colloidal crystal film under compression. (c) RGB and single-wavelength images for representative wavelengths taken under transmission illumination for the gel-immobilized colloidal crystal film under compression.

electrolytes),[56] gas handling,[57] and cellulose processing.[58] The properties of ionic liquids can be adjusted by changing the cation–anion combination. Thus, the use of ionic liquids as solvents for gel-immobilized colloidal crystals will increase the possibility of designing colloidal photonic crystals in which there is no solvent evaporation. We found that hydrophilic ionic liquids could be used as suitable swelling solvents for gel-immobilized colloidal crystals.[59,60]

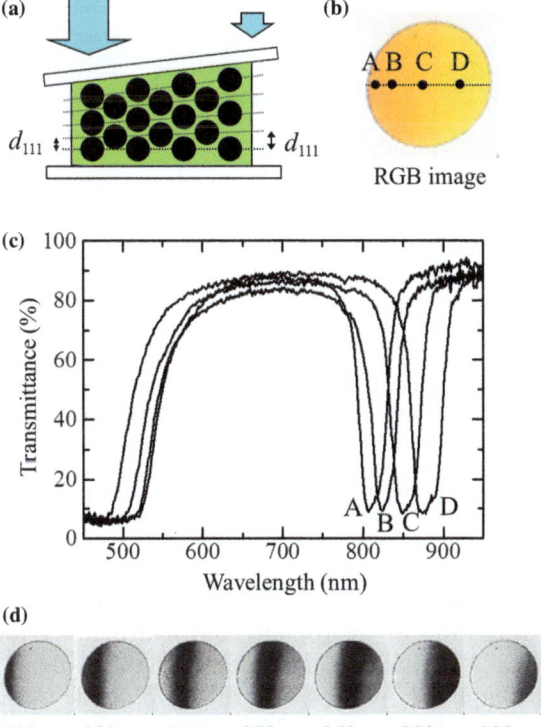

d_{111}

d_{111}

RGB image

800 nm 820 nm 840 nm 850 nm 860 nm 880 nm 900 nm

Figure 6.23 (a) Schematic image of the graded colloidal photonic crystals by inclined compression. (b) RGB image of the gel-immobilized colloidal crystal film under inclined compression. (c) Transmission spectra at points A to D on the inclined compressed film shown in (b). (d) Single-wavelength images for representative wavelengths taken under transmission illumination for the inclined compressed film.

When a gel film is soaked in hydrophobic ionic liquids such as 1,3-diallylimidazolium bis(trifluoromethanesulfonyl)imide, the gel shrinks considerably, similar to the response observed with most organic solvents. As a result, the structure of the colloidal crystal immobilized in the gel is destroyed; the sample turns cloudy, and the Bragg reflection peak in the spectrum disappears. However, when the gel film is soaked in hydrophilic ionic liquids, the crystal structure remains intact. Interestingly, in some hydrophilic ionic liquids, the gel film can swell to an extent greater than that in water. For example, when water was replaced with 1,3-diallylimidazolium bromide, the film swelled to 1.04 times its initial size, leading to a different diffraction color or redshift of the reflection peak in the spectrum, as shown in Figure 6.24.

Furthermore, the swelling volume of the gel can be easily varied by adding a hydrophobic ionic liquid to the hydrophilic one. Thus, the Bragg

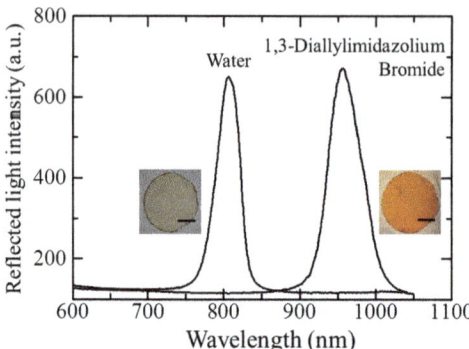

Figure 6.24   Reflection spectra and microscope images of gel-immobilized colloidal
crystal films before and after the replacement of water in the gel
with 1,3-diallylimidazolium bromide. The lengths of the scale bars are
1 mm.
Reprinted with permission from *Macromolecules*, 2011, **44**, 5865.[59]
Copyright 2011 American Chemical Society.

wavelength of the colloidal crystals can be adjusted over a wide range by
varying the mixing ratio of hydrophilic and hydrophobic ionic liquids. As
shown in Figures 6.25(a) and (b), upon decreasing the volume fraction of the
hydrophilic ionic liquid, 1,3-diallylimidazolium bromide to the hydrophobic
ionic liquid, 1,3-diallylimidazolium bis(trifluoromethanesulfonyl)imide, the
Bragg reflection peak shifted to a shorter wavelength over a range of about
400 nm, preserving the uniform crystalline structure. Surprisingly, the
swelling size and the Bragg wavelength showed a linear dependence on the
mixing ratio. In general, however, the swelling–shrinking phenomenon
that depends on the change in the mixing ratio of the solvents is known to
be a type of phase transition, and the gel size (or Bragg wavelength) is known
to be a strongly nonlinear function of the mixing ratio, as already reported
for the water–ethanol system.[40,61–63] Although a theoretical understanding of
this linear dependence poses a serious challenge for future investigation, it is
quite advantageous for adjusting the optical characteristics of the crystals in
practical applications.

Since ionic liquids do not evaporate, even under high vacuum, we can
directly verify the presence of distance-controlled particle arrays in the swollen
gel using ordinary scanning electron microscopy (SEM). For $x_{Br} = 0.2$,
relatively dense arrays of particles were observed (Figure 6.26(a)), whereas for
$x_{Br} = 0.8$, loose arrays of particles were seen (Figure 6.26(b)). From the SEM
images, the distance between the nearest-neighbor particles for $x_{Br} = 0.2$ and
$x_{Br} = 0.8$ was estimated to be 247 ± 15 nm and 335 ± 25 nm, respectively. These
values were in reasonable agreement with those determined using the Bragg
wavelength. This confirmed that uniform expansion or shrinkage of the
periodic arrays could be successfully achieved by changing the mixing ratio of
the two different ionic liquids.

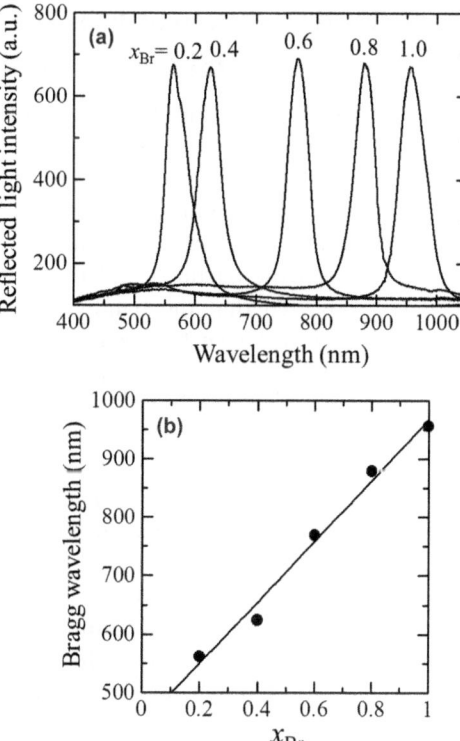

Figure 6.25   (a) Reflection spectra of gel-immobilized colloidal crystal films immersed in a mixed solution of 1,3-diallylimidazolium bromide and 1,3-diallyl-imidazolium bis(trifluoromethanesulfonyl)imide at various mixing ratios. The parameter x_{Br} is the volume fraction of 1,3-diallylimid-azolium bromide measured prior to mixing. (b) Bragg wavelength obtained from (a) as a function of x_{Br}.
Reproduced with permission from *Macromolecules*, 2011, **44**, 5865.[59] Copyright 2011 American Chemical Society.

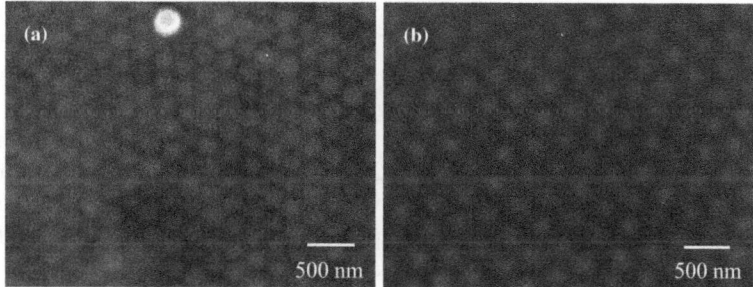

Figure 6.26   Top-view SEM images of gel-immobilized colloidal crystal films with $x_{Br} = 0.2$ (a) and $x_{Br} = 0.8$ (b).
Reproduced with permission from *Macromolecules*, 2011, **44**, 5865.[59] Copyright 2011 American Chemical Society.

6.4 Tuning the Effective Bandwidth in the Gel-Immobilized Colloidal Crystal Film

The optical stop-band in colloidal crystals is characterized by the optical stop-band wavelength and the bandwidth. The wavelength is highly tunable by changing the lattice constant, as described above. However, the control of the bandwidth, which is also important for optical applications, is difficult because the bandwidth for a particular lattice constant is intrinsically determined by the refractive-index contrast between the particles and the surrounding medium. In the opal system, it was reported that the bandwidth could be effectively broadened by introducing a graded structure wherein the superposition of shifted stop-bands effectively yielded a wider stop-band. However, multiple fabrication steps were required to form such graded structures.[64,65] We found that the effective bandwidth in gel-immobilized colloidal crystals could be tuned easily by controlling the photopolymerization conditions for immobilizing the colloidal crystals in the gel network.[66] The dip in the transmission spectrum for the single-crystal-like colloidal crystal film prepared by the flow method can be broadened by simply adjusting the photoirradiation time.

When the light intensity of the irradiation was very low but strong enough to form a self-standing film, degradation of the transmittance characteristics was observed immediately after irradiation, *e.g.*, large shift in the dip wavelength, reduction in the transmittance at the passband wavelength, and appearance of additional satellite dips. By increasing the light intensity, the shape of the spectrum was preserved against irradiation. However, even in this case, the dip expansion occurred during the subsequent equilibration in the dark. Figures 6.27(a) and (b) show temporal changes in the transmission spectra of the colloidal crystal films during equilibration after light irradiation for two representative times, 3 h and 1 h, respectively. In the case of 3 h irradiation, the initial spectrum did not change, at least within the experimental time range, *i.e.* 50 h. By contrast, in the case of 1 h irradiation, the dip at 840 nm gradually expanded during equilibration in the dark after the irradiation was completed, while the spectrum profile, except for the dip region, was preserved, *i.e.* high transmittance at the pass band and the absence of band edge shift at approximately 450 nm were observed. This gives the impression that the optical stop-band was expanded. Figure 6.27(c) shows the time courses of the dip width during the equilibrium process after light irradiation was completed for both cases shown in Figures 6.27(a) and (b). The dip width of the sample irradiated for 3 h remained constant, whereas that of the sample irradiated for 1 h increased with equilibration time, after which it leveled off. The rate of this dip expansion and the stable value of the dip width depend on the irradiation time. Figure 6.27(d) shows the saturation values of the expanded dip width as a function of the light irradiation time after equilibration. For irradiation times exceeding 3 h, the initial dip width was frozen, whereas for irradiation times of less than 3 h, a decrease in the irradiation time resulted in larger saturation values. On the basis of these observations, it was concluded that the final dip width in the colloidal crystal film could be controlled by the irradiation time.

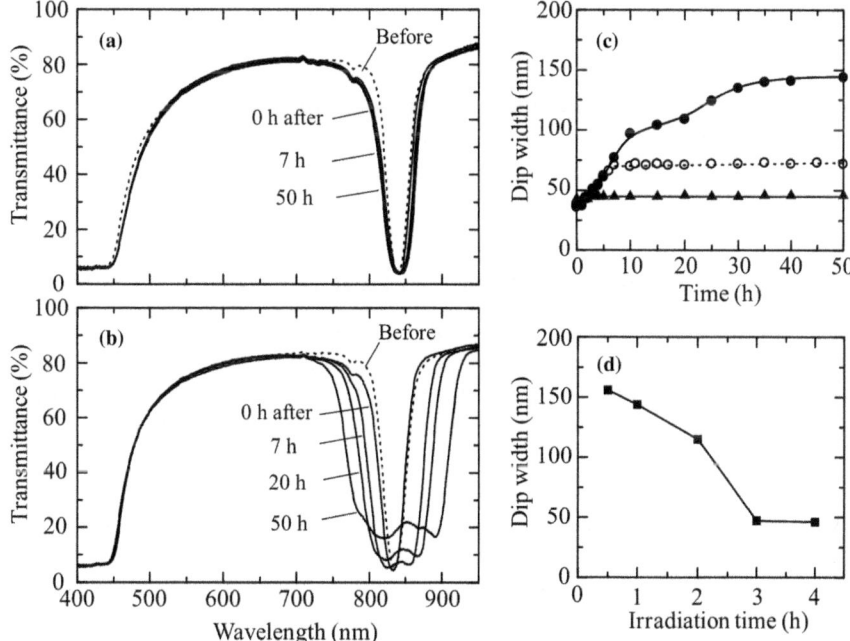

Figure 6.27 Temporal changes in transmission spectra of colloidal crystals after (a) 3 h and (b) 1 h of light irradiation. The dashed line indicates the spectrum before irradiation. (c) Time courses of (111) dip width for 3 h (▲) and 1 h (●) irradiation. The open circle (○) indicates the change in dip width for a multistep case: 1 h of initial irradiation followed by 7 h of retention in the dark, after which additional irradiation for 3 h was carried out. (d) Saturation value for the expanded (111) dip width as a function of light irradiation time.
Reproduced with permission from *Langmuir*, 2007, **23**, 3503.[66] Copyright 2007 American Chemical Society.

In the present material, the dip width could be tuned in the range of 46–156 nm. Another interesting feature in this system was that the dip expansion could be frozen by additional light irradiation. The dashed line in Figure 6.27(c) represents an example in which the sample was first irradiated for 1 h, retained in the dark for 7 h, and irradiated again for 3 h. The expansion of the width could be frozen at 72 nm by the second irradiation; however, it would have increased to 144 nm without the additional irradiation.

The dip expansion observed in this study can be explained if the gel possesses some inhomogeneity. The lattice structure of the colloidal crystals immobilized in the inhomogeneous gel network would become inhomogeneous as a result of subsequent swelling equilibration. This implies that the interplanar spacing of the (111) lattice planes, which determines the dip wavelength, is not constant but fluctuates around a value for the homogeneous case. If the length scale of such an inhomogeneity is smaller than the spatial resolution of the spectral measurement (100 μm) in the lateral direction, the spectrum of the sample would be observed as the superposition of spectra with different dip wavelengths, resulting in an apparent dip expansion. Since an incomplete polymerization reaction causes the inhomogeneity in gels, an increase in the reaction time, *i.e.* irradiation time, would reduce the inhomogeneity. Although the dip expansion due to inhomogeneity is essentially different from the expansion of the optical stop-band in a perfect crystal, it has an equivalent effect on the apparent spectral properties.

6.5   Dry Colloidal Crystal Gel Film without Cracks

The gel-immobilized colloidal crystal film, which includes water, can be converted into a dry film of densely packed colloidal crystals while maintaining excellent optical quality.[67] Since particles are bounded by polymer networks in the gel-immobilized colloidal crystal film, it would seem possible to obtain dry colloidal crystals without destroying the particle arrays by removing the water. However, simply drying the gel film in air destroys the crystalline structure owing to inhomogeneous shrinkage and deformation of the gel, especially at the final stage of the drying process, owing to the capillary force of the final traces of water. To suppress the inhomogeneous shrinkage and deformation of the gel, a circular fragment (diameter: 7 mm; thickness: 0.1 mm) was cut from the wet gel film, and then soaked in aqueous solutions with increasing concentrations (from 0% to 100%, in increments of 1–2 vol%) of ethanol (EtOH), which has poor affinity for the gel-polymer. Since the gel size is a function of EtOH concentration, a gel surrounded by a homogeneous solution with a slightly higher EtOH concentration than the preceding step can be shrunk gradually and homogeneously. After the gel was processed with 100% EtOH solution and shrunk to the densely packed state, it was dried in air. The capillary force problem was expected to be less serious because EtOH has a much lower surface tension than water (EtOH: $23\,\mathrm{mN\,m^{-1}}$, water: $73\,\mathrm{mN\,m^{-1}}$).

Figure 6.28(a) shows a photograph of the resultant dried film. The film was finally contracted to 52% in linear scale, corresponding to a particle concentration that almost achieved the sphere-touching condition, of about 73%. The film showed an even, round shape with a remarkable diffraction color, indicating that isotropic contraction that preserved the crystalline order was achieved. Figure 6.28(b) shows the transmission spectra of the gel film at the initial stage (swelled in water), after replacement with EtOH, and after drying, where the incident light was normal to the film surface. The shift in the dip wavelength was caused by the contraction in the lattice spacing and the change in the refractive index of the medium. All spectra showed a sharp dip at the stop-band and high transmittance at the passband wavelengths, suggesting that the crystallinity of the gel film was preserved throughout the process, not only at the surface but also across its thickness. The regularity of the particle arrays was directly observed through SEM. As shown in Figures 6.29(e) and (f), the particles covered with gels were well ordered and exhibited 6-fold symmetry. It should be noted that the dry colloidal crystal had no cracks at any scale; this

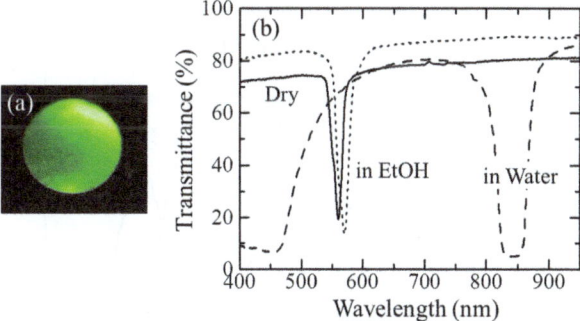

Figure 6.28 (a) Photograph of dried colloidal crystal gel film. The diameter of the film is about 4 mm. (b) Transmission spectra of colloidal crystal gel film swelled in water, shrunk in 100% EtOH, and dried.
Reproduced with permission from *Langmuir*, 2009, **25**, 13315.[67] Copyright 2009 American Chemical Society.

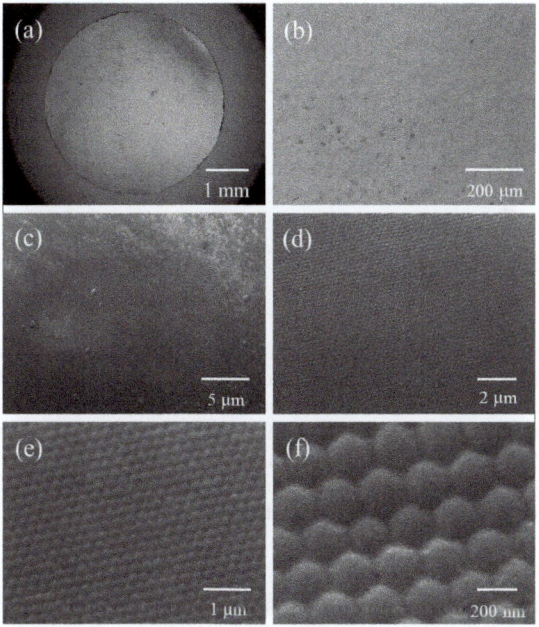

Figure 6.29 Images of dried colloidal crystal gel film at various scale sizes, observed by an optical microscope (a, b) and a scanning electron microscope (c–f). Reproduced with permission from *Langmuir*, 2009, **25**, 13315.[67] Copyright 2009 American Chemical Society.

can be confirmed from the images at various scale sizes observed through an optical microscope (Figures 6.29(a) and (b)) and SEM (Figures 6.29(c)–(f)).

An advantage of this material is that the optical stop-band wavelength of the colloidal crystals can be easily tuned by varying the amount of gelation reagent

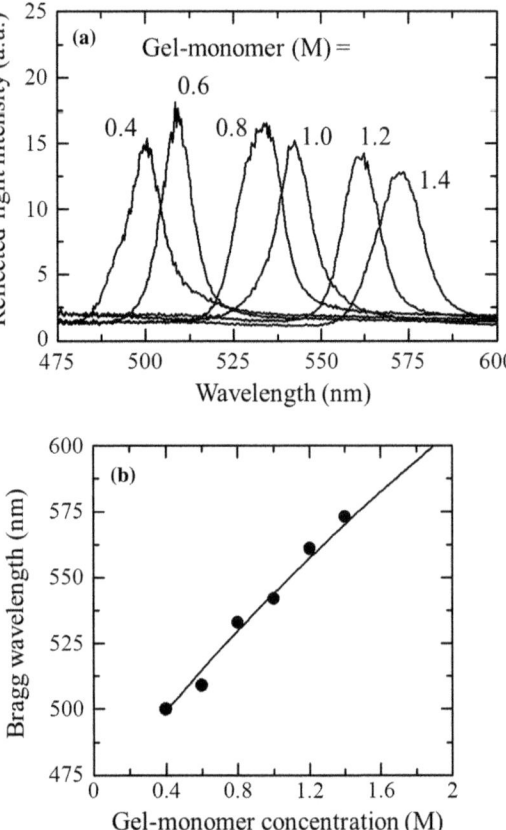

Figure 6.30   (a) Normal reflection spectra of dried colloidal crystal gel films prepared with different gel-monomer concentrations. (b) Plots of the wavelength peaks read from (a) as a function of the gel-monomer concentration. The solid line is calculated using the Bragg equation, assuming that the space among particles is filled with gel-polymer.
Reproduced with permission from *Langmuir*, 2009, **25**, 13315.[67]
Copyright 2009 American Chemical Society.

added in the starting colloidal suspension. Figure 6.30(a) shows the normal reflection spectra of dry colloidal crystal gel films obtained from wet ones prepared with different monomer concentrations, and Figure 6.30(b) shows the plots of the peak wavelength read from Figure 6.30(a) as a function of monomer concentration. The data showed that the stop-band wavelength could be elongated by as much as 14%. The dependence of the stop-band wavelength on the monomer concentration can be explained by assuming that the space between nontouching particles is filled with gel-polymer, as shown by the solid curve in Figure 6.30(b). The adjustability of the stop-band wavelength without changing the particle diameter is a great advantage, especially for industrial applications.

Acknowledgements

Part of this work was carried out at the National Institute for Materials Science (NIMS), hosted by Dr. Tsutomu Sawada, and at Harvard University, hosted by Prof. David A. Weitz. The author also gratefully acknowledges support by the Japan Society for the Promotion of Science [Grant-in-Aid for Young Scientists (A) 22686063], the Mazda Foundation, the Ogasawara Science and Technology Foundation, the Oil & Fat Industry Kaikan Foundation, and the Iketani Science and Technology Foundation.

References

1. K. P. Velikov, C. G. Christova, R. P. A. Dullens and A. van Blaaderen, *Science*, 2002, **296**(106).
2. G. A. Ozin and S. M. Yang, *Adv. Funct. Mater.*, 2011, **11**, 95.
3. F. Garcia-Santamaria, H. T. Miyazaki, A. Urquia, M. Ibisate, M. Belmonte, N. Shinya, F. Meseguer and C. Lopez, *Adv. Mater.*, 2002, **14**, 1144.
4. H. Miguez, F. Meseguer, C. Lopez, A. Mifsud, J. S. Moya and L. Vazquez, *Langmuir*, 1997, **13**, 6009.
5. P. Ni, P. Dong, B. Cheng, X. Li and D. Zhang, *Adv. Mater.*, 2001, **13**, 437.
6. P. Pieranski, *Contemp. Phys.*, 1983, **24**, 25.
7. A. K. Arora and B. V. R. Tata, (eds)., *Ordering and Phase Transitions in Charged Colloids*, Wiley-VCH, New York, 1996.
8. A. P. Gast and W. B. Russel, *Phys. Today*, 1998, **51**, 24.
9. T. Kanai, *J. Ceram. Soc. Jpn.*, 2012, **120**, 87.
10. S. Furumi, H. Fudouzi and T. Sawada, *Laser Photon. Rev.*, 2010, **4**, 205.
11. E. Yablonovitch, *Phys. Rev. Lett.*, 1987, **58**, 2059.
12. S. John, *Phys. Rev. Lett.*, 1987, **58**, 2486.
13. J. D. Joannopoulos, R. D. Meade, and J. N. Winn, *Photonic Crystals*, Princeton University Press, Princeton NJ, 1995.
14. K. Sakoda, *Optical Properties of Photonic Crystals*, Springer-Verlag, Berlin, 2001.
15. K. Ohtaka, *Phys. Rev. B*, 1979, **19**, 5057.
16. S. Noda, K. Tomoda, N. Yamamoto and A. Chutinan, *Science*, 2000, **289**, 604.
17. Y. Akahane, T. Asano, B. Song and S. Noda, *Nature*, 2003, **425**, 944.
18. T. Baba, *Nature Photon.*, 2008, **2**, 465.
19. S. Y. Lin, J. G. Fleming, D. L. Hetherington, B. K. Smith, R. Biswas, K. M. Ho, M. M. Sigalas, W. Zubrzycki, S. R. Kurtz and J. Bur, *Nature*, 1998, **394**, 251.
20. J. E. G. J. Wijnhoven and W. L. Vos, *Science*, 1998, **281**, 802.
21. M. Muller, R. Zentel, T. Maka, S. G. Romanov and C. M. S. Torres, *Adv. Mater.*, 2000, **12**, 1499.
22. H. Miguez, E. Chomski, F. Garcia-Santamaria, M. Ibisate, S. John, C. Lopez, F. Meseguer, J. P. Mondia, G. A. Ozin, O. Toader and H. M. van Driel, *Adv. Mater.*, 2001, **13**, 1634.

23. S. H. Park, D. Qin and Y. Xia, *Adv. Mater.*, 1998, **10**, 1028.
24. S. Matsuo, T. Fujine, K. Fukuda, S. Juodkazis and H. Misawa, *Appl. Phys. Lett.*, 2003, **82**, 4283.
25. P. Jiang, J. F. Bertone, K. S. Hwang and V. L. Colvin, *Chem. Mater.*, 1999, **11**, 2132.
26. Z.-Ze. Gu, A. Fujishima and O. Sato, *Chem. Mater.*, 2002, **14**, 760.
27. P. A. Hiltner and I. M. Krieger, *J. Phys. Chem.*, 1969, **73**, 2386.
28. T. Kanai, T. Sawada and K. Kitamura, *Langmuir*, 2003, **19**, 1984.
29. E. A. Kamenetzky, L. G. Magliocco and H. P. Panzer, *Science*, 1994, **263**, 207.
30. S. H. Foulger, P. Jiang, A. Lattam, D. W. Smith, J. Ballato, D. E. Dausch, S. Grego and B. R. Stoner, *Adv. Mater.*, 2003, **15**, 685.
31. J. Holtz and S. A. Asher, *Nature*, 1997, **389**, 829.
32. H. Fudouzi and Y. Xia, *Adv. Mater.*, 2003, **15**, 892.
33. Y. Iwayama, J. Yamanaka, Y. Takiguchi, M. Takasaka, K. Ito, T. Shinohara, T. Sawada and M. Yonese, *Langmuir*, 2003, **19**, 977.
34. T. Kanai, T. Sawada, A. Toyotama and K. Kitamura, *Adv. Funct. Mater.*, 2005, **15**, 25.
35. T. Kanai, T. Sawada and K. Kitamura, *Chem. Lett.*, 2005, **34**, 904.
36. T. Sawada, Y. Suzuki, A. Toyotama and N. Iyi, *Jpn. J. Appl. Phys.*, 2001, **40**, L1226.
37. B. J. Ackerson, *Nature*, 1979, **281**, 57–60.
38. B. J. Ackerson, J. B. Hayter, N. A. Clark and L. Cotter, *J. Chem. Phys.*, 1986, **84**, 2344.
39. T. Kanai, T. Sawada, I. Maki and K. Kitamura, *Jpn. J. Appl. Phys.*, 2003, **42**, L655.
40. A. Toyotama, T. Kanai, T. Sawada, J. Yamanaka, K. Ito and K. Kitamura, *Langmuir*, 2005, **21**, 10268.
41. T. Kanai, T. Sawada and J. Yamanaka, *J. Ceram. Soc. Jpn.*, 2010, **118**, 370.
42. S.-H. Kim, S.-J. Jeon and S.-M. Yang, *J. Am. Chem. Soc.*, 2008, **130**, 6040.
43. O. D. Velev, A. M. Lenhoff and E. W. Kaler, *Science*, 2000, **287**, 2240.
44. X. Zhao, Y. Cao, F. Ito, H.-H. Chen, K. Nagai, Y.-H. Zhao and Z.-Z. Gu, *Angew. Chem. Int. Ed.*, 2006, **45**, 6835.
45. Y. Zhao, X. Zhao, J. Hu, M. Xu, W. Zhao, L. Sun, C. Zhu, H. Xu and Z. Gu, *Adv. Mater.*, 2009, **21**, 569.
46. J. H. Moon, G.-R. Yi, S.-M. Yang, D. J. Pine and S. B. Park, *Adv. Mater.*, 2004, **16**, 605.
47. S.-H. Kim, J.-M. Lim, W. C. Jeong, D.-G. Choi and S.-M. Yang, *Adv. Mater.*, 2008, **20**, 3211.
48. A. S. Utada, E. Lorenceau, D. R. Link, P. D. Kaplan, H. A. Stone and D. A. Weitz, *Science*, 2005, **308**, 537.
49. R. K. Shah, H. C. Shum, A. C. Rowat, D. Lee, J. J. Agresti, A. S. Utada, L. Y. Chu, J.-W. Kim, A. Fernandez-Nieves, C. J. Martinez and D. A. Weitz, *Mater. Today*, 2008, **11**, 18.
50. H. C. Shum, J.-W. Kim and D. A. Weitz, *J. Am. Chem. Soc.*, 2008, **130**, 9543.

51. T. Kanai, D. Lee, H. C. Shum and D. A. Weitz, *Small*, 2010, **6**, 807.
52. V. Rastogi, S. Melle, O. G. Calderón, A. A. García, M. Marquez and O. D. Velev, *Adv. Mater.*, 2008, **20**, 4263.
53. T. Kanai, D. Lee, H. C. Shum, R. K. Shah and D. A. Weitz, *Adv. Mater.*, 2010, **22**, 4998.
54. B. R. Saunders and B. Vincent, *Adv. Colloid Interf. Sci.*, 1999, **80**, 1.
55. T. Welton, *Chem. Rev.*, 1999, **99**, 2071.
56. R. F. de Souza, J. C. Padilha, R. S. Goncalves and J. Dupont, *Electrochem. Commun.*, 2003, **5**, 728.
57. R. Sheldon, *Chem. Commun.*, 2001, **23**, 2399.
58. R. P. Swatloski, S. K. Spear, J. D. Holbrey and R. D. Rogers, *J. Am. Chem. Soc.*, 2002, **124**, 4974.
59. T. Kanai, S. Yamamoto and T. Sawada, *Macromolecules*, 2011, **44**, 5865.
60. S. Yamamoto, T. Sawada and T. Kanai, *Chem. Lett.*, 2012, **41**, 495.
61. P. J. Flory, *Principles of Polymer Chemistry*, Cornell University Press, Ithaca, 1953.
62. T. Tanaka, D. Fillmore, S.-T. Sun, I. Nishino, G. Swislow and A. Shah, *Phys. Rev. Lett.*, 1980, **45**, 1636.
63. S. Katayama, Y. Hirokawa and T. Tanaka, *Macromolecules*, 1984, **17**, 2641.
64. P. Jiang, G. N. Ostojic, R. Narat, D. M. Mittleman and V. L. Colvin, *Adv. Mater.*, 2001, **13**, 389.
65. J. H. Park, W. S. Choi, H. Y. Koo, J. C. Hong and D. Y. Kim, *Langmuir*, 2006, **22**, 94.
66. T. Kanai, T. Sawada, A. Toyotama, J. Yamanaka and K. Kitamura, *Langmuir*, 2007, **23**, 3503.
67. T. Kanai and T. Sawada, *Langmuir*, 2009, **25**, 13315.

Applications of Stimuli-Sensitive Inverse Opal Gels

YUKIKAZU TAKEOKA

Department of Molecular Design & Engineering, Graduate School of
Engineering, Nagoya University, Furo-cho, Chikusa-ku, Nagoya, 464-8603,
Japan
Email: ytakeoka@apchem.nagoya-u.ac.jp

7.1 Introduction to Stimuli-Sensitive Inverse Opal Gels

Photonic crystals are materials that have a periodic variation in their refractive index on a length scale comparable to the wavelength of light.[1] Analogous to the energy gap in semiconductors, the dielectric periodic structure reveals a stop-band in the spectrum of propagating electromagnetic modes, which is known as the photonic bandgap (PBG). Because the wavelength of a photon is inversely proportional to its energy, the patterned dielectric material will block light with wavelengths in the PBG while allowing other wavelengths to pass freely. These materials exhibit attractive optical properties that are potentially useful for applications in optical fibers, lasers, light-emitting diodes, and reflectors. To fabricate nanostructured materials that have a PBG, both top-down approaches, such as photolithography, and bottom-up approaches, such as the self-assembly method, can be applied. One of the easiest methods for creating inexpensive nanostructured materials is through the self-assembly of submicrometer colloidal particles to prepare a close-packing colloidal crystal (hereafter referred to as a "colloidal crystal"),[2,3] which can have a pseudo-PBG (p-PBG) in the visible region. Colloidal crystals composed of submicrometer colloidal particles of silica (SiO_2) occur in Nature. These naturally occurring

RSC Smart Materials No. 5
Responsive Photonic Nanostructures: Smart Nanoscale Optical Materials
Edited by Yadong Yin
© The Royal Society of Chemistry 2013
Published by the Royal Society of Chemistry, www.rsc.org

colloidal silica crystals are called opals.[4] In fact, opals composed of tiny spheres of SiO_2 are formed over a long period of time, but they are not completely crystalline. However, a portion of the colloidal crystal structure of opals (hereafter referred to as the "opal structure") exhibits vivid colors due to the presence of a p-PBG in the visible region. Iridescence from the opal structure can be observed because the angle-dependent color results from Bragg diffraction. Furthermore, the opal structure can be used as a template to prepare porous structures (hereafter referred to as the "inverse opal structure"), which has a beneficial effect not only on the optical properties but also on the substance permeation properties of other chemicals.[5,6] However, the majority of photonic crystals, including the opal and inverse opal, have specific optical properties that cannot be altered after the crystals have been produced. If the periodicity or the refractive indices of the components of the photonic crystals can be varied by external stimuli, such as electric and magnetic fields, so-called "tunable photonic crystals" can be obtained.[7–13] The use of "soft materials", such as surfactant solutions, liquid crystals, and gels, as the components of photonic crystals enables the energy levels in the p-PBG to be altered by changing the periodicity or the refractive indices. These tunable soft photonic crystals display exceptionally bright and brilliant structural colors, which arise from coherent Bragg optical diffraction and can be tuned across the entire visible region in response to external stimuli.

We have been studying inverse opal gels that exhibit volume changes in response to environmental changes, and these gels were prepared using a colloidal crystal as a template[14–20] (Figure 7.1(a)). Consequently, the obtained environmentally sensitive inverse opal gels exhibit a rapid change in both structural color and volume, and they can be useful for sensors and displays (Figure 7.1(b)). Other research groups also study environmentally sensitive inverse opal gels,[21–25] but the majority of these gels exhibit isotropic volume changes in response to environmental changes. However, inverse opal gels can exhibit anisotropic volume changes because the total amount of elastic energy of such porous gels is small compared with that of homogeneous bulk gels.[26–32] Furthermore, we observed that stimuli-sensitive hydrogel particles confined in an inverse opal hard polymer exhibit a reversible change in the peak intensity of the p-PBG in response to a change in the environment that is different from the optical behavior of inverse opal gels in response to environmental changes.[33–35] This chapter begins with two interesting examples of inverse opals that exhibit an anisotropic volume change in response to environmental changes, and a combined system of environmentally sensitive inverse opal and hydrogel particles is then described.

7.2 Visualization and Spectroscopic Analysis of the Self-Sustaining Peristaltic Motion of Inverse Opal Gels Synchronized with the BZ Reaction

Bolinopsis mikado and *Beroe cucumis* belong to the class of comb jellyfish, which have an organ of locomotion called a comb plate.[36] These jellyfish can

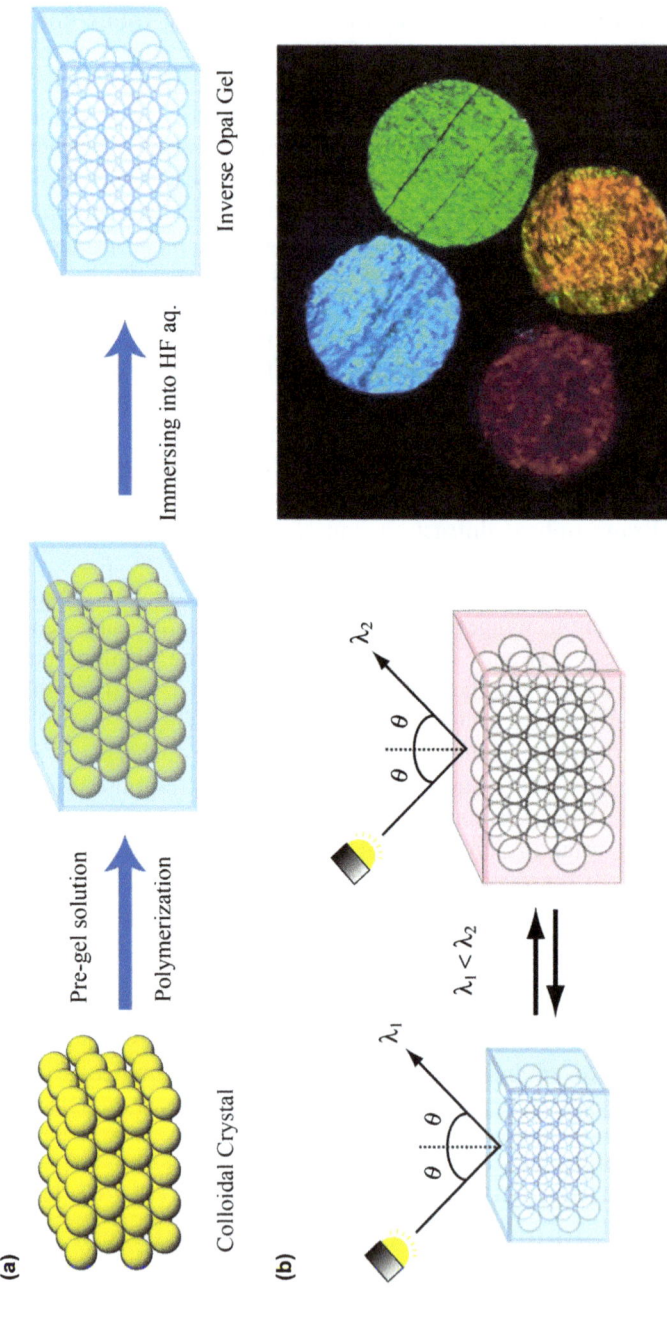

Figure 7.1 (a) Preparation of the inverse opal gel using a colloidal crystal as a template. (b) Left: Structural color change of the inverse opal gel in response to the environment. Right: Photograph of inverse opal gels that reveal different structural colors. Reprinted with permission from Ref. 19. Copyright 2006 American Chemical Society.

move like an airship by beating their comb plate, which is composed of thousands of tiny cilia. The comb plate diffracts certain wavelengths of visible light due to the periodic change in its refractive index on a length scale that is comparable to the wavelength of visible light. Changes in color similar to light show displays can be observed when these jellyfish continuously beat their comb plates (Figure 7.2). Unlike *Aequorea victoria*, which produces color through the green fluorescent protein, comb jellies create their colors with the nanostructures of their comb plates.[37]

There are many other living creatures that can change their body coloration with a variation in body movement, such as cuttlefishes.[38,39] These living creatures can change their body coloration through dynamic variations of the periodic dielectric structures in the cells and organs, which are generated by the conversion of chemical energy within a living organism to kinetic energy. It would appear that these organisms utilize this dynamic optical signaling as a means of communication. The motion and signaling mechanisms of organisms through efficient energy systems will serve as a useful reference for creating the next generation of actuators and displays.

Artificial soft materials that change properties autonomously like jellyfish and cuttlefish have been extensively studied and coupled with nonlinear chemical reactions. Yoshida's group thoroughly studied such a system using a

Figure 7.2 Comb jellyfish (*Bolinopsis mikado*) change color when they continuously beat their comb plate.
This photograph was taken by Y. Takeoka.

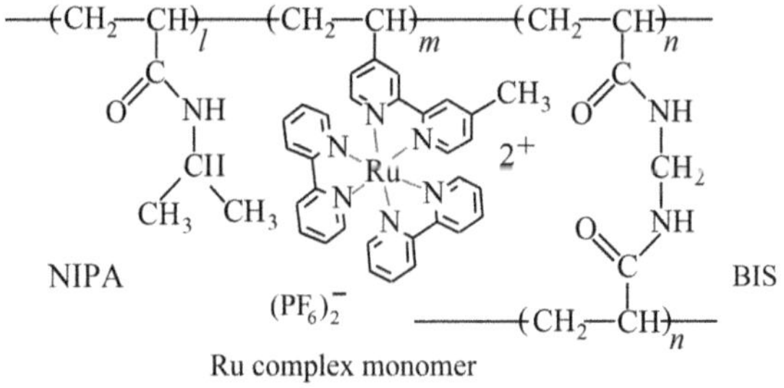

Scheme 7.1 Chemical structure of the poly(NIPA-Ru complex) gel.

gel composed of *N*-isopropyl acrylamide (NIPA), which is a temperature-sensitive monomer; a Ru complex derivative monomer; and *N, N'*-methyl-enebisacrylamide (BIS), which is a crosslinker (namely poly(NIPA-Ru complex) gel)[40–45] (Scheme 7.1).

The poly(NIPA-Ru complex) gel can autonomously change its volume when synchronized with the BZ reaction, which is the most well-known nonlinear chemical reaction.[46] The Ru complex can be a catalyst for the BZ reaction, and this complex can periodically change its oxidation state during the reaction.[47] The Ru complex can change its oxidation state between divalent (under reducing conditions) and trivalent (under oxidizing conditions); the number of counterions in the Ru complex can also be changed depending on the oxidation state of the complex. The volume of the polymer network composed of the polyelectrolyte significantly depends on the inner osmotic pressure of the network, which results from the presence of the counterions and the affinity between the polyelectrolyte chains and the solvent molecules.[48] Because the Ru complex acts as a catalyst for the BZ reaction and periodically changes its oxidation state during the BZ reaction, the polymer network of the gel in which the Ru complex is embedded exhibits a cyclic expansion–contraction change due to the change in osmotic pressure inside the gel. The spatial-temporal pattern of the BZ reaction can be controlled by the species, concentration of the substrate in the BZ reaction solution, and temperature of the reaction; the amplitude and the period of the self-sustaining motion of the poly(NIPA-Ru complex) gel caused by the BZ reaction can also be controlled. Furthermore, the movement pattern of the poly(NIPA-Ru complex) gel induced by the BZ reaction depends on the size and shape of the gel. If the total size of the poly(NIPA-Ru complex) gel is smaller than the wavelength of the chemical wave of the BZ reaction, the redox change occurs homogeneously within the gel without the formation of a pattern.[41] In this case, we can see that the periodic change in the size of the gel is synchronized with the BZ reaction (Figure 7.3(a)).

If we use colloidal poly(NIPA-Ru complex) particles that can change size due to synchronization with the BZ reaction, the flocculation of the colloidal gel particles will also have an effect on the oscillating reaction.[42] We can observe a change in the viscosity of the suspension that contains the colloidal gel particles due to the BZ reaction (Figure 7.3(b)). In the case of a rectangular poly(NIPA-Ru complex) gel, we can observe that the chemical waves of the BZ

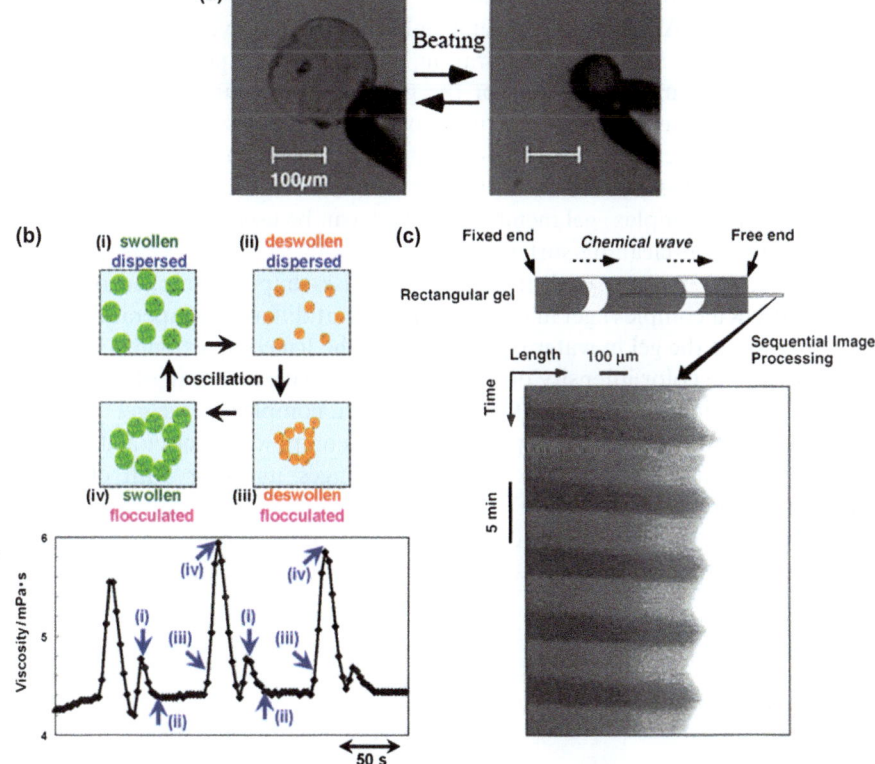

Figure 7.3 (a) Self-oscillation of poly(NIPA-Ru complex) gel resulting from the BZ reaction. (b) Oscillation in the viscosity of a poly(NIPA-Ru complex) gel particle suspension during the BZ reaction. (c) A spatiotemporal pattern of oscillating behavior visualized by image processing for a rectangular poly(NIPA-Ru complex) gel.
Reprinted with permission from Ref. 40. Copyright 1996 American Chemical Society.

reaction travel in the direction of the long axis of the gel (Figure 7.3(c)).[40] Consequently, the overall length of the gel undergoes a cyclic expansion–contraction change as the BZ reaction proceeds. The traveling chemical waves of the BZ reaction in the rectangular gel suggest that we can observe peristaltic motion on the surface of the gel. However, it was difficult to observe the peristaltic motion on the surface of the gel, which was synchronized with the spatiotemporal patterns of the BZ reaction, using an optical microscope because of small changes in the thickness of the gel. One reason for this result is that a phase difference between the chemical and mechanical oscillations is produced from the slow swelling–deswelling response of the gel compared with the speed of the cyclic redox changes.

Because the introduction of porosity into a thin gel membrane is likely to create a rapid response in the swelling-deswelling oscillation during the BZ reaction,[14] we used a colloidal crystal as a template to obtain a periodically ordered poly(NIPA-Ru complex) gel membrane with an inverse opal structure. Another advantage of the incorporation of the periodic porosity is that the dynamic change in the thickness of the gel membrane can be quantitatively determined by observing the structural color of the inverse opal gel, which is based on the reflection from the p-PBG.[29] In this section, I describe how to observe the self-sustained peristaltic motion on the surface of the poly(NIPA-Ru complex) gel membrane, which can be useful for the creation of actuators and self-cleaning surfaces.

Figure 7.4(a) presents optical images of a disk-shaped homogeneous bulk poly(NIPA-Ru complex) gel membrane in water at different temperatures. The orange color of the gel in water is derived from the Ru complex under reducing conditions. The color intensity of the gel increases with increasing temperature because of the increased concentration of the Ru complex in the gel.

Meanwhile, the inverse opal poly(NIPA-Ru complex) gel obtained using a colloidal crystal as a template reversibly changes its color with changing temperature (Figure 7.4(b)).

Figure 7.5(a) presents reflection spectra of the inverse opal gel at different temperatures. Figure 7.5(b) presents the absorption spectrum of the bulk gel

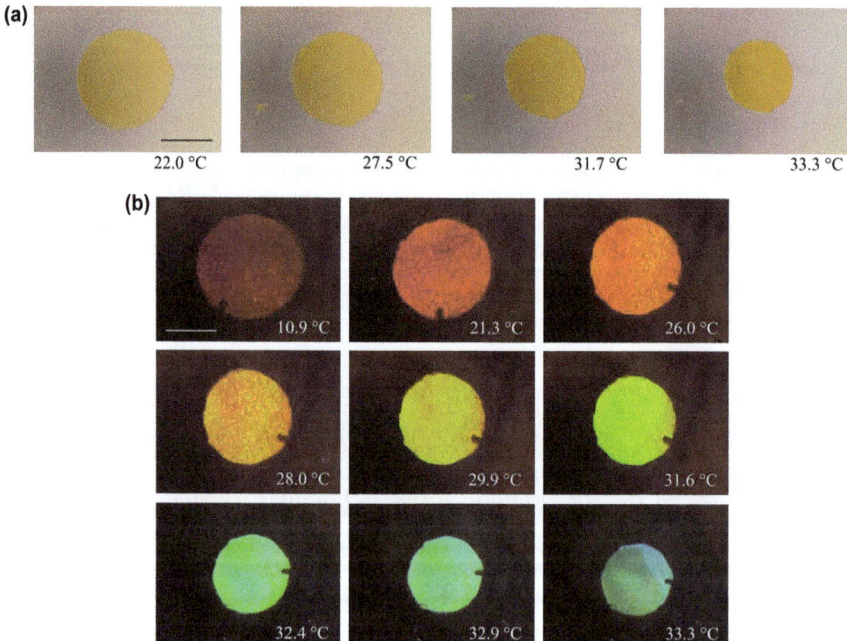

Figure 7.4   (a) Image showing the change in color of a homogeneous bulk poly(NIPA-Ru complex) gel at different temperatures. The scale bar is 1 cm. (b) Image showing the change in color of an inverse opal poly(NIPA-Ru complex) gel at different temperatures. The scale bar is 1 cm.

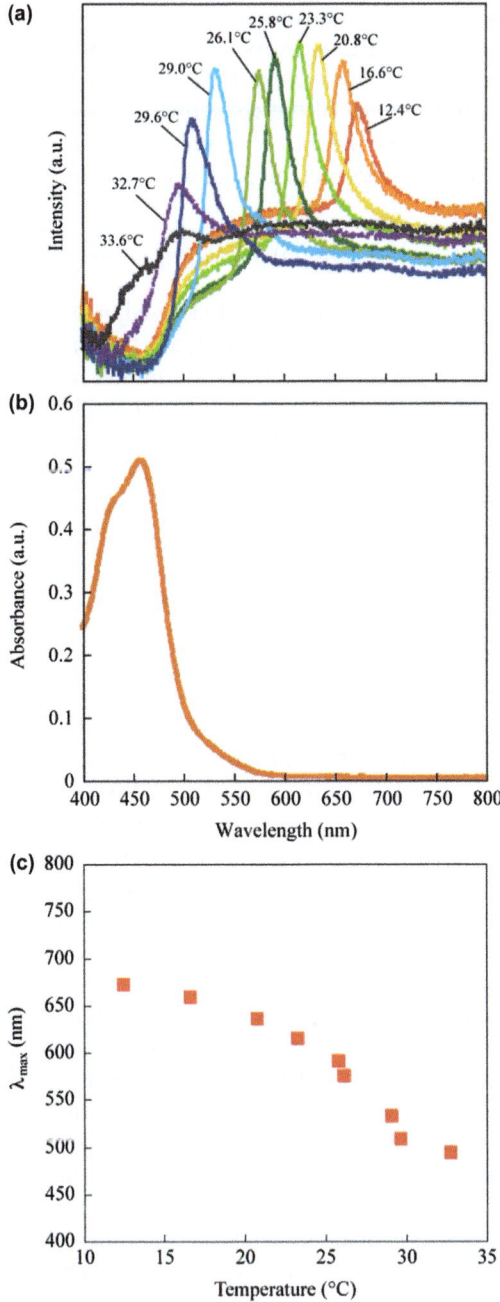

Figure 7.5 (a) Reflection spectra of the inverse opal poly(NIPA-Ru complex) gel at various temperatures in water. (b) UV absorption spectrum of the homogeneous bulk poly(NIPA-Ru complex) gel in water. (c) Temperature dependence of λ_{max} of the reflection spectra shown in (a).

membrane. A sharp reflection peak can be observed that depends on the temperature of the water in these inverse opal gel spectra. The position of the sharp peak is shifted to lower wavelengths with increasing water temperature, which reflects the color change of the gel shown in Figure 7.4(b).

If the porous gel maintains the fine structure of the precursor colloidal crystal and the volume change is isotropic, the peak wavelength can be estimated using the following equation,[29]

$$\lambda_{max} = 1.633(d/m)(D/D_0)(n_a^2 - \sin^2\theta)^{1/2} \qquad (7.1)$$

where d is the diameter of the colloidal particle used, m is the order of the Bragg reflection and is normally 1, D/D_0 (D and D_0 are the characteristic sizes of the gel in the equilibrium state at a certain condition and in the preparative state, respectively) is defined as the degree of equilibrium swelling of the gel, n_a is the average refractive index of the inverse opal gel, and θ is the angle measured from the normal to the plane of the gel. The peak at 460 nm in the absorption spectrum in Figure 7.5(b) arises from the metal–ligand charge-transfer transition that occurs in the reduced form of the Ru complex.[27] When the reflection peak from the inverse opal gel overlaps with the absorption spectrum of the Ru complex, the intensity of the reflection peak decreases. For example, at temperatures greater than 34 °C, almost no peak is observed in the reflection spectrum (Figure 7.5(a)). Accordingly, the observation of structural color in the gel at temperatures greater than 34 °C is difficult. However, the presence and the position of the reflection peak at temperatures less than 34 °C can be verified from the reflection spectrum. Figure 7.5(c) shows the plot of λ_{max} of the reflection spectra as a function of the water temperature. Because the volume of this gel becomes smaller with increasing temperature, we can observe the correlation between λ_{max} and the degree of equilibrium swelling of the gel. To quantitatively evaluate the peristaltic motion of the gel coupled with the BZ reaction, we must know the quantitative relationship between λ_{max} and the degree of equilibrium swelling of the gel. Here, I clarify the relationship between these parameters.

At approximately 32 °C, the poly(NIPA-Ru complex) gel undergoes a reversible volume change between a hydrated swollen state at lower temperatures and a dehydrated collapsed state at higher temperatures, which reflects the nature of polyNIPA (Figure 7.6(a)). Based on eqn (7.1), the changes in both the degree of swelling (D/D_0) and n_a are parameters in the determination of the value of λ_{max}. Therefore, the change in the refractive indices of a bulk poly(NIPA-Ru complex) gel that does not have pores and water at various temperatures was measured using a refractometer to estimate the value of n_a at various temperatures. Figure 7.6b shows the change in the refractive indices of the bulk poly(NIPA-Ru complex) gel, water, and the inverse opal gel calculated using eqn (7.2), as shown below, at different temperatures. The refractive index of the bulk gel is approximately 1.338 at 10 °C, which is close to the refractive index of pure water because the gel at 10 °C is composed of more than 90% water. From 10 to 32 °C, the refractive index of the bulk

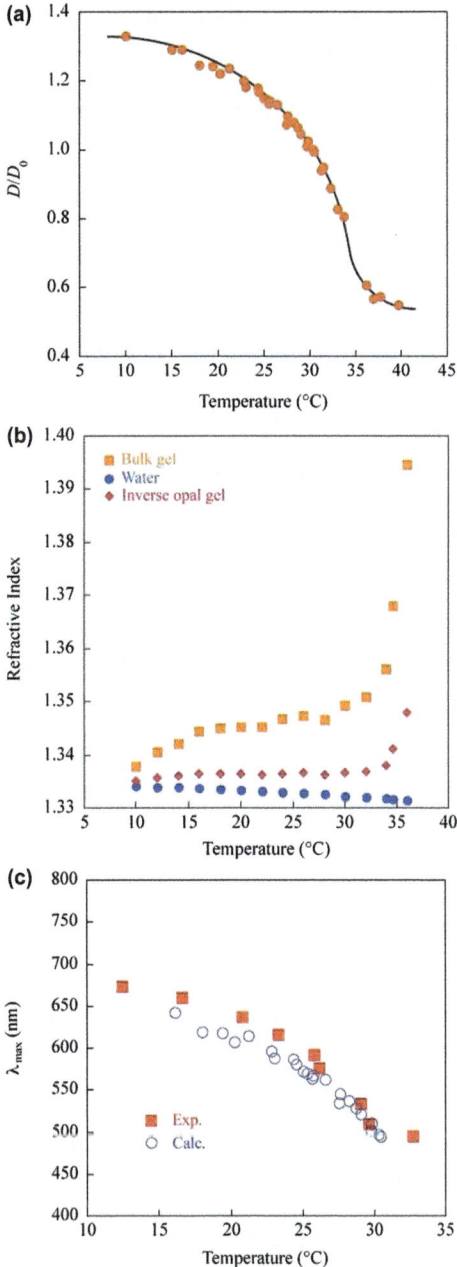

Figure 7.6 (a) Swelling ratios of the homogeneous bulk poly(NIPA-Ru complex) gel are plotted as a function of temperature. (b) Refractive indices of bulk poly(NIPA-Ru complex) gel, water, and the inverse opal poly(NIPA-Ru complex) gel in water are plotted as a function of temperature. (c) The calculated λ_{max} is compared with the experimental data in Figure 7.5(c).

gel gradually increases and then rather sharply increases at temperatures greater than 32 °C. Meanwhile, the refractive index of water gradually decreases as the temperature increases. The value of n_a is calculated as a weighted sum of the refractive indices of the spherical portion and the gap portion as follows:

$$n_a^2 = \sum n_i^2 \phi_i \qquad (7.2)$$

where n_i and ϕ_i are the refractive index and the volume fraction of each portion, respectively. The calculated refractive index of the inverse opal gel slightly varies from 10 to 32 °C, but the value sharply increases at temperatures greater than 32 °C. However, the change in n_a in the inverse opal gel from 10 to 32 °C is only 0.002. Even from 32 to 36 °C, the change in n_a is only approximately 0.01. Therefore, these are negligible changes for λ_{max} compared with the change in the degree of swelling (Figure 7.6(a)). The values of λ_{max} estimated using $d = 230$ nm, $n_a = 1.336$, and D/D_0, which are values that were experimentally obtained with increasing temperature (Figure 7.6(a)), are consistent with the value of λ_{max} obtained experimentally with increasing temperature (Figure 7.6(c)).

Because the aqueous solution we used for the BZ reaction contains malonic acid as a reactive substrate, $NaBrO_3$ as an oxidizing agent, and nitric acid to keep the solution acidic, the degree of swelling of the poly(NIPA-Ru complex) gel can change under the influence of these solutes, especially the electrolytes. One reason for this behavior is that the degree of swelling of the polyelectrolyte gel decreases with increasing electrolyte concentration in a solution because the inner osmotic pressure of the polyelectrolyte gel can be affected by the presence of electrolytes.[48] Therefore, the swelling behavior of the poly(NIPA-Ru complex) gel was observed under conditions where the concentration of the electrolytes was approximately the same as the concentration of electrolytes in the BZ reaction solution. In addition, we need to know the effect of the oxidation state of the Ru complex attached to the polymer network on the degree of swelling of the poly(NIPA-Ru complex) gel, which directly influences the value of λ_{max}. Figure 7.7(a) shows the degrees of swelling of the poly(NIPA-Ru complex) gels with oxidized and reduced Ru complexes at different temperatures. These oxidation states of the Ru complex were controlled using a reducing agent and an oxidizing agent. These gels underwent continuous volume changes with increasing temperature. The degrees of swelling of the poly(NIPA-Ru complex) gel with the oxidized Ru complex were larger than those of the poly(NIPA-Ru complex) gel with the reduced Ru complex at all temperatures. The increase in the inner osmotic pressure of the poly(NIPA-Ru complex) gel caused by the increase in the number of counterions in the Ru complex caused the increase in the degree of swelling. Figure 7.7(b) shows the plot of experimentally obtained λ_{max} values of these inverse opal gels versus temperature. As expected, we observed that the values of λ_{max} for these gels decreased with increasing temperature. Additionally, the values of λ_{max} for the gels with the oxidized state of the Ru complex are always greater than those for the gels with the reduced state of the Ru complex.

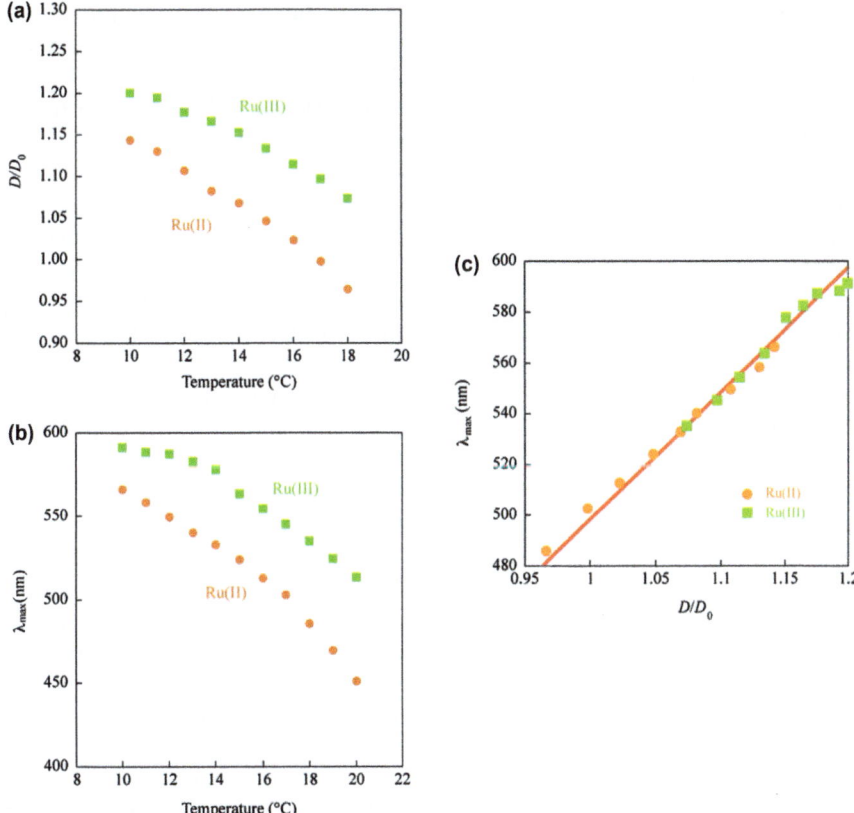

Figure 7.7   (a) Swelling ratios of the homogeneous bulk poly(NIPA-Ru complex) gels with oxidized or reduced Ru complexes under the condition in which the concentration of electrolytes is approximately the same as the electrolyte concentration in the BZ reaction solution, plotted as a function of temperature. (b) The value of λ_{max} of the reflection spectra of the inverse opal poly(NIPA-Ru complex) gels with oxidized or reduced Ru complexes under the condition where the concentration of electrolytes is approximately the same as the electrolyte concentration of the BZ reaction solution, plotted as a function of temperature. (c) The plot of λ_{max} versus the swelling ratio of the poly(NIPA-Ru complex) gels.

Figure 7.7(c) shows a plot of the experimental peak position versus the degree of swelling of these porous gels. A strong linear correlation between λ_{max} and D/D_0 was observed in both porous gels, which implies that the theory and experiment are in agreement and that the degree of swelling is dominant over the value of λ_{max}. The red line was theoretically obtained using $d = 230$ nm, $n_a = 1.336$, and $\theta = 0$ and was in good agreement with the experimentally obtained plots. Therefore, we can estimate the value of D/D_0 of the porous gel by observing the value of λ_{max} in the reflection spectra of the porous gel during the BZ reaction.

We used an optical microscope to observe the dynamic change in the pigmented color from the homogeneous bulk disk-shaped poly(NIPA-Ru complex) gel during the BZ reaction at 25 °C. The steadily evolving, faint green patterned portion that appeared in the orange-colored bulk gel is due to the oxidation of the Ru complex during the BZ reaction[47] (Figure 7.8(a)).

The structural color change in the disk-shaped inverse opal poly(NIPA-Ru complex) gel during the BZ reaction was also observed at various temperatures using an optical microscope. Figure 7.8(b) shows the changes in the structural colors over time at different temperatures.[27] The time at which concentric rings developed when the Ru complex was in an oxidized state during the BZ reaction was indicated with a white "X" mark at each temperature, and this time was selected as the starting time of structural color change. A concentric colorful ring caused by the swelling of the inverse opal gel due to the oxidation of the Ru complex during the BZ reaction, which is shown with a dotted line, spreads out from the white "X" mark at each temperature. The concentric developing change in structural color indicates a peristaltic motion occurring on the surface of the gel, which resembles the intestinal motility and motion of an earthworm. The changes in the reflection spectra during the BZ reaction were monitored at different temperatures. A reflection probe used to observe the spectra was fixed in the prescribed position above the inverse opal gel. Periodic alterations in the reflection spectra were observed during the BZ reaction, indicating that the swollen portions of the gel passed by the probe in a regular sequence[26] (Figure 7.8(c)). The oscillatory redox reaction during the BZ reaction was also simultaneously observed by measuring the optical transmittance of light at 670 nm. As a characteristic absorption band of the oxidized Ru complex only appeared at 670 nm in the UV/Vis spectra and was not observed for the reduced Ru complex (Figure 7.9(a)), the oscillating redox reaction can be easily followed spectrophotometrically by measuring the optical transmittance of light at 670 nm. Figure 7.9(b) shows the periodic changes in λ_{max} that were observed from the reflection spectra and in the optical transmittance of light at 670 nm from the inverse opal gel at 13 °C. The mechanical oscillation of the gel autonomously occurs with the same period as that of the redox oscillations; the chemical and mechanical oscillations are synchronized without a phase difference.

Because the value of D/D_0 during the BZ reaction can be estimated using the spectroscopically observed value of λ_{max}, the variable thickness, L, of the gel membrane can be calculated using the spatiotemporally determined values of λ_{max}, d, and n_a using the following equation:[27,28]

$$L = L_0(D/D_0) = L_0 m\lambda_{max} / 1.633dn_a \qquad (7.3)$$

where L_0 is the preparative thickness of the gel membrane. The potential variation in $\Delta\lambda_{max}$ assessed from the results shown in Figure 7.7(b) is 43 nm at 13 °C. The variation will be achieved only in the case where the Ru complex changes between the completely oxidized state and the completely reduced state during the BZ reaction. As the gel membrane used in this experiment is 500 μm thick, the potential change in the thickness of the membrane is 43 μm at 13 °C,

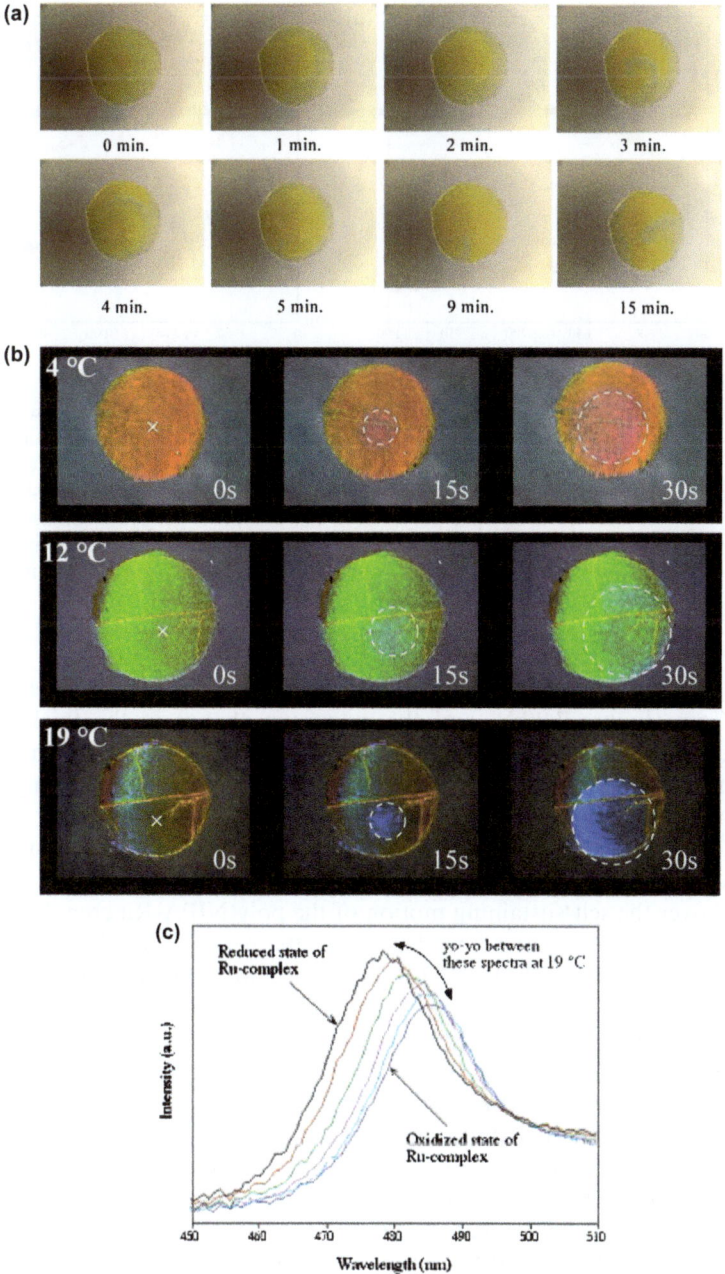

Figure 7.8 (a) Spatiotemporal color patterns of oscillating behavior for a disk-shaped bulk poly(NIPA-Ru complex) gel during the BZ reaction. (b) Spatiotemporal structural color changes for a disk-shaped inverse opal poly(NIPA-Ru complex) gel during the BZ reaction at different temperatures. (c) Change in the reflection spectra of the disk inverse opal poly(NIPA-Ru complex) gel during the BZ reaction at 19 °C.
Reprinted with permission from Ref. 26. Copyright 2006 American Chemical Society.

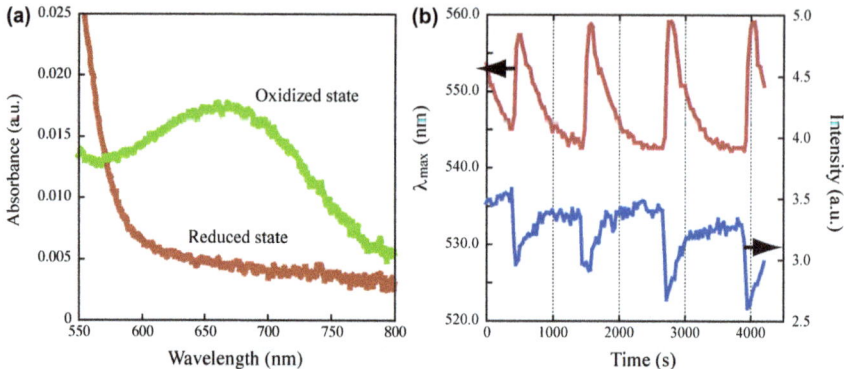

Figure 7.9 (a) UV/Vis spectra of Ru(III) (oxidized state, aqueous solution of Ce(SO$_4$)$_2$ (5 mM) and HNO$_3$ (916 mM)) and Ru(II) (reduced state, aqueous solution of Ce$_2$(SO4)$_3$ (5 mM) and HNO$_3$ (906 mM)) of the bulk poly(NIPA-Ru complex) gel. (b) Oscillating behavior of the structurally colored inverse opal poly(NIPA-Ru complex) gel synchronized with the BZ reaction. Temporal changes in the value of λ_{max} and the intensity of transmitted weak light through the inverse opal poly(NIPA-Ru complex) gel were observed at 13 °C.

which was estimated using eqn (7.3). However, the maximum change in $\Delta\lambda_{max}$ was 17 nm (Figure 7.9(b)), and the maximum change in the thickness was 17 μm. This result must arise because the Ru complex cannot change between the completely oxidized state and the completely reduced state during the BZ reaction.

The cyclic rhythm can be tuned by the external concentration of each substrate in the BZ reaction solution.[28] By considering this behavior, chemical control over the self-sustaining motion of the poly(NIPA-Ru complex) gel can be achieved. It is common knowledge that the period is very sensitive to the concentration of NaBrO$_3$, [NaBrO$_3$], and we observed the [NaBrO$_3$] dependence on the period and amplitude of the self-sustaining peristaltic motion derived from the reflection spectra.[28] Figure 7.10 shows the periodic changes in the thickness of the inverse opal gel membrane during the BZ reaction with various [NaBrO$_3$] values.

The period of the peristaltic motion decreased with increasing [NaBrO$_3$]. Meanwhile, the amplitude was almost independent of [NaBrO$_3$] at lower concentrations, whereas it was significantly decreased at higher [NaBrO$_3$] values. We interpret this phenomenon as follows: as the swelling–deswelling change cannot follow the BZ reaction with high frequency at higher [NaBrO$_3$] values, the amplitude of the peristaltic motion notably decreased. As described above, we have shown that the self-sustaining peristaltic motion on the surface of the inverse opal poly(NIPA-Ru complex) gel can be observed with the naked eye and quantitatively determined through measurements of reflection spectra. Because the motion can be controlled by other stimuli, such as light[27,28] and temperature,[26] this system may be useful in the development of new

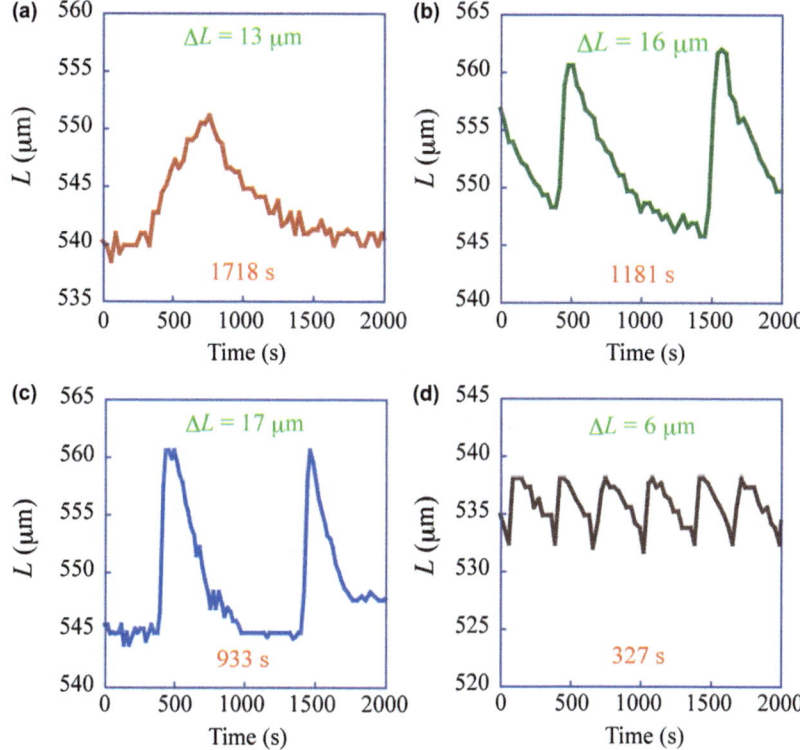

Figure 7.10 Temporal changes in the thickness of the inverse opal gel during the BZ reaction at 13 °C with different initial compositions of the reaction solution as follows: [malonic acid] = 0.0625 M, [HNO₃] = 0.890 M, and [NaBrO₃] for (a) 28 mM, (b) 42 mM, (c) 60 mM, and (d) 84 mM solutions.

micromachines capable of imitating biological functions, future lab-on-a-chip systems, and functional surfaces that transport fluids.

7.3 A Light-Sensitive Inverse Opal Gel that Exhibits Rapid Two-State Switching between Two Arbitrary Structural Colors

In this section, I introduce a photosensitive inverse opal hydrogel that exhibits a reversible, light-triggered rapid change between two arbitrary volume states at a certain temperature. The hydrogel composed of NIPA; azobenzene monomer; 4-acryloylaminoazobenzene (AAB), which changes its dipole moment upon the photoisomerization of the azobenzene group; and BIS can change its volume in response to temperature variations and light irradiation (Scheme 7.2).[31,49]

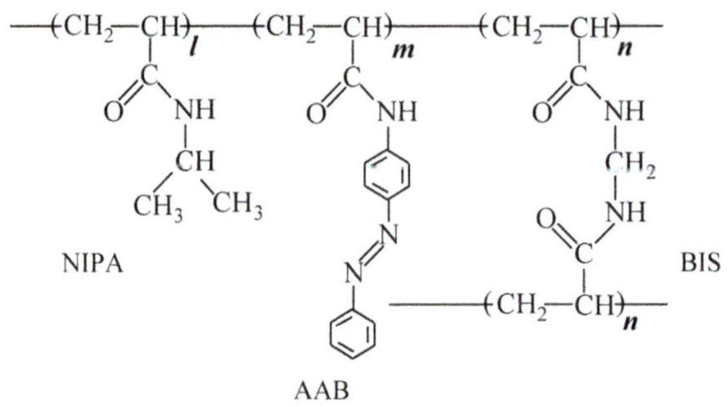

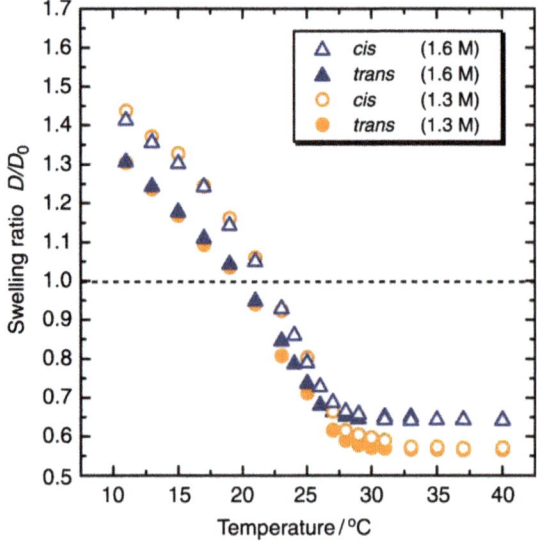

Scheme 7.2   Chemical structure of the poly(NIPA-AAB) gel.

Figure 7.11   The degree of swelling of a cylindrical poly(NIPA-AAB) gel in water in the dark and upon irradiation with UV light (366 nm), plotted as a function of temperature. The total monomer amounts of pregel solutions used to prepare the gels were 1.3 and 1.6 M, and the amount of AAB was 3 mol% in these solutions.

For example, poly(NIPA-AAB) gels with 3 mol% of AAB in a total monomer amount of 1.3 or 1.6 mol/L in the pregel solution undergo a reversible, drastic change in volume at approximately 27 °C under dark conditions, where the majority of the azobenzene group of AAB is in the *trans* state[31] (Figure 7.11).

We know that the hydrogel composed of only NIPA and BIS undergoes an abrupt volume change at approximately 32 °C.[50] Compared with this hydrogel, because the *trans* state of the azobenzene group in AAB may have no dipole moment, the lower critical solution temperature (LCST) of the poly(NIPA-AAB) gel decreases due to the hydrophobicity of AAB. Azobenzene derivatives are well-known photochromic compounds that exhibit two different states, namely the *trans* form and the *cis* form, which can be reversibly interconverted using a light stimulus. The *trans* state of azobenzene is thermally stable under dark conditions but can be isomerized to the *cis* state by applying ultraviolet light, which corresponds to the energy gap of the $\pi-\pi^*$ transition. Because the *cis* state has a larger dipole moment than the *trans* state, the volume of the hydrogel with an azobenzene derivative increases by applying ultraviolet light (Figure 7.11). Meanwhile, the *cis* state can be recovered to the *trans* state with blue light, which is equivalent to the $n-\pi^*$ transition. Therefore, the volume of the hydrogel with an azobenzene derivative decreases by applying blue light. As described above, a reversible change in the dipole moment of the azobenzene moiety can be useful for inducing a volume change in the poly(NIPA-AAB) gel.[51]

We introduced porosity into the poly(NIPA-AAB) gel by using a colloidal crystal as a template.[31] The obtained inverse opal poly(NIPA-AAB) gel maintains the fine structure of the precursor colloidal crystal and exhibits brilliant structural color resulting from Bragg reflections. Because inverse opal gels can change their volume isotropically under the LCST, the value of λ_{max} diffracted from the inverse opal gel can be expressed by eqn (7.1) in the previous section. Because the volume of the poly(NIPA-AAB) gel becomes smaller as the temperature increases, the value of λ_{max} decreases with increasing temperature (Figure 7.12).

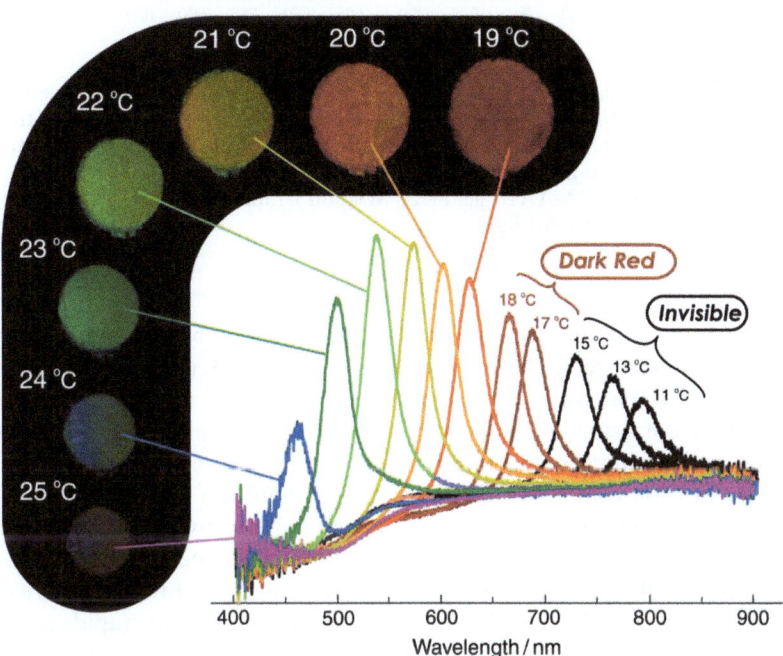

Figure 7.12 Spectroscopic characterization of the inverse opal gel. Photographs and reflection spectra of the inverse opal poly(NIPA-AAB) gel in the dark at various temperatures.
Reprinted with permission from Ref. 31. Copyright 2007 Wiley-VCH.

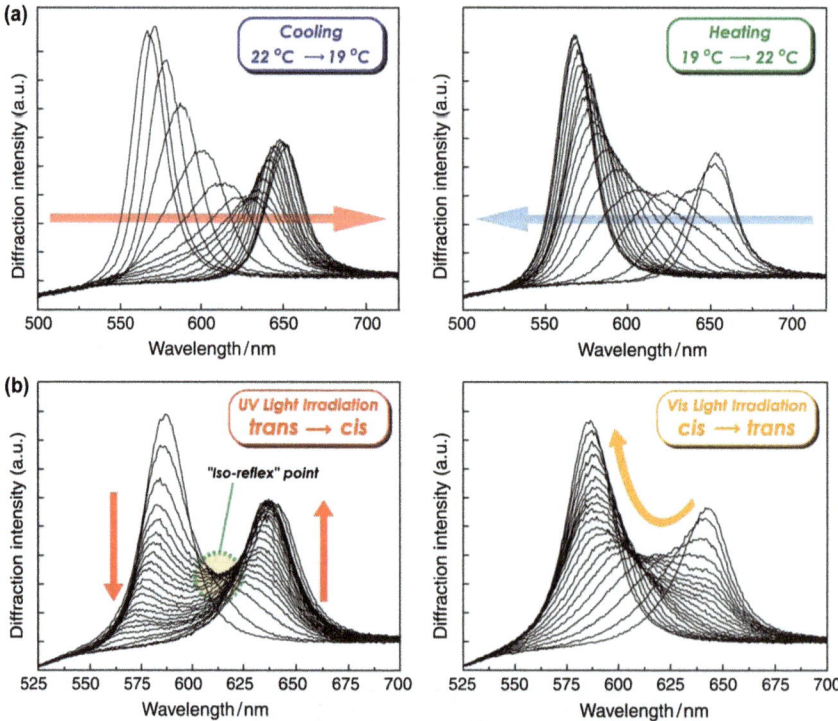

Figure 7.13   (a) Changes in the reflection spectra with time for the inverse opal
poly(NIPA-AAB) gel in the dark upon decrease in temperature from 22
to 19 °C (left) and upon increase in temperature from 19 to 22 °C (right).
(b) Changes in the reflection spectra with time for the inverse opal
poly(NIPA-AAB) gel at 21 °C upon irradiation with 366-nm (left) and
437-nm (right) light.

A rapid and continuous shift in λ_{max} was induced by temperature jumps
between two given temperatures lower than the LCST (Figure 7.13(a)). This
result indicates that the inverse opal gel exhibits a fast and isotropic volume
change that corresponds to the rapid temperature change. The decrease in
intensity and the increase in width that occur at the half-maximum of the peaks
upon the abrupt volume change must be attributed to the distortion of the
inverse opal structure. The distortion may be induced by the difference in the
size of the various components of the inverse opal gel because the relaxation
time of the swelling and shrinking of a gel is proportional to the square of its
characteristic size. However, the inverse opal gel exhibits rapid two-state
switching between two arbitrary structural colors that is in synch with the
trans–cis photoisomerization of the azobenzene moiety as a result of irradiation
with light at a certain temperature. Figure 7.13(b) shows the kinetic alteration
in the reflection spectra upon irradiation with UV or visible light. An isosbestic
point was observed in both cases shown in Figure 7.13(b) *en route* to new

equilibrium conditions. Isosbestic behavior indicates that there are two distinct swelling states during the alterations.

To explain these results, we present simple sketches of the alterations in Figures 7.14(a) and 7.14(b). The continuous shifting of λ_{max} is attributed to the isotropic change in the volume of the gel upon a sharp temperature increase (Figure 7.14(a)). Upon exposure to UV, part of the azobenzene moiety must isomerize to the *cis* state. The incident UV light is primarily absorbed by the azobenzene moiety in the vicinity of the surface of the gel on the side near the light source because of the strong absorption of light at approximately 366 nm by the azobenzene moiety. Consequently, the incident UV light gradually reaches the azobenzene moieties that are in the *trans* state in the interior of the gel membrane. Therefore, upon exposure to UV light, a portion of the gel where the azobenzene moiety exists in the *cis* form will absorb water, and this swollen portion must then coexist with the remaining smaller portion of the gel

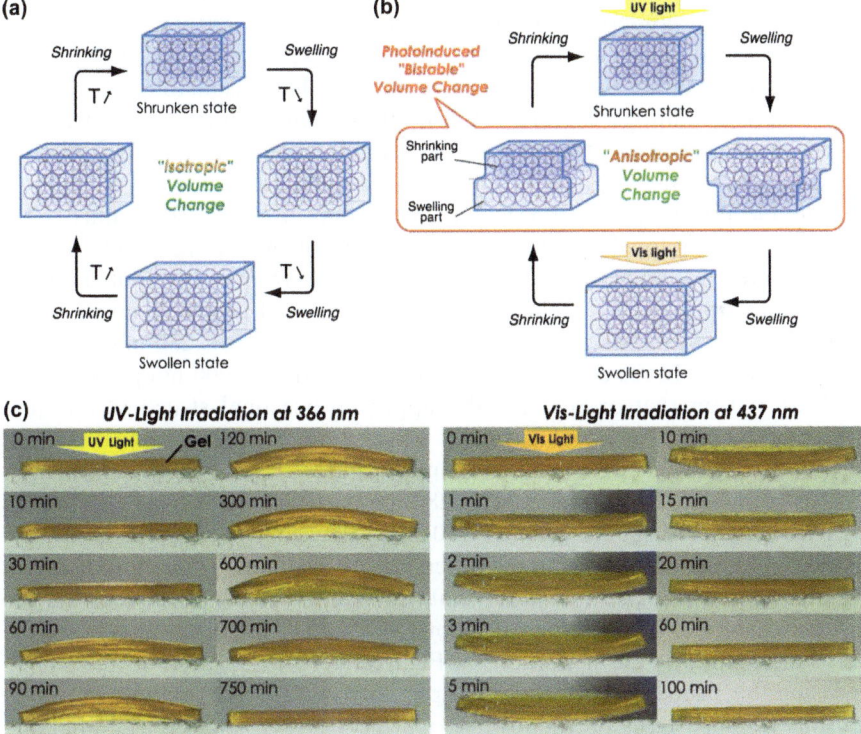

Figure 7.14   Models of the temporal volume change mechanisms of the inverse opal poly(NIPA-AAB) gel driven by (a) a temperature jump and (b) light irradiation. (c) Shape variation behavior of the homogeneous bulk poly(NIPA-AAB) gel driven by light irradiation with 366-nm (left) and 437-nm (right) light was observed on the side of the disk-shaped gel membrane.

in its initial state (Figure 7.14(b)). If we use a homogeneous bulk gel that is composed of NIPA, AAB, and BIS, the bulk gel bends away from the light source to reduce the elastic energy generated by the deformation[52] (Figure 7.14(c)). The total amount of elastic energy in the bent gel is less than that in the gel with the coexistence of two different swollen states. Therefore, the bending deformation of the gel is caused by the reduction in the total amount of elastic energy in the gel. We know that the total amount of elastic energy is proportional to the net volume of a gel.[53] As the net volume of the inverse opal gel is considerably less than that of the bulk gel, the total amount of elastic energy in the inverse opal gel with the coexistence of two different swollen states is also considerably less than that in the bulk gel. Consequently, the coexistence of two different swollen states in the inverse opal gel can be facilitated by irradiation with light in a temperature-controlled environment.

To confirm our above hypothesis, we prepared a double inverse opal gel that reflected two notably different wavelengths of light using a bilayer colloidal crystal; one side was prepared using a colloidal crystal composed of SiO_2 particles that were 280 nm in diameter, and the other side was prepared using a colloidal crystal composed of SiO_2 particles that were 220 nm in diameter[52] (Figure 7.15(a)). We can observe two distinct peaks at 560 and 710 nm in the reflection spectrum of the double inverse opal gel that was maintained at 15 °C under dark conditions (Figure 7.15(b)). If the temperature of the water is changed, the positions of both peaks are simultaneously shifted according to the isotropic volume change. If our hypothesis is correct, upon irradiation of the double porous gel with UV light from one direction, there must be a response time lag for the change in the positions of these peaks. Thus, we can observe a response time change in the reflection spectrum of the double inverse opal gel by applying UV light to double inverse opal gels with upper layers consisting of membranes prepared with either 220- or 280-nm particles (Figure 7.15(c)). In the case of membranes prepared with 200-nm particles, the peak at 560 nm that arose from the upper inverse opal structure was first replaced by a new peak at a higher wavelength, and the peak at 710 nm subsequently changed in a similar manner (Figure 7.15(c), left). In contrast, in the case of membranes prepared with 280-nm particles, the two-state switching was first observed in the peak at 710 nm, followed by the peak at 560 nm (Figure 7.15(c), right). Thus, these observations verify our prediction shown in Figure 7.14(b). Furthermore, we confirmed the two-state switching in the structural color of the inverse opal gel by applying UV light from the side of the gel, which supports our hypothesis[52] (Figure 7.15(d)).

Using this property of the inverse opal poly(NIPA-AAB) gel, we can draw various color patterns on the inverse opal poly(NIPA-AAB) gel by applying light. For example, if the inverse opal gel is stored in the dark for a certain amount of time and then exposed to UV light through a photomask, the irradiated portion will increase in volume and will undergo a change in structural color.[31] Because the swollen portion has little influence on the nonirradiated portion in terms of volume, patterns of color can be generated on the gel membrane (Figure 7.16).

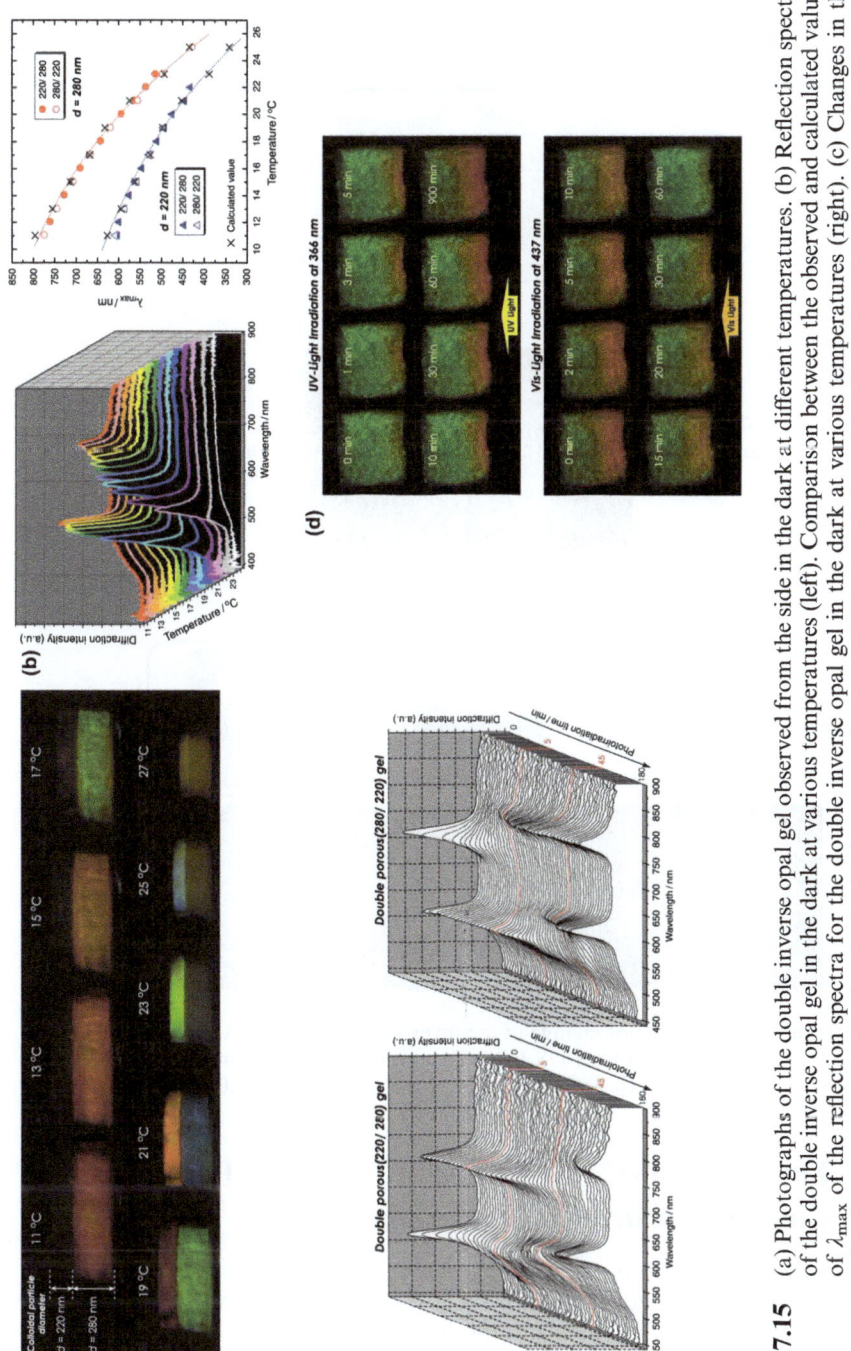

Figure 7.15 (a) Photographs of the double inverse opal gel observed from the side in the dark at different temperatures. (b) Reflection spectra of the double inverse opal gel in the dark at various temperatures (left). Comparison between the observed and calculated values of λ_{max} of the reflection spectra for the double inverse opal gel in the dark at various temperatures (right). (c) Changes in the reflection spectra with time for the double inverse opal gel at 15 °C by irradiation with 366-nm light from the normal to the plane of the gel membrane. The upper layer is the inverse opal membrane prepared with 220-nm particles (left), and the upper layer is the inverse opal membrane prepared with 280-nm particles (right). (d) Photographs of the structural color changes with photoirradiation from the side of the inverse opal poly(NIPA-AAB) gel induced by (upper) *trans–cis* and (lower) *cis–trans* photoisomerization of azobenzene moieties in the gel.

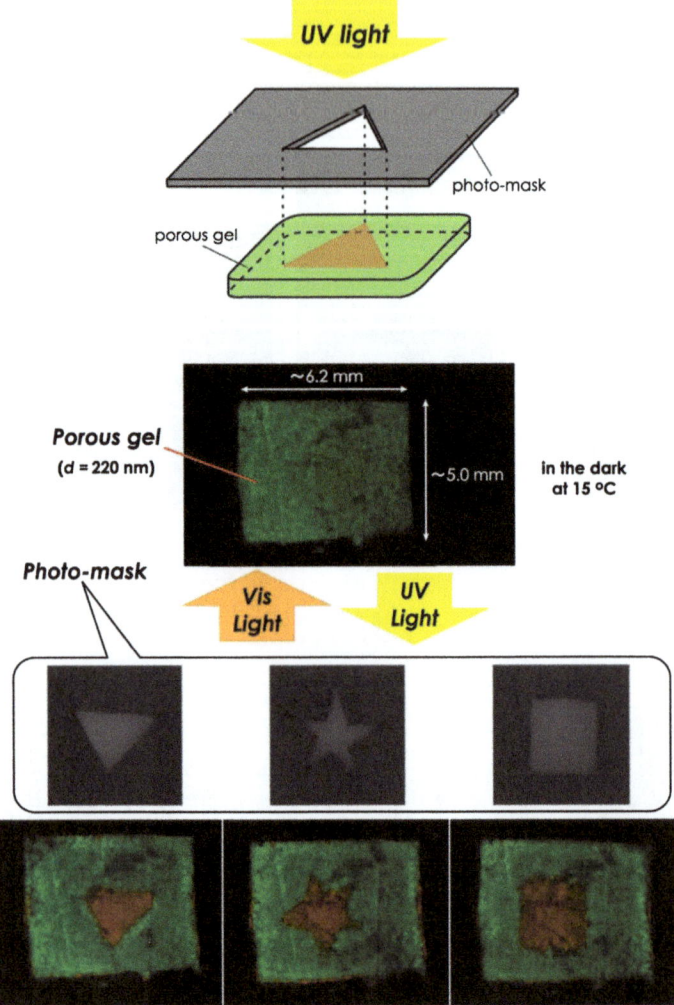

Figure 7.16   Structural color change in the inverse opal poly(NIPA-AAB) gel caused by irradiation with UV light through the corresponding photomasks. Reprinted with permission from Ref. 31. Copyright 2007 Wiley-VCH.

7.4   Tunable Full-Color Material from Gel Particles Confined in Inverse Opal Gels

Tunable photonic crystals may have possible applications in future visual systems, such as the paper-like color displays that have attracted attention over the years. Most recently, some groups have independently developed electrically tunable soft photonic crystals that may be applicable for flexible electronic paper, multicolor display technologies. These newly created tunable soft photonic crystals exhibit hue changes in the entire visible region that are voltage tunable without the use of any other complicated systems that are used in existing displays. However, the display of subtle shades of color cannot be accomplished using these tunable soft photonic crystals because these systems do not have a mechanism by which to control the brightness and saturation of color. A new method is necessary to manipulate the delicate change in coloration required to produce paper-like, full-color displays in the future.

Here, I provide details of the recently developed dual-tunable soft photonic crystal[35] that exhibits independent control of the hue and brightness or saturation of the structural colors as a full-color material. We have created display media and sensing materials that show multicolored images based on the tunable p-PBG using stimuli-responsive inverse opal-type porous hydrogels obtained using a colloidal crystal of silica colloidal particles as a template (Figure 7.17).[16,18,29,32] By changing the number of target molecules, applying light, and tuning the voltage and pH applied to the stimuli-sensitive inverse opal gels, and thereby the variation of the degree of swelling of the gels, the spacing between the submicrometer pores can be altered, which changes the displayed color. These results demonstrate that it is possible to rapidly tune the structural color hue of the inverse opal gels, similar to the tuning of colors by the iridophores of fish and chameleons.

Additionally, we have studied other tunable soft photonic crystals where the peak intensity of the reflection spectra changes with a change in the volume of gel particles confined in an inverse opal hard polymer (Figures 7.18(a) and (b)), whereas no change in the peak position was observed.[33] When the pores of the inverse opal hard polymer are filled with the gel particles, the normal-incidence reflection spectra in the visible region show distinct peaks due to diffraction from the periodic layers composed of the hard polymer and gel. As the gel particles undergo a collapsed transition, they become stickier on the cavity wall of the inverse opal hard polymer. This behavior can induce a disturbance in the ordered array of the gel particles and may cause the formation of many layers of rough surfaces in the inverse opal hard polymer (Figure 7.18(c)).

In other words, the gel array in the inverse opal hard polymer undergoes a reversible order–disorder transition depending on the variation of the size of the gel. This situation leads to stronger incoherent scattering of light from the gel particles, which results in a decrease in the coherent Bragg scattering when the gels are in a collapsed state. Consequently, the brightness or saturation of the structural color from this system can be controlled by external stimuli. Whether we can control the brightness or saturation of the structural color depends on whether the background color of the photonic crystal is black or white. However, there have been few reports on independently and extensively tuning both the hue and brightness or saturation of the structural color. To accomplish the aim of this study, we prepared a new soft photonic crystal that combined the above two different types of tunable photonic crystals (Figure 7.19).[35]

The methodology associated with our new concept for a dual-tunable photonic crystal (DTPC) is as follows. First, an inverse opal gel that is responsive to stimulus A is prepared using a colloidal crystal as a template.

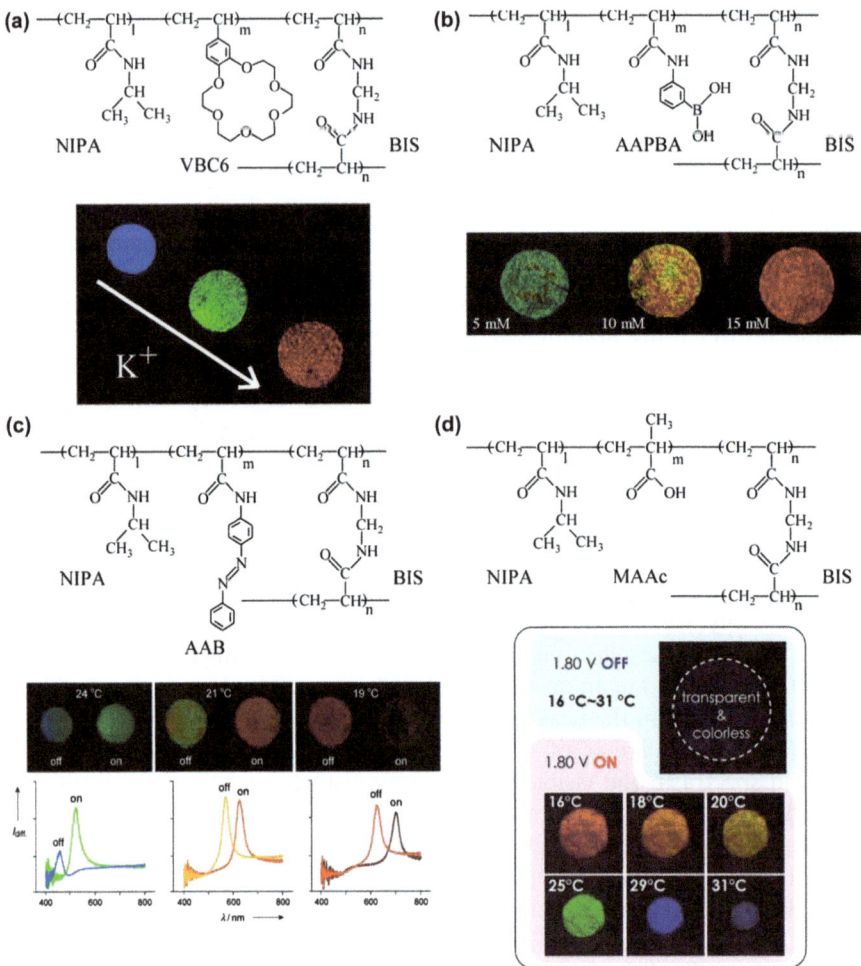

Figure 7.17 Various applications of stimuli-responsive inverse opal gels in sensors and displays. (a) Ion response, (b) glucose response, (c) light response, and (d) electric-field response.
Reprinted with permission from Ref. 30. Copyright 2007 Wiley-VCH.

Subsequently, gel particles that are sensitive to stimulus B are prepared in the porous regions of the inverse opal gel. The important conditions for the gel particles confined in the inverse opal gel system are that the inverse opal gel has no reaction to stimulus B and that the gel particles are unresponsive to stimulus A. Consequently, these gel particles that are confined to the inverse opal gel system will change the hue of the structural color in response to stimulus A, whereas the variation in the brightness or the saturation of the structural color can be evoked by stimulus B. In the present work, we selected pH as stimulus A and temperature as stimulus B to prepare a prototype DTPC that exhibits tunable color. The resultant DTPC prepared from thermosensitive gel particles

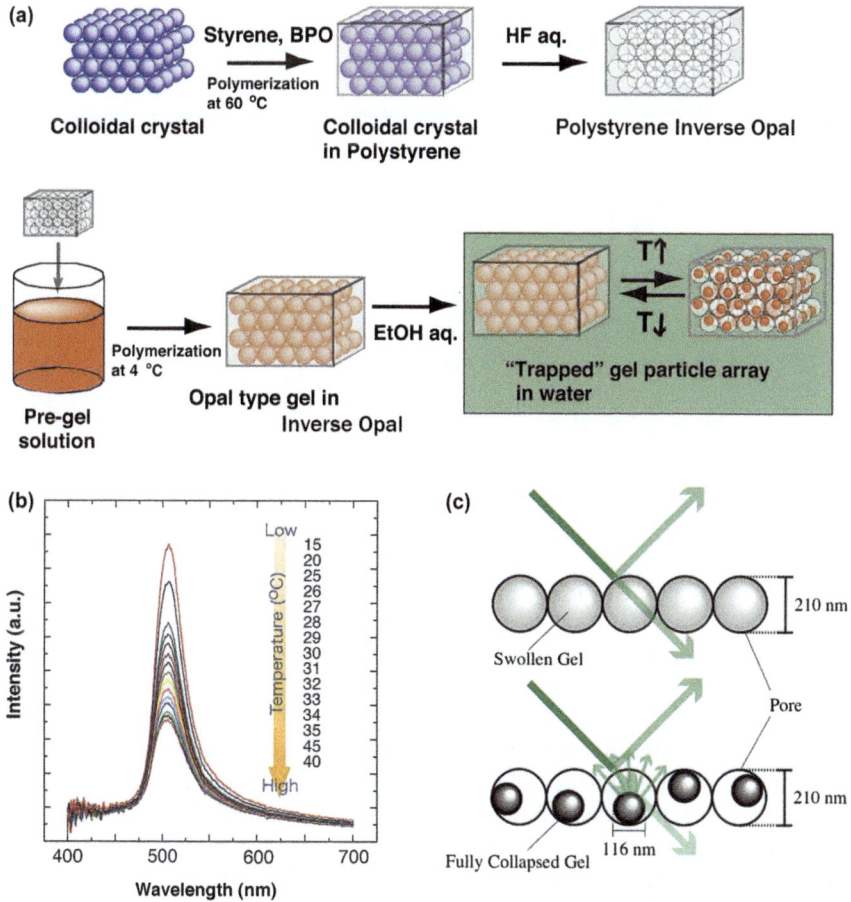

Figure 7.18 (a) Preparation of a "trapped" gel particle array in an inverse opal polymer. (b) Temperature dependence of the reflection spectra of the trapped gel particle array in water. (c) Size and situation of hydrogel particles in pores of an inverse opal polymer as a function of temperature.

Reprinted with permission from Ref. 33. Copyright 2006 American Chemical Society.

confined in the pH-sensitive inverse opal gel system independently displays variable structural color in terms of the hue and the brightness or saturation, which depend on the pH and temperature, respectively.

A colloidal crystal composed of monodispersed spherical particles of amorphous silica with a diameter of 233 nm was used as a template to prepare a pH-sensitive inverse opal gel. Because the experimentally obtained peak position ($\lambda_{max} = 482$ nm) of the reflection spectrum from the colloidal crystal coincides with the theoretical prediction ($\lambda_{max} = 495$ nm), the colloidal crystal has a tightly controlled numbers of layers that are oriented with their fcc (111)

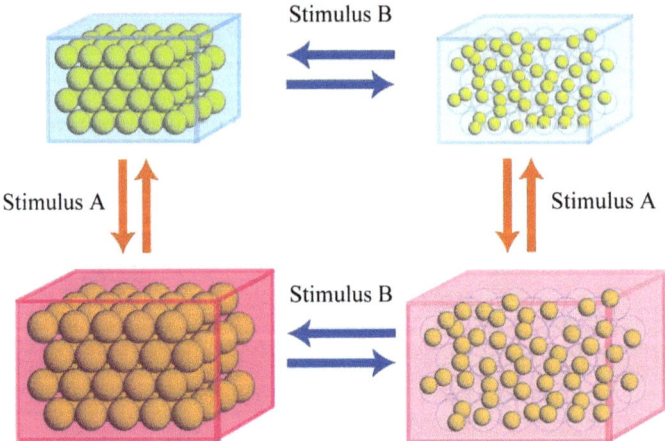

Figure 7.19 Concept of dual-tunable photonic crystals composed of gel particles confined in an inverse opal gel. The inverse opal gel can reversibly change its volume depending on the intensity of stimulus A, whereas the volume of the gel particles can be varied according to the intensity of stimulus B. Reprinted with permission from Ref. 35. Copyright 2009 Wiley-VCH.

axes parallel to a glass substrate.[29] We can estimate the crystalline quality from the normalized stop-bandwidth, $\Delta\lambda/\lambda_{max}$, where $\Delta\lambda$ is the width at the half-maximum of the peak. The experimental value of $\Delta\lambda/\lambda_{max}$ is 0.075 for the crystal, whereas the theoretically calculated value of $\Delta\lambda/\lambda_{max}$ along the (111) direction of a close-packed fcc air-filled SiO_2 colloidal crystal is 0.049.[54] Although there is a discrepancy between these values, this crystal can be regarded as being a fine crystal. The sharp hexagonal peaks in the two-dimensional Fourier transform of the SEM image also confirm the presence of long-range crystalline order. The thickness of the crystal is approximately 0.5 mm, and the number of layers is approximately 2700 for a fcc (111) crystal.

We used 2-hydroxyethyl methacrylate (HEMA) and acrylic acid (AAc) to obtain a pH-sensitive inverse opal gel. HEMA is a water-soluble, nonionic monomer and is essentially insensitive to pH. This monomer is a very well-known component of hydrogels that are useful for a variety of medical applications, such as soft contact lenses. AAc is a pH-sensitive, weak acid monomer with a pKa of 4.25 at 25 °C. Therefore, hydrogels based on HEMA and AAc exhibit marked volume changes in response to changing external pH values in aqueous solutions. The HEMA-AAc gel prepared using ethylene glycol dimethacrylate (EGDMA) as a crosslinker behaves as a neutral polymer network in a pH 2 buffer solution because the pKa values of AAc copolymerized in polymers become greater than 4.25 due to the electrostatic interaction caused by the neighbouring AAc monomers located on a polymer chain. In a pH 7 buffer solution, the majority of the AAc monomers in the HEMA-AAc gel must be in an essentially dissociated state based on the above-mentioned pKa value. Consequently, the electrostatic repulsions of the equally charged chains and the osmotic pressure of the counterions due to the high degree of ionization of the AAc groups in the polymer network force the HEMA-AAc gel to expand, which results in relatively large degrees of swelling. However, in electrolyte solutions, such as buffer solutions, excess cations and anions screen the charges on the polymer chains and eliminate the osmotic imbalance, which greatly decreases the swelling power of the polyelectrolyte gels. In the buffer solutions with an ionic strength of 0.1 used in this project, however, the electrostatic repulsion and the osmotic pressure caused by the ionization of the AAc groups sufficiently contributed to the increase in the degree of swelling of the HEMA-AAc gel. Therefore, the degree of swelling of the HEMA-AAc gel can be tuned within pH levels ranging from 2 to 7 in buffer solutions with an ionic strength of less than 0.1.

Figure 7.20 shows typical top- and side-view SEM images of the inverse opal gel composed of HEMA, AAc, and EGDMA obtained using the colloidal crystal as a template. This dried polymer network has clearly retained the three-dimensional crystalline order of the colloidal crystal template and exhibits ordered close packing of air spheres. In general, it is difficult to observe such a highly ordered array of pores in the inverse opal gels composed of weakly crosslinked gels in SEM images because the porous structure cannot be retained, even upon freeze drying. In contrast, this inverse opal gel is quite robust because the highly crosslinked network structure must be formed by not only the chemical crosslinkage but also the physical crosslinkage due to the entanglement of the polymers. The porous structure can be retained and observed in SEM images.

The representative colored hydrogels presented in Figures 7.21(a) and 7.21(b) are the inverse opal gels created from buffer solutions of pH 2 and pH 7, respectively, at 15 °C.

The reflected colors are caused by the Bragg diffraction of visible light by the array of aqueous solution spheres. As the pH increased, a distinct color variation from green (in pH 2) to orange (in pH 7) was observed. This color change was primarily induced by the change in the volume caused by the degree

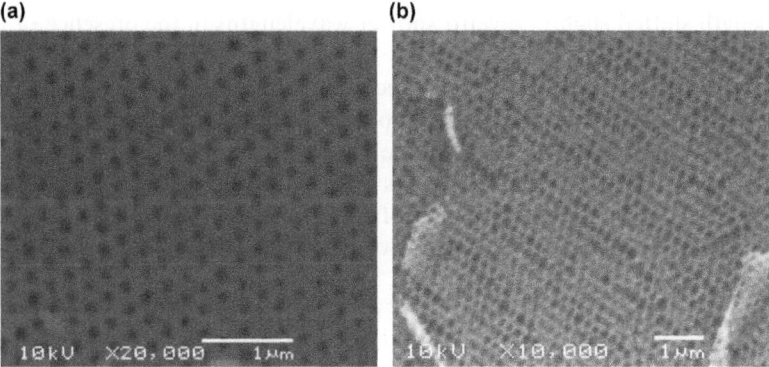

(a)   **(b)**

Figure 7.20   SEM images showing a (a) top view and (b) side view of the dried inverse opal gel.

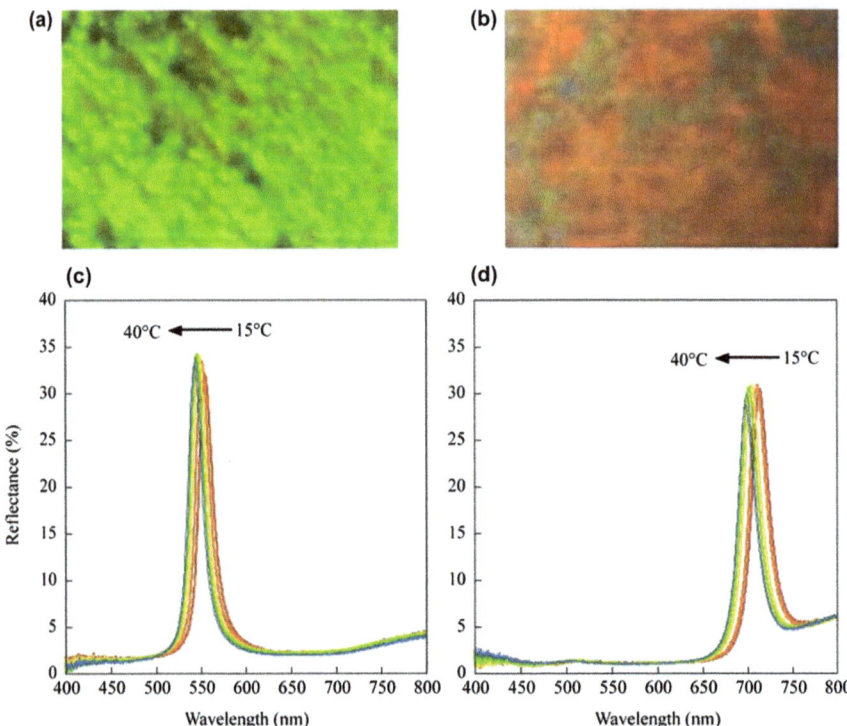

Figure 7.21 Optical photographs of the inverse opal gel in (a) pH 2 and (b) pH 7 buffer solutions at 15 °C. Reflectance spectra of the inverse opal gel in (c) pH 2 and (d) pH 7 buffer solutions at different temperatures.

of dissociation of the carboxylic acids in the polymer network because the ionic strengths of these buffer solutions have the same value. Figures 7.21(c) and 7.21(d) present the reflection spectra of the inverse opal gel in these buffer solutions at different temperatures. As the temperature increased, the reflected wavelength shifted slightly toward shorter wavelengths in the presence of both solutions (Figure 7.22(a)).

These shifts are due to the slight decrease in the lattice spacing of the fcc structure that occurs with the collapse of the gel because the pure polyHEMA gel exhibits a slight change in its volume in water, depending on temperature.[55] The reflectance of the spectra changes little with temperature in both cases (Figure 7.22(b)). The peak wavelength from the p-PBG of the inverse opal gel under normal incidence to the plane of the gel can be estimated using the following equation:[29]

$$\lambda_{\max} = 1.633(d/m)(D/D_0)n_a \tag{7.4}$$

In eqn (7.4), θ is 0 in eqn (7.1) in the second section. I have already reported that the change in n_a for the inverse opal-type periodic porous gel was observed

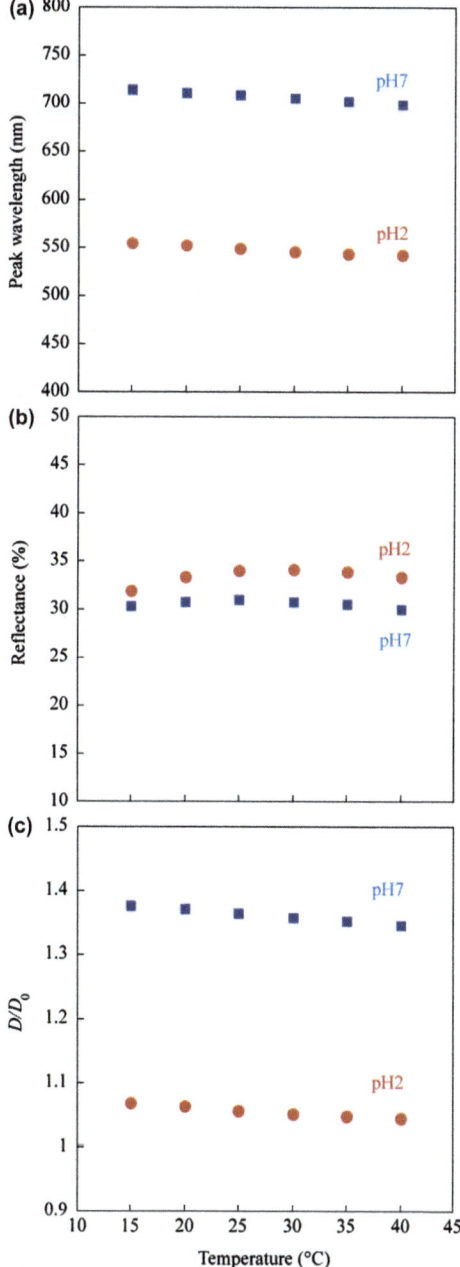

Figure 7.22 Dependence of temperature on (a) the peak wavelength, (b) reflectance of the spectra from the inverse opal gel, and (c) the swelling degree of the inverse opal in different pH buffer solutions.
Reprinted with permission from Ref. 35. Copyright 2009 Wiley-VCH.

to be extremely small when the gel undergoes volume changes; therefore, we can treat the value of n_a as a constant under varying degrees of D/D_0.[29] Therefore, the value of D/D_0 in certain situations can be estimated using the spectroscopically observed value of λ_{max}. Figure 7.22(c) presents the temperature dependence of the calculated value of D/D_0 based on the temperature dependence of the experimentally obtained value of λ_{max} (Figure 7.22(a)). Here, we use the values of each parameter to calculate the value of D/D_0 as follows: d, 233 nm; m, 1; refractive index of water, 1.334; refractive index of the inverse opal gel portion, 1.435; volume fraction of water spheres, 0.74; and volume fraction of the inverse opal gel portion, 0.26. Therefore, the calculated value of n_a is 1.36 based on eqn (7.2). The change in the degree of swelling of the inverse opal gel in each pH environment is less than 0.03 when the temperature changes from 15 to 40 °C, whereas the change in D/D_0 with a variation in pH from 7 to 2 at 15 °C is approximately 0.31; the size variations of this pH-sensitive inverse opal gel with temperature are considerably small.

The DTPC system was composed of thermosensitive gel particles composed of a thermosensitive monomer, N-isopropylacrylamide (NIPA), confined in the pH-sensitive inverse opal gel. Figure 7.23 presents a typical SEM image of the top view of the DTPC. A patterned concavo-convex surface, which is distinctly different from the surfaces of the colloidal crystal and the inverse opal gel, can be observed in the SEM image.

The photographs in Figures 7.24(a) and 7.24(b) show the difference in color of the DTPC in pH 2 and pH 7 buffer solutions, respectively, at 15 °C.

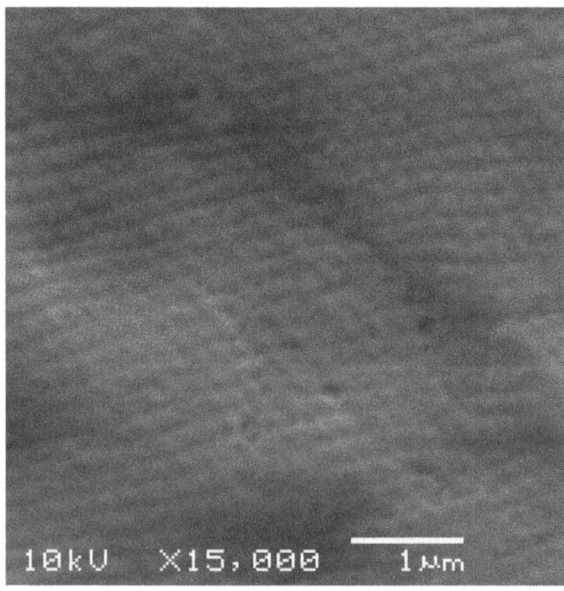

Figure 7.23   SEM image showing a top view of the dried DTPC.

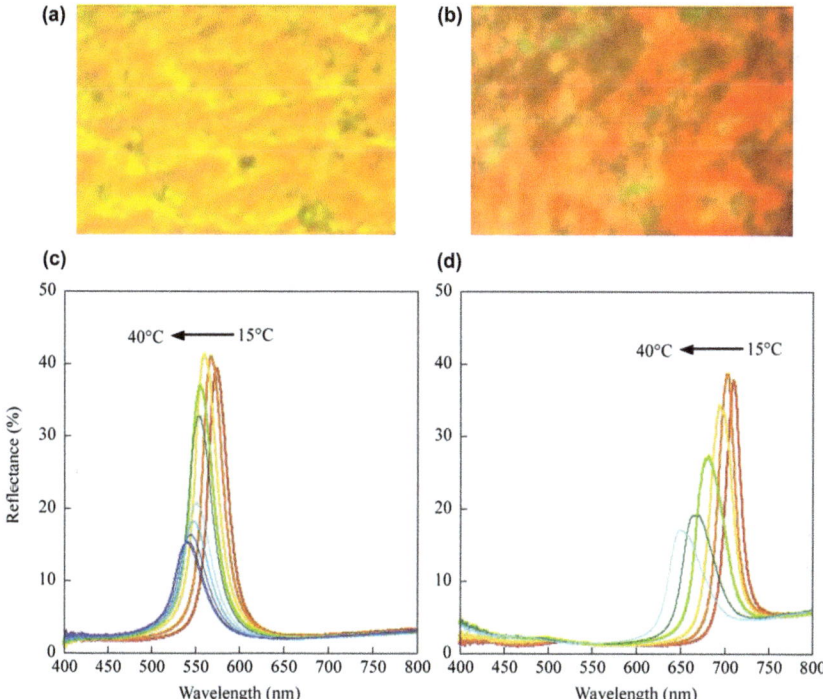

Figure 7.24   Optical photographs of the DTPC in (a) pH 2 and (b) pH 7 buffer solutions at 15 °C. Reflectance spectra of the DTPC in (c) pH 2 and (d) pH 7 buffer solutions at different temperatures.

This color change essentially reflects the pH sensitivity of the inverse opal gel. A visible change in the color of the DTPC at pH 2 was observed compared with the color of the inverse opal gel, whereas the DTPC at pH 7 was orange without any obvious difference in color compared with the color of the inverse opal gel. However, the peak positions of the reflection spectra decreased moderately with increasing temperature in both cases (Figure 7.24(c), 7.24(d), and 7.25(a)). These changes were amplified by the contraction of the NIPA gel particles.

The most interesting aspect of the DTPC system is that not only the peak position but also the peak intensity can be tuned by external stimuli. In the reflection spectra of the DTPC in pH 2 and pH 7 buffer solutions at different temperatures, which are shown in Figures 7.24(c) and 7.24(d), respectively, the peak intensity of the Bragg reflection from the p-PBG increased slightly in the initial state with increasing solution temperature in both cases (Figure 7.25(b)).

The fine structure of the inverse opal may be distorted depending on the combination of the swelling degrees of both the inverse opal gel and the gel particles. At present, it is difficult for us to quantitatively explain the initial increase in the peak intensity. In contrast, the peak intensity considerably

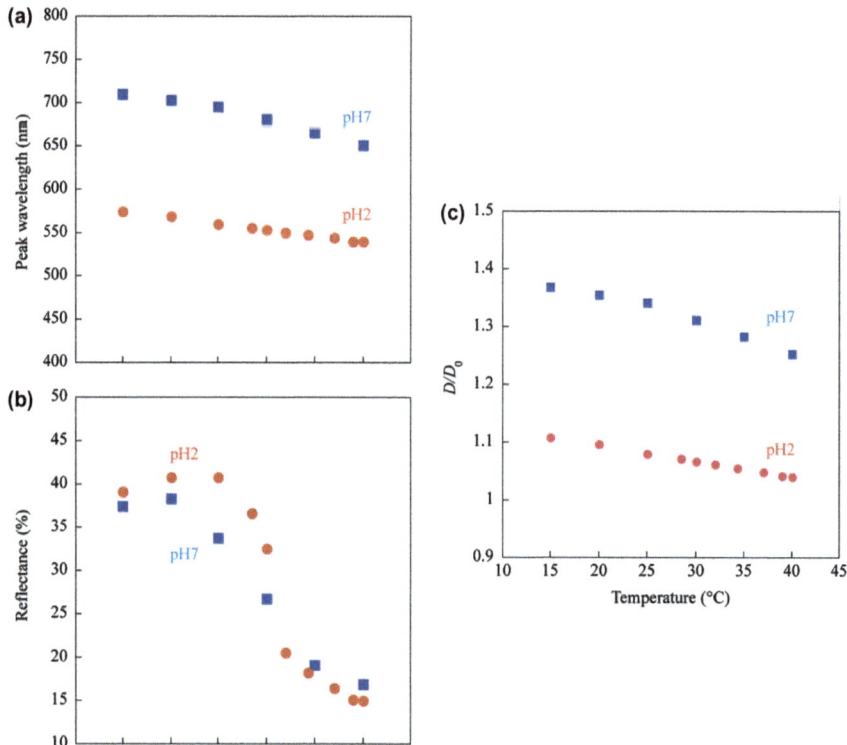

Figure 7.25   Temperature dependence of (a) the peak wavelength, (b) reflectance of the spectra of the DTPC, and (c) swelling degree of the inverse opal portion of the DTPC in different pH buffer solutions. Here, we use a value of n_a of 1.37, which is slightly larger than that in Figure 7.22(c) because there are NIPA gel particles in the system (refractive index of NIPA gel at 25 °C is 1.365). The value of D/D_0 in the preparative state (in ethanol at 25 °C) is 1.27, as measured by the overall size of the DTPC based on the size of the inverse opal (in ethanol at 25 °C).
Reprinted with permission from Ref. 35. Copyright 2009 Wiley-VCH.

decreased at temperatures greater than 25 °C. The decrease in the reflectance can be explained as follows. The NIPA gel exhibits the same change in size at different pH levels, demonstrating that this thermosensitive neutral NIPA gel is not pH-sensitive (Figure 7.26).

Note that the NIPA gel particles prepared in the DTPC system are designed to exhibit a swelling ratio of almost unity at a fully swollen state in the buffer solutions by controlling the amount of the monomer and the crosslinker in the pregel solution.[17,29,56] This swelling ratio is important for the fabrication of the DTPC system. As the increase in temperature above 25 °C causes the abrupt collapse of the NIPA gel particles,[50] the gel particles can move freely and scatter throughout the pores of the inverse opal gel. This process may induce a disturbance in the ordered array of the gel particles and may form many layers

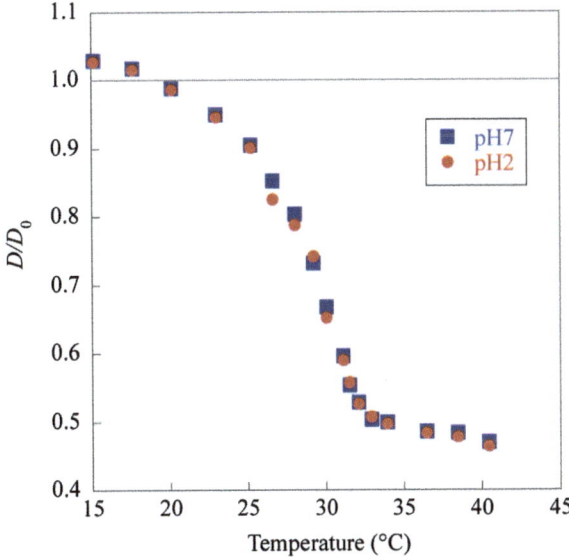

Figure 7.26   The degree of swelling, D/D_0, of the NIPA gel as a function of temperature in different pH buffer solutions is shown, where D denotes the gel diameter in the equilibrium state in certain situations and D_0 is the gel diameter upon synthesis.

of rough surfaces in the inverse opal gel. As explained above, this situation leads to stronger incoherent scattering of light from the gel particles, which results in a decrease in the coherent Bragg scattering at higher temperatures. The change in brightness or saturation of the structural color of the DTPC with temperature was also confirmed with the naked eye. Unfortunately, however, we could not capture this change in photographs.

Furthermore, these spectra in Figures 7.24(c) and 7.24(d) indicate that the positions of the p-PBG moderately change with temperature (Figure 7.25(a)). The variations in the swelling ratio of the DTPC, which were determined through calculations using eqn (7.4), were confirmed for both cases (Figure 7.25(c)); the rate of change in the swelling ratio *vs.* temperature for the DTPC is appreciably larger than that for the inverse opal gel. To describe the spectral behavior, one should account for the temperature dependence of the fine structure of the DTPC system. The experimentally obtained $\Delta\lambda/\lambda_{max}$ ratio may help explain this temperature dependence.[29] Figure 7.27 shows the measured $\Delta\lambda/\lambda_{max}$ of the inverse opal gel and the DTPC as a function of temperature.

The experimental value of $\Delta\lambda/\lambda_{max}$ for the DTPC increased as the temperature increased, whereas that for the inverse opal gel was temperature-independent. The changes in $\Delta\lambda/\lambda_{max}$ for the DTPC become considerably more pronounced at higher temperatures, regardless of the pH of the solution. We can interpret this phenomenon as follows. The thermosensitive gel particles are

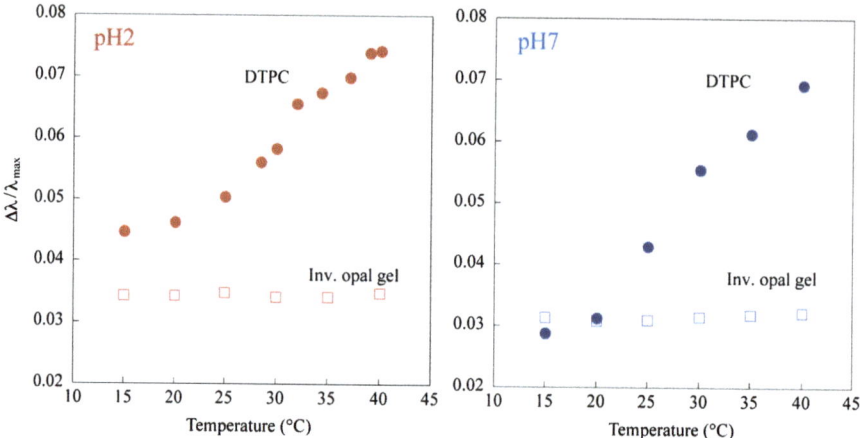

Figure 7.27 Temperature dependence of $\Delta\lambda/\lambda_{max}$ obtained from the reflection spectra of the inverse opal gel and the DTPC at (a) pH 2 and (b) pH 7.

swollen at lower temperatures than the LCST and undergo a decrease in volume with increasing temperature. If all of the gel particles are isolated in each pore in the inverse opal gel, the volume change of the particles may not have a significant effect on the shape and volume of the inverse opal gel unless there are strong interactions between the surface of the particles and the inner walls of the pores. During the preparatory procedure of the DTPC, however, a partially interlaced area between the HEMA-AAc polymer network and the NIPA polymer network must have been formed in the DTPC because the inverse opal gel was slightly swollen in the pregel solution used to create the NIPA gel. Additionally, the isolation procedure of the gel particles could be imperfect because of the material softness of the inverse opal gel. These conditions may induce the deformation of the fine porous structure of the inverse opal gel portion of the DTPC with the volume change of the NIPA gel particles, especially at higher temperatures. Therefore, a steep increase in $\Delta\lambda/\lambda_{max}$ would occur at higher temperatures.

Figure 7.28 shows the temperature dependence of the diameters of the pores in the inverse opal gel and of the gel particles at pH 2 and pH 7.

The pore sizes at each pH are calculated by the value of D/D_0 shown in Figure 7.25(c) and the diameter of the silica particle, 233 nm, which can be used as the value of D_0. Thus, the calculated value of D is the diameter of the pores. The size of the gel particles in the DTPC can be determined by considering that the volume of a cylindrical NIPA gel depends on temperature at different pH levels in these buffer solutions, as shown in Figure 7.26 and Figure 7.25c. The size of the gel particles in the DTPC below 25 °C at pH 2 is regarded to be the same as the pore size because the size of the gel particle cannot be larger than the pore size. Consequently, in a pH 2 buffer solution, the pores of the inverse opal gel are occupied by the gel particles at temperatures less than 25 °C, whereas there is space between the pores and the gel particles at temperatures

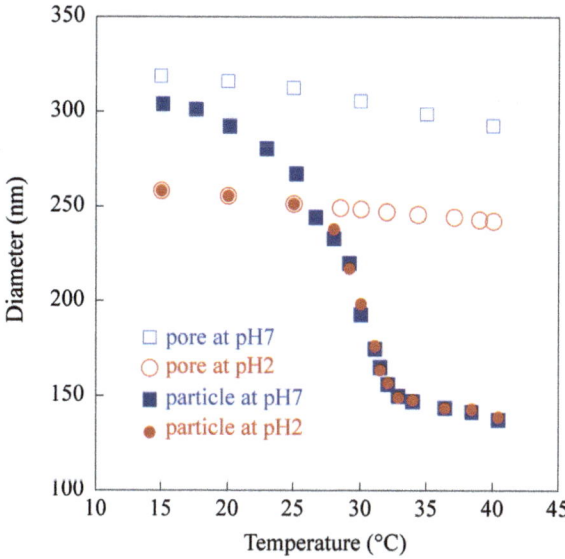

Figure 7.28 Temperature dependence of the diameter of the NIPA gel particles and the pore size of the inverse opal gel portion of the DTPC at different pH levels.

greater than 25 °C. In contrast, the diameter of the gel particles is always smaller than the pore size of the inverse opal gel at pH 7. Based on the increase in the value of $\Delta\lambda/\lambda_{max}$ shown in Figure 7.27(a), the NIPA particles must increase the size of the pores of the inverse opal gel, which produces considerable distortion of the DTPC.

In this investigation, the dual tuning of the p-PBG structure in the thermosensitive gel particles confined in the pH-sensitive inverse opal gel using temperature and pH as external stimuli was examined from a full-color material standpoint. The position of the p-PBG can be thermally tuned, whereas the intensity of the p-PBG is dramatically changed by pH (Figure 7.29).

Consequently, the hue from the DTPC was remarkably changed by the environmental pH. Furthermore, the change in the brightness or saturation of the structural color of the DTPC with temperature was also confirmed with the naked eye, but we could not capture this change in optical photographs. The reversible, independent and extensive switching of the hue and the brightness or saturation of the structural color of the DTPC could be achieved using different external stimuli. Although there is still room for improvement, such as in the angle dependence on color and the response speed for applications such as displays and switches, this dual-tunable soft photonic crystal corresponds to a conceptually new approach to the fabrication of full-color display materials. As we have already developed a method to fabricate structural colored materials without angle dependence,[57–59] the DTPC can be used to construct full-color, paper-like displays in the near future.

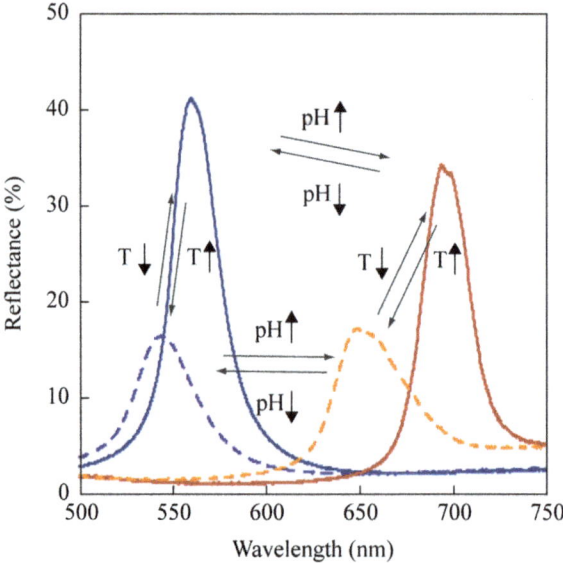

Figure 7.29   Reflection spectra of the DTPC at different temperatures and in buffer solutions of different pH values. Left solid line: pH 2, 15 °C; right solid line: pH 7, 15 °C; left dashed line: pH 2, 40 °C; right dashed line: pH 7, 40 °C. Reprinted with permission from Ref. 35. Copyright 2009 Wiley-VCH.

7.5   Conclusions

In this chapter, I described some applications of stimuli-sensitive inverse opal gels that can exhibit anisotropic volume changes and of a system that combines stimuli-sensitive hydrogel particles with a stimuli-sensitive inverse opal gel. As described above, these stimuli-sensitive soft materials exhibit unique properties and are promising candidates for the development of sensing and color materials.

Acknowledgements

This work was supported by a Grant-in-Aid for Scientific Research (No. 22107012) in the innovative area of "Fusion Materials" (Area no. 2206) and a Grant-in-Aid for Scientific Research (No. 23245047) from the Ministry of Education, Culture, Sports, Science and Technology, as well as by the Industrial Technology Research Grant Program in 2008 (Grant 08C46502d) of the New Energy and Industrial Technology Development Organization (NEDO) of Japan. The author is also grateful to Mr. S. Shinohara, Dr. K. Matsubara, and Mr. M. Honda for their contributions to this study. This chapter was written by using their master theses.

References

1. K. Busch, S. Lölkes and R. B. Wehrspohn, *Photonic Crystals* (ed. H. Föll), Wiley-VCH, Weinheim, 2004.
2. A. S. Dimitrov and K. Nagayama, *Langmuir*, 1996, **12**, 1303–1311.
3. P. Jiang, J. F. Bertone, K. S. Hwang and V. L. Colvin, *Chem. Mater.*, 1999, **11**, 2132–2140.
4. P. J. Darragh, A. J. Gaskin, B. C. Terrell and J. V. Sanders, *Nature*, 1966, **209**.
5. A. A. Zakhidov, R. H. Baughman, Z. Iqbal, C. Cui, I. Khayrullin, S. O. Dantas, J. Marti and V. G. Ralchenko, *Science*, 1998, **282**, 897–901.
6. P. Jiang, K. S. Hwang, D. M. Mittleman, J. F. Bertone and V. L. Colvin, *J. Am. Chem. Soc.*, 1999, **121**, 11630–11637.
7. M. Harun-Ur-Rashid, T. Seki and Y. Takeoka, *Chem. Record*, 2009, **9**, 87–105.
8. F. Marlow, P. Muldarisnur R. Sharifi R. Brinkmann and C. Mendive, *Angew. Chem.-Int. Ed.*, 2009, **48**, 6212–6233.
9. Y. Zhao, X. Zhao and Z. Gu, *Adv. Funct. Mater.*, 2010, **20**, 2970–2988.
10. C. I. Aguirre, E. Reguera and A. Stein, *Adv. Funct. Mater.*, 2010, **20**, 2565–2578.
11. J. P. Ge and Y. D. Yin, *Angew. Chem.-Int. Ed.*, 2011, **50**, 1492–1522.
12. J. X. Wang, Y. Z. Zhang, S. T. Wang, Y. L. Song and L. Jiang, *Acc. Chem. Res.*, 2011, **44**, 405–415.
13. S.-H. Kim, S. Y. Lee, S.-M. Yang and G.-R. Yi, *NPG Asia Mater*, 2011, **3**, 25–33.
14. Y. Takeoka and M. Watanabe, *Langmuir*, 2002, **18**, 5977–5980.
15. Y. Takeoka and M. Watanabe, *Adv. Mater.*, 2003, **15**, 199–201.
16. D. Nakayama, Y. Takeoka, M. Watanabe and K. Kataoka, *Angew. Chem.-Int. Ed.*, 2003, **42**, 4197–4200.
17. Y. Takeoka and M. Watanabe, *Langmuir*, 2003, **19**, 9104–9106.
18. H. Saito, Y. Takeoka and M. Watanabe, *Chem. Commun.*, 2003, 2126–2127.
19. Y. Takeoka and T. Seki, *Macromolecules*, 2007, **40**, 5513–5518.
20. H. Murayama, A. Bin Imran, S. Nagano, T. Seki, M. Kidowaki, K. Ito and Y. Takeoka, *Macromolecules*, 2008, **41**, 1808–1814.
21. Y.-J. Lee and P. V. Braun, *Adv. Mater.*, 2003, **15**, 563–566.
22. Y. J. Lee, S. A. Pruzinsky and P. V. Braun, *Langmuir*, 2004, **20**, 3096–3106.
23. R. A. Barry and P. Wiltzius, *Langmuir*, 2006, **22**, 1369–1374.
24. J. Wang and Y. Han, *Langmuir*, 2009, **25**, 1855–1864.
25. E. T. Tian, J. X. Wang, Y. M. Zheng, Y. L. Song, L. Jiang and D. B. Zhu, *J. Mater. Chem.*, 2008, **18**, 1116–1122.
26. Y. Takeoka, M. Watanabe and R. Yoshida, *J. Am. Chem. Soc.*, 2003, **125**, 13320–13321.
27. S. Shinohara, T. Seki, T. Sakai, R. Yoshida and Y. Takeoka, *Angew. Chem.-Int. Ed.*, 2008, **47**, 9039–9043.

28. S. Shinohara, T. Seki, T. Sakai, R. Yoshida and Y. Takeoka, *Chem. Commun.*, 2008, 4735–4737.
29. Y. Takeoka and T. Seki, *Langmuir*, 2006, **22**, 10223–10232.
30. K. Ueno, K. Matsubara, M. Watanabe and Y. Takeoka, *Adv. Mater.*, 2007, **19**, 2807–2812.
31. K. Matsubara, M. Watanabe and Y. Takeoka, *Angew. Chem.-Int. Ed.*, 2007, **46**, 1688–1692.
32. K. Ueno, J. Sakamoto, Y. Takeoka and M. Watanabe, *J. Mater. Chem.*, 2009, **19**, 4778–4783.
33. M. Kumoda, M. Watanabe and Y. Takeoka, *Langmuir*, 2006, **22**, 4403–4407.
34. M. Honda, K. Kataoka, T. Seki and Y. Takeoka, *Langmuir*, 2009, **25**, 8349–8356.
35. M. Honda, T. Seki and Y. Takeoka, *Adv. Mater.*, 2009, **21**, 1801–1804.
36. http://is2.sss.fukushima-u.ac.jp/fks-db//txt/20024.008/pdf/00002.pdf.
37. http://www.nobelprize.org/nobel_prizes/chemistry/laureates/2008/shimomura-lecture.html.
38. L. M. Mäthger, T. F. T. Collins and P. A. Lima, *J. Exp. Biol.*, 2004, **207**, 1759–1769.
39. L. M. Mäthger, M. F. Land, U. E. Siebeck and N. J. Marshall, *J. Exp. Biol.*, 2003, **203**, 3607–3613.
40. R. Yoshida, T. Takahashi, T. Yamaguchi and H. Ichijo, *J. Am. Chem. Soc.*, 1996, **1181**, 5134–5135.
41. R. Yoshida, E. Kokufuta and T. Yamaguchi, *CHAOS*, 1999, **9**, 260–266.
42. D. Suzuki, H. Taniguchi and R. Yoshida, *J. Am. Chem. Soc.*, 2009, **131**, 12058–12059.
43. S. Maeda, Y. Hara, T. Sakai, R. Yoshida and S. Hashimoto, *Adv. Mater.*, 2007, **19**, 3480–3484.
44. S. Maeda, Y. Hara, R. Yoshida and S. Hashimoto, *Angew. Chem.-Int. Ed.*, 2008, **47**, 6690–6693.
45. R. Yoshida, *Adv. Mater.*, 2010, **22**, 3463–3483.
46. A. N. Zaikin and A. M. Zhabotinsky, *Nature*, 1970, **225**, 535–537.
47. L. Kuhnert, *Nature*, 1986, **319**, 393–394.
48. I. Ohmine and T. Tanaka, *J. Chem. Phys.*, 1982, **77**, 5725–5729.
49. M. Irie, *Adv. Polym. Sci.*, 1993, **110**, 49–65.
50. Y. Hirokawa and T. Tanaka, *J. Chem. Phys.*, 1984, **81**, 6379–6380.
51. M. Kamenjicki, I. K. Lednev and S. A. Asher, *J. Phys. Chem. B*, 2004, **108**, 12637–15639.
52. Dr. Kazuki Matsubara's Master Thesis.
53. T. Tomari and M. Doi, *Macromolecules*, 1995, **28**, 8334–8343.
54. I. I. Tahran and G. H. Watson, *Phys. Rev. B*, 1996, **54**, 7593–7597.
55. H. M. Nizam and A. W. M. EI-Naggar, *J. Appl. Polym. Sci.*, 2005, **95**, 1105–1115.
56. C. Alvarez-Lorenzo, O. Guney, T. Oya, Y. Sakai, M. Kobayashi, T. Enoki, Y. Takeoka, T. Ishibashi, K. Kuroda, K. Tanaka, G. Q. Wang,

A. Y. Grosberg, S. Masamune and T. Tanaka, *Macromolecules*, 2000, **33**, 8693–8697.

57. Y. Takeoka, M. Honda, T. Seki, M. Ishii and H. Nakamura, *ACS Appl. Mater. Interf*, 2009, **1**, 982–986.
58. M. Harun-Ur-Rashid, A. Bin Imran, T. Seki, M. Ishi, H. Nakamura and Y. Takeoka, *Chemphyschem*, 2010, **11**, 579–583.
59. Y. Gotoh, H. Suzuki, N. Kumano, T. Seki and Y. Takeoka, *New J. Chem.*, 2012 **DOI:** 10.1039/C2NJ40368D, Letter.

CHAPTER 8

Bioinspired Fabrication of Colloidal Photonic Crystals with Controllable Optical Properties and Wettability

FENGYU LI, JINGXIA WANG AND YANLIN SONG*

Institute of Chemistry, Chinese Academy of Sciences, Beijing, 100190, China
*Email: ylsong@iccas.ac.cn

8.1 Introduction

Photonic crystals (PCs), constituted of periodic arrangement materials with different dielectric constant, can effectively modulate light manipulation.[1–7] This manipulating property is similar to the electrical semiconductor, which may bring about a revolution of light communication, is selected as one of the ten scientific progresses in the journal of Science in 1999 and 2006. In fact, biological PC[8–17] exists widely in Nature, such as the vivid color of the butterfly's ring, peacock's feather, and the colorful shell of some beetles. These biological PCs not only demonstrate the bright structure color owing to the Bragg diffraction effect of their periodic microstructure, but also have some special properties in order to meet or survive various living environments, for example, special wettability and tough mechanical strength. These special properties will provide important insight for the fabrication of novel function materials.

RSC Smart Materials No. 5
Responsive Photonic Nanostructures: Smart Nanoscale Optical Materials
Edited by Yadong Yin
© The Royal Society of Chemistry 2013
Published by the Royal Society of Chemistry, www.rsc.org

8.1.1 Functional Biological PCs

In Nature, many bewildered biological surface is made of PCs,[8,9,11] which have many attractive properties, such as iridescent color, special wettability and high mechanical strength. Wherein, the wings of butterflies (especially Morphos) is one of the typical examples. Their vivid color originates from the Bragg diffraction effect of their periodic structure. These bright colors are favorable for their information communication, identification, *etc.*[11,13] What is important, special wettability is crucial for their swift movement in the rain or special environment. As shown in Figure 8.1, typically, a butterfly wing[12] shows a special superhydrophobicity and directional adhesion, which makes it easy for shedding water droplets on its surface in the high-humidity environment, and benefits its flight in the rain. The similar phenomenon exists in other birds' wings, such as the feathers of peacocks or ducks. All of this special wettability of creature surface is derived from the living requirement under special environments. While the hydrophilic property is also necessary for some other living environment, for example, the wings of *tmesisternus isabellae* and the hydrophilic pearl.[13] The hydrophilicity of pearl mainly is originated from the continuously absorbing hydrophilic organic component for its continuous growth. The hydrophilic *tmesisternus isabellae* is due to its hydrophilic materials surface, which can vary its structure color with environmental humidity. The wing swells after absorbing moist air with an environmental humidity increase, its color will change and redshift due to the replacement of air by water and its swollen volume. In contrast, the structure color returns as the humidity falls due to the evaporation of adsorbed water. It is found that the change of structure color corresponds to the change of the soil's color under

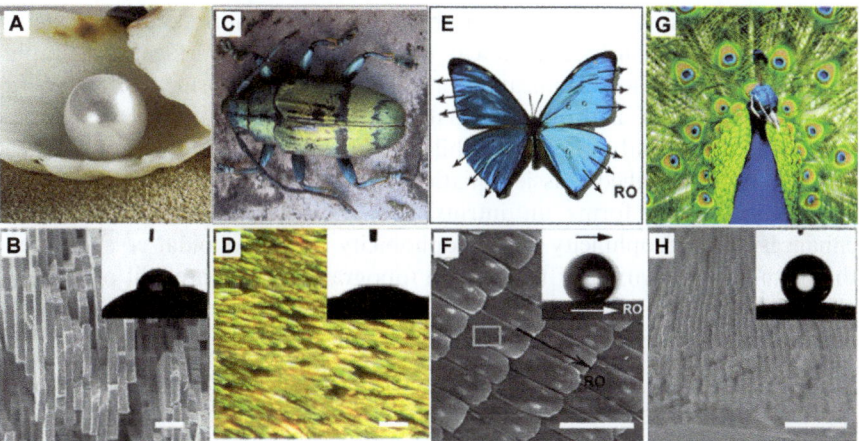

Figure 8.1 Typical biological PCs with controllable wettability. (A, B) Hydrophilic pearl, (C, D) Hydrophilic longhorn beetles *tmesisternus isabellae*, (E, F) Superhydrophobic butterfly wings, (G, H) Superhydrophobic peacock. Reproduced with permission. Copyright 2011 The American Chemical Society.

rain or sunny conditions. This obvious color change of *tmesisternus isabellae* could mainly be attributed to the self-protection from a natural enemy after long-time evolution. It is worth noting that biological PCs demonstrate remarkably high mechanical strength, which enables it to maintain its iridescent structure color and special wettability even in strong winds and pouring rain. The strong mechanical properties could mainly be originated from the tough linkage interaction among periodic structures. These special properties of biological PCs will afford significant insight for the fabrication of high-performance functional optical materials. In the following text, we will summarize typical biofabrication of colloidal PCs with special properties.

8.2 Bioinspired Fabrication of Functional PCs

Inspired by the special functionality of biological PCs, colloidal PCs with advanced properties have been designed and fabricated to meet various application requirements. In the following sections, we put the emphasis on the fabrications of colloidal PCs with controllable wettability, high mechanical strength and anisotropic structure.

8.2.1 Fabrication of PCs with Controllable Wettability

Wettability of solid surface plays an important role in the various applications,[18–29] including industrial production, agriculture and daily life. The special wettability of colloidal PCs can endow materials with novel functions such as self-cleaning properties for optical devices, the fully wetting and spreading of the precursor to infiltrate into the opal template,[22] the full diffusion of a catalyst into a reaction system,[23] or the quick detection of liquid in microfluidic systems.[24] The wettability of a solid surface can be mainly controlled by its surface chemical composition and surface topographic structure.[18–20] Surface chemical composition determines surface free energy (*i.e.* hydrophilic or hydrophobic materials), while the surface topographic structure can amplify the hydrophilicity or hydrophobicity based on the Wenzel[18] and modified Cassie equation.[19] For colloidal PCs, their periodic latex arrangement forms an intrinsic rough structure, which definitively enhances the hydrophilicity or hydrophobicity of the colloidal crystals, while the further modification of the surface topographic structure will produce an effective influence on the dynamic wetting behaviors. In our work, we demonstrate the modulation of the static or dynamic wettability of the colloidal crystals by varying the surface chemical composition or latex morphology.[17,25] Wherein, the static wettability indicates the apparent contact angle of the substrate, which could be controlled by its surface chemical composition, while the dynamic wettability denotes the sliding angle of the oil or water droplet on the substrate, it could be modified by varying the surface morphology of latex particles.

Generally speaking, the modulation of the static wettability mainly is realized by controlling its surface chemical composition, while the roughness structure derived from the periodic arrangement of the latex particles can amplify its hydrophilicity or hydrophobicity based on the modified Cassies's equation. The pioneering report on the superhydrophobic colloidal PCs is demonstrated by Gu *et al.*[16] They combined the self-cleaning properties of the lotus leaf and the vivid structure color of the butterfly wing into one sample by just coating low-energy materials over the surface of colloidal crystals, which presents a facile fabrication of colloidal crystals with controlled wettability. Subsequently, continuous research developed for the fabrication of colloidal crystals with special wettability by coating the responsive materials onto the surface of colloidal crystals; the as-prepared colloidal PCs could reversibly change its wettability by an external stimulus. However, one problem is encountered for the frequent use of the as-prepared sample, the special wettability is fragile as the coated material is prone to be peeled off during use. To solve these problems, we developed a novel approach to fabricate the colloidal crystals with special wettability by well-designed latex particles.[30–36] The wettability can be easily modified based on the change of the surface chemical composition or roughness structure of the latex particles. In this case, the resultant film can maintain good stability even if the latex particles of the above layer are peeled off, this is due to the complete indepth wettability for the whole film rather than just upper layer (Figure 8.2). This will provide an effective revolution of developing a novel and functional film by well-designed latex particles. Simultaneously, the wettability of the colloidal PCs can also be controlled by various actions of external fields, such as light, electronic, temperature or pH only if the latex particles are suitably treated by chemical modification.

8.2.1.1 PCs with Controllable Static Wettability

Static wettability of films reflects the static wetting or spreading state of the liquid droplet lying on the substrate, which is characterized by the apparent contact angle of the film, is a preliminary parameter related to the wetting behavior of the solid surface. The static wettability of the latex surface can be effectively modified by its surface chemical composition. The water contact angle of the film can be changed from superhydrophobic to superhydrophilic on modifying its surface chemical composition combining its intrinsic periodic structure. In a typical study, the latex particles with special core-shell morphology[30] are synthesized by emulsion polymerization, where hydrophobic polystyrene (PS) is latex core, while the hydrophilic polyacrylic acid (PAA) is the latex shell. When the latex particles are assembled at lower temperature (lower than 20 °C), the hydrophilic PAA latex shell will keep itself around the latex surface, which results in the hydrophilicity of the latex surface. On combining the amplification effect of periodic structure with the hydrophilic properties of latex surface, the as-prepared colloidal PCs show superhydrophilicity.[31–32] In contrast, the superhydrophobic colloidal PCs can be easily obtained at higher assembly temperature (for example, higher than 80 °C), this can be due to the phase separation of the polymer segment upon the latex surface based on the principle of minimum surface energy. When the latex

(a)

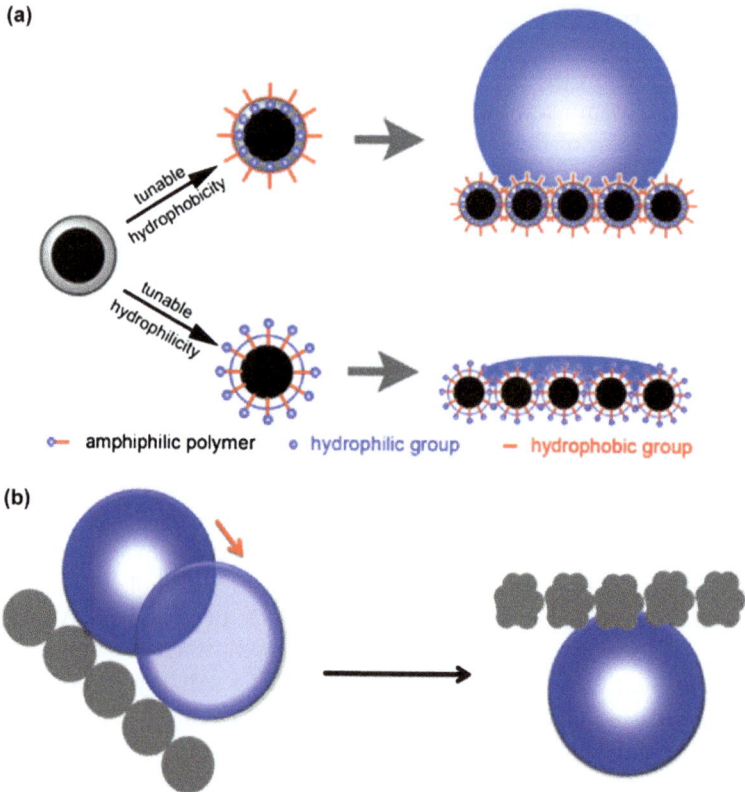

(b)

Figure 8.2 Schematic illustration of colloidal PCs with controllable wettability obtained from well-designed amphiphilic latex particles.
Reproduced with permission. Copyright 2011 The American Chemical Society.

particles are assembled at temperatures higher than that of the glass transition of the polymer, the polymer segments tend to change the conformation. As a result, the hydrophilic group PAA, will shrink and shield from the latex surface, while the hydrophobic polymer segment PS will extend toward the latex exterior. This results in a change of the latex surface from hydrophilicity to hydrophobicity. The resultant superhydrophobicity film is a combination of an amplification effect of the periodic arrangement to the hydrophobic surface.

In this case, the key issue of the wettability transition is the phase separation[37–40] of the polymer segment at the desired temperature, the temperature of glass transition of the polymer (T_g) is an effective parameter to control the phase separation. A complete phase separation occurs only if the temperature is close to or above the T_g. Accordingly, modifying the T_g of the polymer segment can finely control the wettability transition temperature of the resultant films. It makes possible the achievement of colloidal PCs with suitable wettability at desired temperature. Figure 8.3 shows a typical example of controlling the wettability transition temperature by modifying the T_g of the

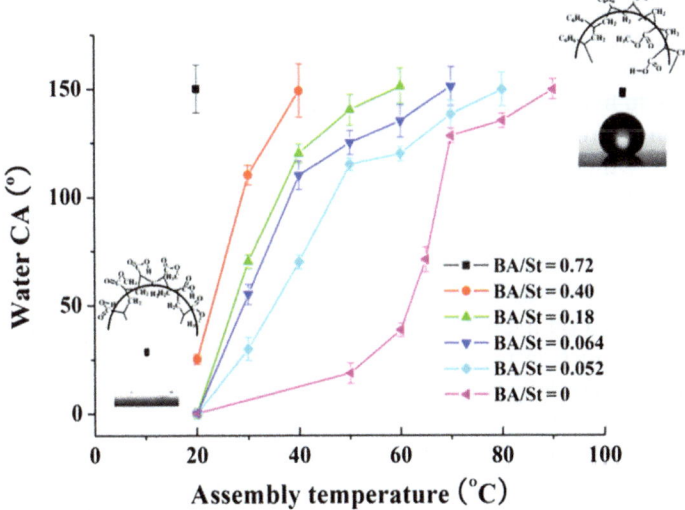

Figure 8.3 Fine control of the wettability transition temperature by varying the ratio of butyl acrylate (BA)/St. The insert images indicate the water droplet shape on the different substrate and the possible surface chemical composition of latex particles respectively.
Reproduced with permission. Copyright 2007 Wiley-VCH Verlag GmbH & Co. KGaA.

polymer segment based on changing the ratio of soft or hard polymer segments in latex particles, which could effectively change the wettability transition temperature from 20, 40, 60, 80 to 90 °C by continuously increasing the ratio of the hard segment in the polymer chain.[32] This is greatly beneficial for the extended applications of polymer colloidal PCs in special environments (Figure 8.3).

8.2.1.2 Colloidal Crystals with Reversible Wettability

This aspect has potential importance for the films with reversible wettability, where, the wettability can be reversibly controlled from hydrophilic to hydrophobic, or from superhydrophobic to superhydrophilic. The films with reversible wettability can show promising application in smart mcirofluidic switches or other advanced optical devices. Colloidal PCs with reversible wettability can be realized by modifying the latex surface by responsive materials,[26,34] such as light-responsivity, electronic-responsivity or pH-responsitivity. Typically, the colloidal PCs assembled from polystyrene latex particles show superhydrophobicity due to the hydrophobic surface of PS spheres and the relative amplification effect of the periodic structure on the wettability. It is interesting that the films can vary its wettability from super-hydrophobic to highly hydrophilicity upon UV irradiation. This wettability change can be attributed that the irradiation process results in a distinct change of surface chemical composition, *i.e.* UV irradiation breaks the bonds of the polystyrene chain and results in the formation of carboxyl groups on the latex surface, which lead to a dramatic change in the surface properties from hydrophilic to hydrophobic. As a result, the wettability of colloidal crystals can easily change from highly hydrophobic to highly hydrophilic upon UV irradiation. Otherwise, introducing the responsive material (for example, electronic-responsive polypyrrole)[34] into the interstice of latex particles and subsequent calcinations could remove the template and also achieve the inverse opal PCs with reversible wettability. The wettability of the inverse opal structure can be reversibly controlled from highly hydrophobic to highly hydrophilic based on the protonation or deprotonation reaction resulting from oxidation or reduction processes. The process changes the surface properties from hydrophobic to hydrophilic, resulting in the change of the wettability of films from superhydrophobic to highly hydrophilic. Meanwhile, this process changes reversibly the optical properties and conductivity of the film, owing to the simultaneous change of the volume or refractive index of the materials (Figure 8.4). Other responsive colloidal crystals with controllable wettability change have been realized by introducing the responsive material, such as, pH-responsive, heat-responsive, *etc.* This wettability change will produce an important influence on the application of colloidal crystals to extended smart displays and microfluidic systems.

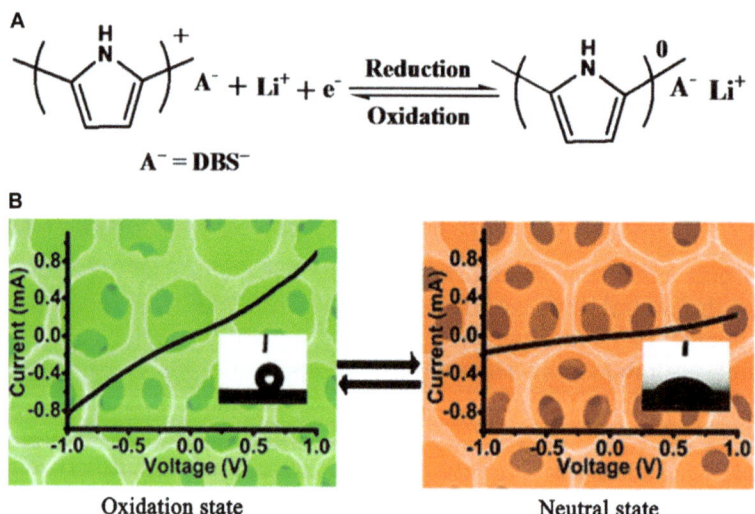

Figure 8.4   (A) The changing chemical structure of polypyrrole under the oxidization/reduction states, (B) The reversible transition of structure color, conductivity and wettability of the polypyrrole inverse opal. Reproduced with permission. Copyright 2008 The American Chemical Society.

8.2.1.3   Colloidal Crystals with Controllable Dynamic Wettability

It is known that the static wettability of the colloidal crystals can easily be controlled by modifying its surface chemical composition. However, in real applications, the detailed wetting process, such as the pining or rolling of droplet during evaporation process, is especially important to understand the printing resolution, or dynamic kinetic of swelling or shrinkage of hydrogel systems. The dynamic wettability[41–43] of the film surface mainly depends on the micro-/nanocomplicate structure of the latex surface, such as a lotus leaf, which has extremely excellent water-repellent properties owing to its special micro-/nanocomplicate structure. These nanostructures cause air pockets among the interstice of the microstructure, the presence of hydrophobic air pockets makes it difficult for liquid droplets to contact fully onto the film's surface, leading to a change of the three-phase contact line (TCL) of the liquid droplet from continuous to discontinuous. As a result, the wetting state of droplets on the film's surface changes from a stable Wenzel state to an unstable Cassie state, and the droplet easily rolls away from the interface of substrate. *Vice versa*, the presence of hydrophilic microstructure makes the droplet wet fully onto the substrate surface, the droplet will pin onto the surface tightly. Accordingly, the dynamic wettability of colloidal crystals can be easily modulated by altering its micro-/nanostructure. In our laboratory, we have successfully fabricated colloidal crystals with controllable dynamic wettability by modifying the microstructure of the latex particles.

The latex particles with increased rough structure were first synthesized by emulsion polymerization, where the latex surface of particles becomes rougher with increasing charging times of the initiator. It was found that the rougher the latex particle is, the higher the water contact angle for the colloidal crystals assembled from these particles, and the higher the adhesion force toward the droplet for the colloidal crystals. This experiment provides a preliminary confirmation that the rough structure is responsible for the enhancement of the sliding behavior of water droplets on the film's surface. The rough structure plays a similar role in the underwater wettability of the colloidal crystals. However, a rougher latex surface is required for modifying the underwater dynamic wettability of the colloidal crystals, since the inevitable underwater swelling of the latex particles will decrease the roughness of the films. In this experiment, we further enlarged the difference of latex morphology by synthesis of latex particles with shape changing from spherical, cauliflower-like to single-cavity particles,[44,45] and these latex particles were assembled at an air/water interface.[46] As a result, the as-prepared colloidal crystals show a distinct morphology for different latex particles, which affects the wetting states of TCL of the latex droplet on these substrates. The film from spherical latex particles shows only the periodic arrangement of the latex spheres, which favors the continuous TCL for oil droplets on the latex surface. While the film from cauliflower-like particles shows the additional microstructure from the micro-nipple of the latex surface, which results in a discontinuous TCL for oil

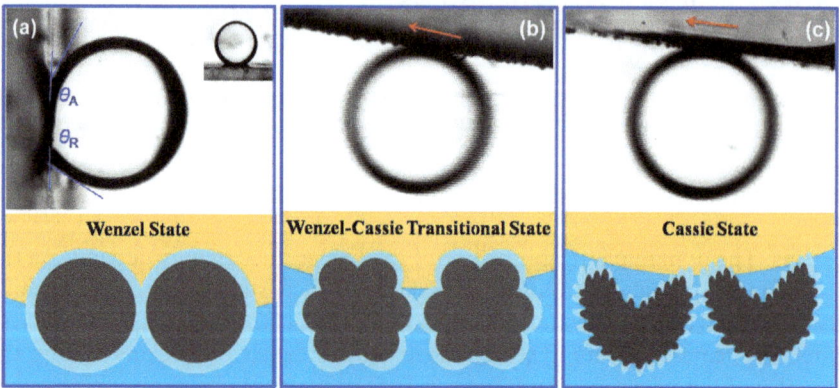

Figure 8.5 The underwater dynamic wettability change of the colloidal crystals assembled from different latex particles, (a) sphere (b) cauliflower-like and (c) single-cavity latex particles.
Reproduced with permission. Copyright 2011 Wiley-VCH Verlag GmbH & Co. KGaA.

droplets. However, colloidal crystals assembled from single-cavity latex particles with more than 1/3 cavity structure, form a special dot contact mode for the oil droplet on the surface. These different morphologies and distinct TCL determine the resultant wetting state of the droplet on a surface. As a result, the adhesive force of the oil droplet on colloidal crystals decreases with the varying morphology of the latex particles. The colloidal crystals from spherical particles show the highest adhesive force owing to the stable TCL and continuous large contact area, while the films from cauliflower-like particles show decreased adhesive force due to the discontinuous TCL and discontinuous contact of oil droplets on the substrate. Colloidal crystals from single-cavity particle show the smallest adhesive force due to the discontinuous point contact mode of TCL, and the oil droplet can easily roll away on the film's surface (Figure 8.5). The colloidal crystals with controllable underwater adhesive force will produce an important influence on the fabrication of functional adhesive materials, and will have important applications in oil release, drug release, *etc.*

8.2.1.4 Applications of Colloidal Crystals with Controllable Wettability

The colloidal PCs with special wettability demonstrate promising applications in various fields, which is mainly inspired by the amazing use of biological PCs with special wettability. In Nature, the special wettability of biological PCs is necessary for their environmental adaptability. For example, the special superhydrophobicity of butterfly wings can shield them from rainfall.[12] Thus, rain droplets can swiftly roll off with a slight tremble. Simultaneously, dust particles on the surface are easily removed. While superhydrophobicity of a mosquito's complex eyes itself produces good vision in humid environments due to the easily rolling of the fog droplet from the surface.[14] Also, longhorn

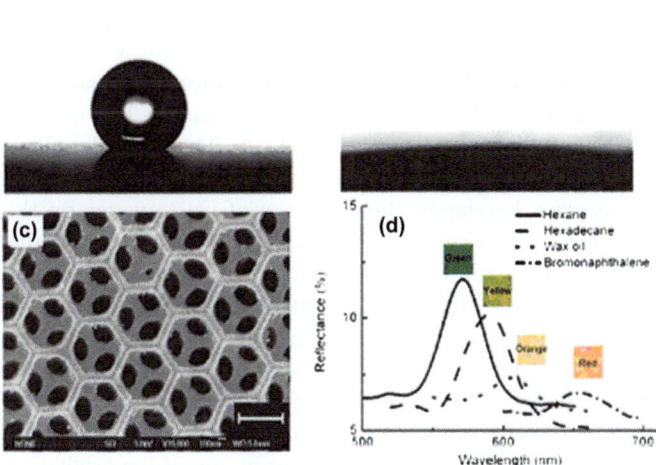

Figure 8.6   Oil-sensitive phenolic resin inverse opal. (a) and (b) is the wetting state of water or oil on the substrate surface, (c) is an SEM image of the inverse opal, (d) is the reflection spectra or color change of the film responding to different oils.
Reproduced with permission. Copyright 2008 Wiley-VCH Verlag GmbH & Co. KGaA; and Copyright 2008 The Royal Society of Chemistry.

beetles show an ingenious example with harmony of PCs and special wettability.[15] On its hydrophilic elytra an amazing camouflage behavior was observed and the color of the pattern region changes from yellow to red when the environmental humidity increased, as shown in Figure 8.6. This is mainly due to water adsorption/desorption of the hydrophilic pattern region under different humidity, which results in the change of their periodic structure. Such a color change can keep consistent with the soil's color in rainy or sunny conditions in order to protect it from its natural enemies. Understanding these functions of biological PCs will open a new way for the design and synthesis of novel functional materials.

In fact, more applications are developed for colloidal crystals with special wettability in different solution systems, owing to their special application in sensing, detecting or catalytic systems. In one of our studies, the as-prepared colloidal crystals was used as an effective humidity-detecting sensor[47] based on the distinct color change, which can easily monitor the humidity change from 0, 20, 40, 60 or even 100% based on the structure color changing from transparent, violet, blue, green or even red. It is very effective to develop naked-eye detector for humidity changes. In this case, the colloidal crystals are fabricated by infiltrating the hydrophilic polyacrylamide into the interstice of the latex particles. In these processes, the hydrophilicity of the colloidal crystals is a key parameter, which ensures the full wetting and spreading of the water droplet upon the films, and the crosslinkage degrees and the extended process of polyacrylamide affects the color change of the colloidal crystals after adsorbing different amounts of water. It could be expected that the specific wettability of the colloidal crystals will enable its use in special sensing applications. Based on a similar idea, the oil-sensitive[35,36] colloidal crystals were fabricated by infiltrating oil-sensitive phenolic resin into the interstice of the latex particles and subsequently dissolving the latex template by toluene. The as-prepared films show superhydrophobicity and superolephilicity. The superolephilicity makes it possible for use as an oil detector to adsorb oil, while the superhydrophobicity provides a necessary barrier to the influence of waste water on the detection result. As expected, the as-prepared films can effectively monitor the change of the kinds of oil by its distinct color or optical signal. It should be noted that the as-prepared film shows highly oil-sensitive properties, it even can detect different oil with refractive index differences of just 0.02.[36]

Furthermore, the colloidal crystals can also be used as detecting elements for some dynamic systems. For example, a self-oscillation system is well known due to its special use for the biological mimics of the human muscle, heart or minirobots.[48] However, the detailed information about the oscillation process is generally judged from the length change of the system. It is expected to monitor the oscillating process by color or optical signal change. Introducing the colloidal crystal into an oscillating system may provide a more convenient way for the fine judgment of the oscillation parameter. We presented a typical example to monitor pH oscillation process by using of colloidal crystals,[49] which is shown in Figure 8.7. Here, the pH-responsive colloidal crystals were synthesized and introduced into the system. By suitably modifying the content of the acid group and crosslinked degree of the pH-responsive PC hydrogel, we can easily monitor the oscillation process by the distinct color change and optical signal. It should be noted that excellent solvent resistance is necessary for the as-applied colloidal crystals.[50] Monitoring of hte dynamic process by using colloidal crystals is especially important for the real applications of PCs in extended systems.

8.2.2 Anisotropic PCs from Special Latex Particles

The specific light-manipulation properties of PCs make them widely applicable in various optical devices, sensing, *etc.* The optical properties of PCs mainly originate from the periodic arrangement of monodisperse latex particles, which can effectively modulate the light transmittance route and direction based on the Bragg diffraction equation. Recently, anisotropic PCs have been desired as photonic materials because of the enhanced light control, provided by the anisotropy of the building blocks,[51–57] which was predicted that since anisotropic building blocks can lift the symmetry-induced degeneracy of photonic bands. Anisotropic colloidal PCs also have wide applications in microlens arrays and antireflective coatings, complex colloidal structure, *etc.* However, it is difficult to achieve homogeneous orientation for anisotropic latex particles. Some additional approaches have been developing for the well-ordered latex assembly. Typically, Ding *et al.* attained iridescent ellipsoidal CC self-assembled from Fe_2O_3/SiO_2 core-shell particle in a magnetic field.[58,59] Liddell and coworkers[60,61] fabricated PCs from building blocks of

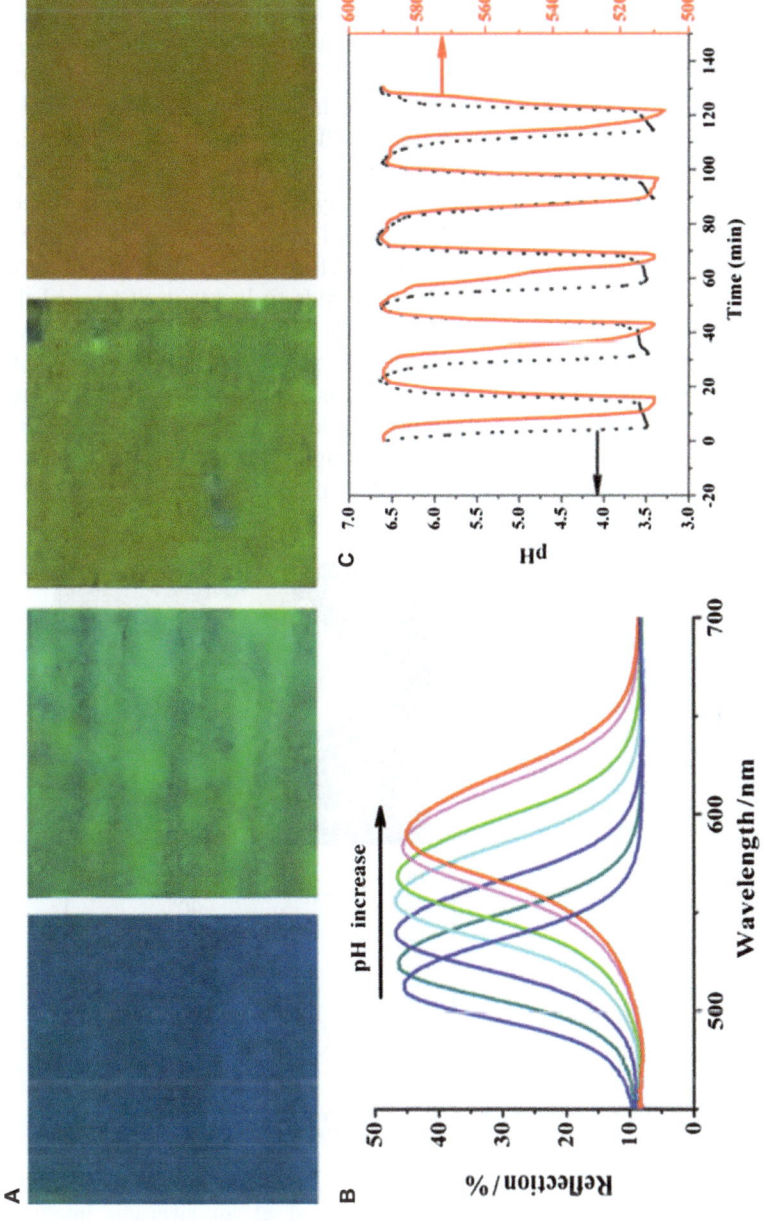

Figure 8.7 The application of the responsive PCs in self-oscillation system. The varying structure color (a) or reflection spectra (b) of the film with increasing pH of the system, (c) the comparison of the reflection spectra change and the pH oscillation system. Reproduced with permission. Copyright 2009 Wiley-VCH Verlag GmbH & Co. KGaA.

dimer-based and mushroom cap-based particles *via* a convectively self-assembly approach. To develop more effective assembly routes for anisotropic latex particles, we developed a novel assembly means of anisotropic latex particles in the air/water interface based on the wettability difference of the two sides of mushroom-shaped latex particles.[45] The mushroom-shaped latex particles were synthesized by emulsion polymerization. The special morphology was achieved by introducing a definite amount of DVD into the polymerization system at the suitable time. It is interesting that the as-synthesized single-cavity latex particles showed the distinct wettability difference between the two sides of latex particles: the sphere side is hydrophilic due to the richness of hydrophilic carboxyl groups, and the other side is hydrophobic due to the introduction of DVD onto the cavity side. This special wettability difference made it possible to assemble the latex particle in the interface of the air/water interface. Here, the hydrophilic sphere side of particles tends to locate itself toward the water side, while the hydrophobic cavity side set itself toward the air side. As-prepared colloidal crystals show special optical properties, there is obviously less change in its stop-band when varying its observing angle, which indicates the special effect from the uneven structure, could be obtained. This assembly of anisotropic latex particles, by using the wetting difference, opens up a novel way for fabrication of novel anisotropic colloidal crystals (see Figure 8.8).

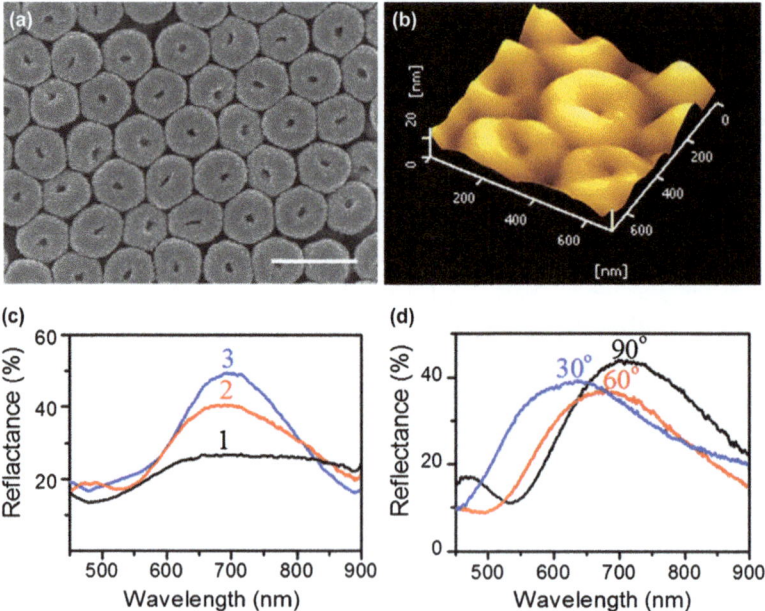

Figure 8.8 The anisotropic colloidal PCs assembled at an air/water interface. (a) SEM and (b) AFM of as-prepared colloidal PCs, reflection spectra of the films with varying thickness (c) and observing angle (d).
Reproduced with permission. Copyright 2010 Wiley-VCH Verlag GmbH & Co. KGaA.

8.2.3 Fabrication of Colloidal PCs with High Mechanical Strength

Colloidal crystals are periodic structures of materials with different refractive index, and there is a discontinuous structure from latex particles and air according to the means of assembly. Here, the latex particle is arranged in a point-contact mode, the open structure among contacting latex particles and resulting from the air matrix will impair the mechanical properties,[62,63] which reduces the mechanical strength of the structure for the conventional applications. Accordingly, various approaches have been developed to improve the latex linkage and mechanical strength of films. Typical examples include post-treatment of the colloidal crystals by an additional calcinations process.[64,65] Similar to the tough biological PCs from crosslinked structure, introducing the crosslinked system[66–76] into the latex interstices presents another effective approach for improving the mechanical strength. Meanwhile, the solvent resistance of the film is improved greatly, which could withstand the solvent for more than 2 h. Further improvement of the mechanical strength of the films requires the use of the high-strength materials.[77] In one of our experiments, the precursor of SiC was selected as the infiltrating material, which has the highest mechanical strength, solvent resistance, and high-temperature resistance. The material was infiltrated into the interstices of the latex particles accompanied with subsequent curing and calcination to remove the template. As-prepared SiC inverse opal PCs show extremely high mechanical strength, it can withstand the common solvents including strong acid (HCl), strong base (NaOH), and THF and toluene, which demonstrates the colloidal crystals can be used in extreme-temperature or harsh-solvent environments.

Although the mechanical strength of the colloidal crystals can be improved by changing the kinds of materials, there remains an unintended crack formation in the self-assembly process.[78,79] This has severely hindered the fabrication and application of large-scale colloidal PCs in advanced optical devices. Generally, cracking is mainly caused by the tensile stress generated in the self-assembly process, which arises from the shrinkage of the colloidal spheres during the final drying process and the constraint of this shrinkage by the rigid substrate. Based on this understanding, various approaches have been devised to eliminate the tensile stress and avoid cracking. First, preventing the colloidal spheres from shrinking by preshrinking them before assembly is an effective approach to eliminate the tensile stress and avoid cracking.[80–83]

Secondly, using the liquid substrate (molten Ga)[84] or patterned substrate[85] allows the homogeneous shrinkage of the colloidal spheres, thus avoiding the stress production during the assembly process. Finally, the preferential selection of the low stacking density crystal structure[86] is favorable for the elimination of cracking as it releases the stress from the interstices between the lattice faces. Although various achievements have been demonstrate for the fabrication of crack-free colloidal PCs, the as-fabricated crack-free colloidal PCs may degrade the optical properties due to the crack region may be connected by the pure polymer, or the crack-elimination process will restrict the well-ordered latex assembly. To solve this problem, we present a facile strategy for large-area crack-free single-crystalline PCs by the combined effects of polymerization-assisted assembly and substrate deformation.[87] The coassembling monomers infiltrate and polymerize in the interstices of the colloidal spheres, which could reduce the shrinkage of the colloidal spheres and lower the generated tensile stress. The infiltrated polymer also strengthens the long-range interactions of colloidal spheres based on its elasticity, otherwise, the timely transformation of the flexible substrate releases the residual stress. Elastic polymer infiltration and substrate deformation results in an integration of micrometer-sized single-crystal domains into centimeter-scale crack-free single-crystalline PCs. The obtained PCs show an improved optical property due to the large-area well-ordered single-crystalline structure. This facile approach to large-area single-crystalline PCs is promising for the practical applications of PCs in high-quality optical integrated devices, optical waveguides, *etc.*

To seek the simpler approach for the high-quality PCs, we further select a low-adhesive superhydrophobic substrate as assembly substrate, since that can allow the latex particle to shrink freely, which is beneficial to the fabrication of crack-free colloidal PCs. In Figure 8.9, as expected, the as-prepared colloidal PCs show a perfectly well-ordered arrangement, large-scale crack elimination, more close-packed structure and sufficient thickness. The colloidal PCs show a narrow stop-band with a full width at half-maximums (FWHM) of 12 nm. These properties profit from low-adhesive superhydrophobic substrate. The latex suspension on this substrate is in the Cassie regime and undergoes an obvious receding of the solid–liquid–gas three-phase contact line (TCL) during the drying process. The receding of TCL causes a timely release of the stress induced from latex shrinkage, leading to complete crack elimination and a more close-packed assembly structure. Furthermore, the simultaneous nucleation and crystallization of latex particles on the outermost layer of the

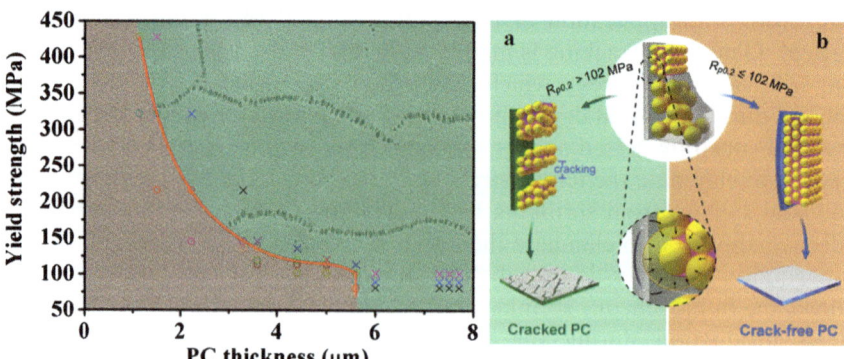

Figure 8.9   Fabrication of single-crystal photonic crystals by combined the effect of the flexible substrate and polymer infiltration. (Left) the critical thickness for the film without crack; (Right) the schematic for the formation of crack-free single-crystalline PCs.
Reproduced with permission. Copyright 2012 The Nature Publisher Group.

spreading liquid film produce a perfect assembly structure spanning the whole PCs. This facile fabrication of colloidal PCs with narrow stop-bands will be of great importance not only for the development of PC-based optical devices, but also for the extensive creation of various advanced crystal structure of other materials.[88]

8.2.4 Large-Scale Fabrication of Colloidal PCs by Spray Coating or Inkjet Printing

Colloidal PCS have demonstrated important application in various fields, but the large-scale fabrication of colloidal PCs remains a problem for real applications. To overcome the problem, we developed an ultrafast fabrication of colloidal crystals by spray coating;[89] it can easily fabricate the colloidal crystals with area of about $7 \, cm^2$ in several seconds. In this experiment, the key issue is the ultrafast assembly of latex particles, which have been effectively dissolved by the application of monodispersed latex particles with soft poly(acrylic acid) shell and hard poly(styrene) core structure. Here, the strong adhesion force among the soft latex shell accelerates the well-ordered latex assembly, and the hydrogen bonding among the carboxyl group of latex surface also contributes to the well-ordered assembly. Meanwhile, the monodispersed latex particles synthesized in our laboratory also contribute to the well-ordered latex assembly. The fabrication of colloidal crystals by spray coating opens a potential door for the fabrication of functional colloidal crystals.

With extended application of colloidal PCs in various systems, patterned PC has become a necessary step for its advanced use. Fabrication of patterned PCs by a common inkjet printer affords an effective approach for the specific applications.[90–96] We developed a facile fabrication of patterned colloidal PCs by a common printer,[94,96] which will greatly extend the applications of PCs in various applications. In this study, the special wettability of the printing substrate is especially crucial, which affects greatly the wetting/spreading of the latex droplet and well-ordered latex assembly on the substrate. As a result, the colloidal crystals with bright structure color and distinct patterns have been successfully fabricated using inkjet printing. However, different colors of the films mainly depend on the different ink made of latex suspension with different latex dimension for the printing system. It was expected that the colloidal crystal with different colors can be directly obtained by just using one kind of latex suspension. Fortunately, this idea is successfully realized by applying responsive latex particles,[95,96] which can show different structure color under varying exterior environments. Yin and coworkers fabricated the pattern PCs by screen printing, where one kind of magnetic latex particle was used.[95] The latex assembly can be easily realized by magnetic force, while the different structure color can be obtained by modifying the magnetic force of the assembly process, which can modify the distance between latex particles and change the color of the film. Recently, we extended the approach to the fabrication of color patterns by a pattern printer; the different color can be simultaneously displayed by using the same latex particles,[96] as shown in

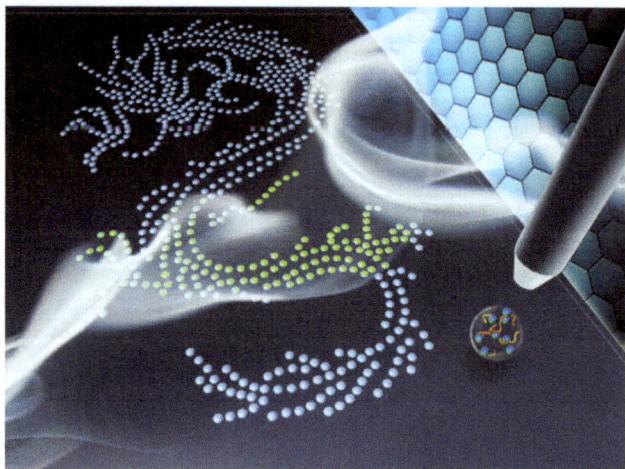

Figure 8.10   Fabrication of responsive pattern PCs from inkjet printing, where, the ink is the aqueous solution of latex particle and responsive monomer. Reproduced with permission. Copyright 2012 The Royal Society of Chemistry.

Figure 8.10. This facile fabrication of pattern PCs by printing provides a simple approach for the extended application of PCs in various displays.

8.3   Application of Photonic Crystals in Various Optic Devices

Photonic crystals (PCs) are attractive optical materials for controlling and manipulating the flow of light. The photon stop-band, light-wavelength selectivity and slow photon effect of PCs materials supply the excellent model and platform to fabricate and carry out: total reflection, light waveguide, superprism, light-beam bending,[97,98] spontaneous emission modification,[99–106] light-emitting diodes[107,108] and fluorescence detection, *etc*. Basic on the outstanding properties of PCs in light adjustment, the new materials with PCs structures draw great attention in research and application fields. In the following sections, we will present several cases of PCs material applications in high-performance photoluminent, ultrasensitive detecting, error-defending optical data storage and high-efficiency photocatalysts. Also, we will see the new aspects created by PCs on utilizing facile PCs nanotechnique processing to improve vastly current and traditional photoactive materials and optical devices.

8.3.1   High-Performance Photoluminescent Devices

8.3.1.1   Fluorescence Enhancement

The high-performance light emission keeps attracting research because of its applications in illumination, detection, display and advanced optical devices.

The light-propagation inhibition of a stop-band can modify significantly the emission characteristics of embedded optically active materials (dyes, polymers, semiconductors, *etc.*) at a certain wavelength.[99–106] PCs and optically active materials integration could fabricate the high-performance photoluminescent devices. For example, the fluorescence intensity is related to the radiation rates of light-emitting materials,[109] electromagnetic field,[110] and Bragg reflection. With fluorescence emission and PCs stop-band matching, employing PC surfaces as a Bragg mirror, the solid-state fluorescence emission could be enhanced drastically.

The photonic crystal films were prepared by self-assembly of colloidal crystals using a vertical deposition method. On depositing organic dyes on PC surfaces, we can investigate the fluorescence surface emission enhancement effect on PC surfaces (Bragg mirror) compared with glass substrates (transparent) and aluminum film (regular mirror). Figure 8.11 illustrates simply the fluorescence enhancement mechanism by reflection of PCs. Light with a wavelength that overlaps the stop-band of PCs is suppressed by the PCs, *i.e.* there is no way for the emission to enter into the PCs. So, the light will be reflected back in the direction of the stop-band. However, light incident on glass without PCs will transmit to the substrate because of its low reflectivity. Figure 8.11(a) shows that a light source emitted from the surface of PC film would be prohibited from transmitting through the surface and reflected back in the direction of the stop-band and enhanced when its wavelength overlaps the stop-band of the PCs. As for Al film, the fluorescence can be reflected to some extent because of its reflection characteristics. The fluorescence data (Figure 8.12(a)) demonstrate that the PC surface is a more efficient and selective reflection film than an Al film. In contrast, as illustrated in Figure 8.11(b), the light would transmit and be lost if emitted from the glass surface. Hence, the fluorescence of RB on glass is the weakest. In addition, Figure 8.12(b) also shows that the reflection of PCs film is selective (light whose wavelength overlaps the photonic stop-band can be reflected), while that of Al film is not. Therefore, the PCs film is more efficient and selective than an Al film as a reflection mirror.

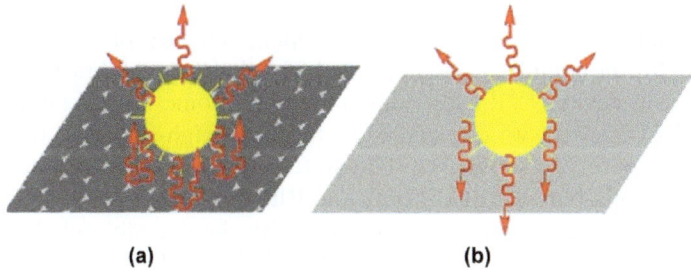

(a)   **(b)**

Figure 8.11   Schematic illustration of a light emission source on the surface of a PC film (a) and a glass substrate without PCs (b).
Reproduced with permission. Copyright 2007 The Royal Society of Chemistry.

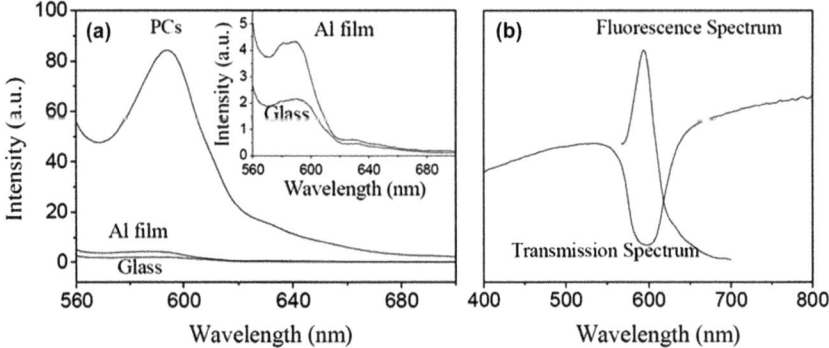

Figure 8.12   (a) The fluorescence spectra of 20-nm thick RB deposited on yellow PCs, aluminum film and glass substrate ($\lambda_{ex} = 550$ nm). Here the fluorescence spectra were collected in the stop-band direction of PCs. The inset shows the fluorescence spectra of RB (20 nm) on aluminum film and glass. (b) The transmission spectrum of yellow PCs and the fluorescence spectrum of RB deposited on the PCs.
Reproduced with permission. Copyright 2007 The Royal Society of Chemistry.

Solid-state fluorescence of organic dyes has been enhanced drastically by PCs as the excitation and/or emission wavelengths of the organic dyes are in the range of the photonic stop-bands of the matched PCs, which demonstrates that the PCs will be more efficient and selective reflection mirrors than aluminum films. It could be extended to other light-emitting materials and corresponding PCs. A promising application of the strategy could lead to a new generation of optoelectronic and lighting devices.[111]

8.3.1.2   Brightlight-Emitting Diodes (LEDs)

High-efficiency and brightlight solid-state lighting devices are considered as one of the most promising next-generation lighting devices in response to the great need for reducing energy consumption, especially for the light-emitting diodes (LEDs).[112,113] Immense efforts have been made in bright white-lighting luminous films, including the complex doping schemes or combinations of materials, for the complications associated with design and fabrication of solid-state lighting devices.[114–116] By combining a photonic crystal luminous film with an LED device, we can expect to fabricate high-efficiency bright LED lamps. As Figure 8.13 shows, the solid-state lighting device was obtained by a combination of a blue-light LED bulb and Rhodamine 6G doped PC film. The luminous film obtains from the PC doping with the dye molecule Rhodamine 6G. Illuminated by the commercial blue-light LED, the PC-structured luminous film gives rise to bright yellow light. The combination of the yellow and blue light can give rise to bright white light with CIE (0.33, 0.38), which agrees well with that of the pure white CIE (0.33, 0.33) (Figure 8.13(a)). The PC-structured luminous film shows brilliant yellow due to the Bragg reflection

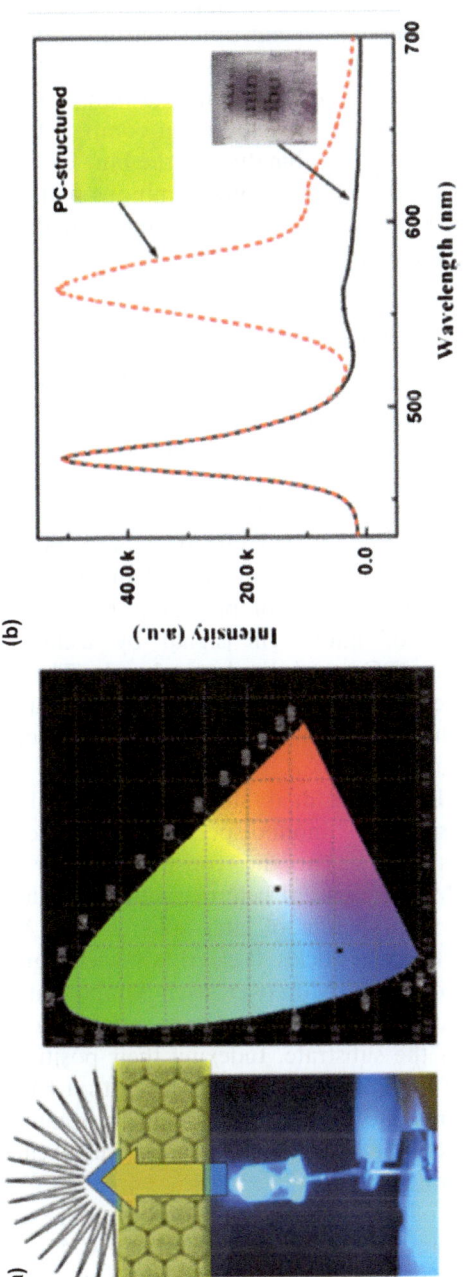

Figure 8.13   (a) The scheme of a highly efficient LED system combining a LED and a PC-structured R6G film, and its CIE chromaticity point, compared with that of the conventional planar R6G film. (b) The emission spectra of the active devices. Reproduced with permission. Copyright 2009 Springer-Verlag GmbH.

caused by its well-ordered periodic structure (see the upper inset in Figure 8.13(b)). A control sample is transparent and has a pink color due to the doping R6Gmolecules (see the lower upper inset in Figure 8.13(b)). Figure 8.13(b) shows the combination emission spectrum of the blue-light LED and the illuminated PC-structured luminous film (dotted line). It is clearly observed that the emission intensity of the luminous film with the PC structure is dramatically enhanced by a factor of over ten compared with that of the usual planar luminous film. The combination of the blue emission of the LED and the yellow emission of the PC-structured luminous film leads to a bright white light. Its total emission intensity is 2.36 times bigger than that of the usual LED system containing a planar luminous film.[117]

8.3.1.3 Hierarchical Fluorescence Enhancement

Fluorescence enhancement based on photonic crystals (PCs) has opened up a new horizon for the fluorescence-based detection.[118,119] Near the stop-band of PCs, light propagates at a reduced group velocity owing to resonant Bragg scattering, which can enhance the optical gain leading to stimulated emission, as well as amplify the excitation of incident light.[120–123] Although a significant amount of research has focused on taking advantage of the benefits of the stop-band of PCs for fluorescence enhancement, the extent of amplification is still limited.[124–126] In the following, we will present a facile approach on the remarkable enhancement of fluorescent signal by using heterostructure colloidal photonic crystals (PCs) with dual stop-bands. The intensity of the fluorescent medium in heterostructure PCs with dual stop-bands overlapping the excitation wavelength and the emission wavelength of fluorescent medium can produce further enhancements.

The heterostructure PCs with dual stop-bands are fabricated by the successive vertical depositions of monodisperse latex spheres with different diameters onto glass substrates at the constant temperature humidity. Figure 8.14 presents SEM side views of heterostructure PC films composed of different self-assembly concentrations of latex spheres with 260-nm particles deposited on top of the 210-nm particles. It clearly shows the growth of one PC over another with good ordering. Moreover, each component stack is a face-centered cubic (fcc) crystal film of latex spheres with (111) planes oriented parallel to the surface of the substrate. Indexing their positions allows us to estimate the thickness of the PC films. The PC 260 with a 645-nm stop-band will be used for the fluorescence emission matching to organic dye Nile Red ($\lambda_{em} = 620$ nm), and PC 210 (stop-band $= 516$ nm) for the excitation matching of Nile Red ($\lambda_{ex} = 545$ nm).

To evaluate the effects of stop-band positions on fluorescence emission, PC 180, PC 210 and PC 260 samples with the stop-band at 410, 516, 645 nm and the PC 210-260 with dual stop-bands were fabricated from latex spheres with different diameters. All PC samples have the same thickness of 7 μm. The reflection spectra of all the colloidal PCs, complemented with the excitation laser and the fluorescence emission of Nile Red are shown in Figure 8.15(a).

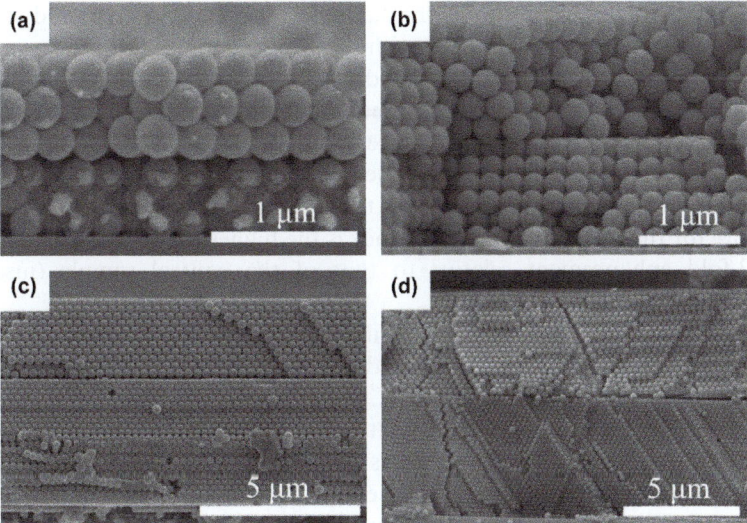

Figure 8.14   SEM side views of heterostructure PC films made of different self-assembly concentrations of poly(St–MMA–AA) spheres with 260-nm particles deposited on top of the 210-nm particles.
Reproduced with permission. Copyright 2011 from Elsevier B. V.

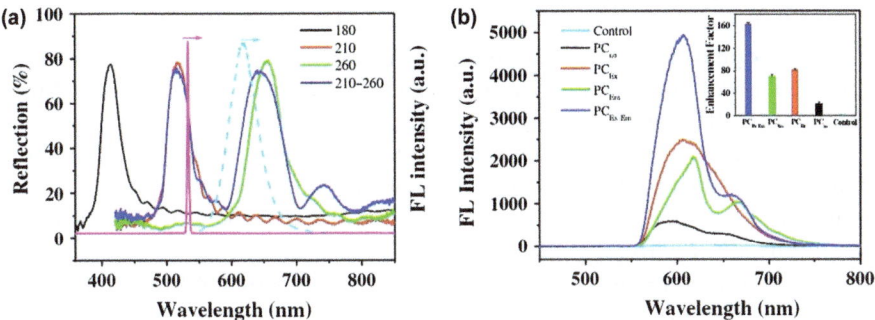

Figure 8.15   (a) Reflection spectra of PCs with different stop-bands and their relationship with the excitation laser ($k = 532$ nm, magenta solid line) and emission of Nile Red ($k = 620$ nm, cyan dashed line). (b) Fluorescence spectra of Nile Red absorbed on the control sample and PCs with colloidal particle diameters of 180 nm (PC_{no}), 210 nm (PC_{Ex}), 260 nm (PC_{Em}) and 210down–260up nm (PC_{Ex-Em}). The inset shows the enhancement factor of fluorescence intensity on PCs with different stop-bands. The factor is the intensity ratio of fluorescent medium on PCs relative to that on the control sample.
Reproduced with permission. Copyright 2011 From Elsevier B. V.

Here, the laser with 532 nm wavelength is used as the excitation light and the emission of Nile Red is centered at 620 nm. For PC 210 with stop-band at 516 nm (PC_{ex}), its band edge overlaps the excitation laser. For PC 260 with stop-band at 645 nm (PC_{em}), the blue band edge overlaps the emission of the

Nile Red. For PC 210–260 with dual stop-bands with 516 and 645 nm (PC_{ex-em}), the stop-bands coincide with the overlapping of both the excitation laser and the emission of the Nile Red. The stop-band at 410 nm of PC 180 (PC_{no}) overlaps neither the excitation nor emission of Nile Red. Figure 8.15(b) shows different fluorescence enhancements on PC films infiltrated with Nile Red. The fluorescence intensities of detections on all colloidal PC substrates are higher than that on the control sample. The fluorescence intensity for the PC with dual stop-bands is the highest of all samples, showing a 162-fold enhancement relative to that of the control sample (insert in Figure 8.15(b)).

In detail, PC_{no} has a fluorescence enhancement of 21.8 times relative to the control sample, which can be attributed to an enrichment effect from the large surface area of the PC structure. PC_{ex} coincides with the excitation laser, leading to increased local energy density, which strongly excites fluorophores. The excitation enhancement alone is about 3.7-fold. When the emission wavelength of Nile Red significantly overlaps the blue band edge of PC_{em}, there is a 68.9-fold enhancement at maximum emission wavelength of Nile Red on the PC compared with that on the control sample. The fluorescence enhancement is mainly attributed to the combination of large surface area and enhanced extraction (ca. 3.2-fold) of light generated in PC_{em}, which can serve as the dielectric cavity and act as a local resonance mode for the emission propagation. Photons can couple to the overlapping local resonance mode and Bragg scatter out of the structure, thereby greatly reducing the amount of light trapped as guided modes. Moreover, the high density of states near the stop-band enhances the coupling of spontaneously emitted photons.[127,128] More importantly, there is an enormous enhancement of 62-fold in PC_{ex-em}, because its stop-bands overlap both the excitation laser and the emission of the Nile Red, which can produce the doubly resonant enhancement. The excitation, emission, or possibly both at the same time can be enhanced and yield the composite effect when combined.

The photon stop-band of photonic crystal could generate the selective and high-efficiency Bragg reflection and scattering. Rationally corresponding photonic crystals and light-emitting materials, with the fluorescence emission or excitation matching, will yield a bright and high-efficiency fluorescent signal release. This gives one approach to improve and increase the current fluorescent dye and materials without redesigning and resynthesizing the new advanced dye or materials. In this way, we can fabricate high-performance sensors and develop sensitive fluorescence-based detection.[129]

8.3.2 Ultrasensitive Detecting

In current detection analysis methods, fluorescence detection analysis has the great advantage on nontouch, undamaged and multichannel complex environments detection and analysis.[118,119] The high sensitivity is the most important property for high-performance detection. PC materials could carry out the light emission enhancement and fluorescence amplification.[111,124–126,129]

The PC materials applications in fluorescence detection analysis will perform the ultrasensitive detection with a big amplification of the fluorescence signal. Beyond that, we will give some deep discussions on fluorescence resonance energy transfer (FRET)[130,131] application of PC materials in ultrasensitive detection.

8.3.2.1 Amplifying Trace TNT Detection on Inverse Opal Photonic Crystal

After the 9/11 attacks, in the practical ultrasensitive detection, there is no aspect being paid more attention than trace explosives. Consequently, a great variety of methods have been proposed for the improvement of explosive detection sensitivity. After designing and synthesizing the new sensor compound, nanoscientists considered construction of macroporous sensing substrates for improving TNT detection sensitivity, utilizing the large surface areas and porous structure to facilitate the diffusion of the trace TNT vapor and reduce the self-aggregation of fluorescencepolymers.[132–140] Combining macroporous sensing substrates and the optical behavior of PC materials, we can build inverse opal photonic crystal, (Figure 8.16) and improve the TNT detection sensing by not only utilizing the large surface areas of the inverse opal structure, but also the special light-manipulation properties, especially, the slow-photon effect of PC.

A prototypical fluorescent conjugate polymer, namely poly[2-methoxy-5-(2-ethylhexyloxy)-p-phenylenevinylene] (MEH-PPV) is selected as a sensitive chemosensing material to capture TNT electron donor MEH-PPV and the electron acceptor TNT.[141] A silica inverse opal PC is fabricated by a sacrificial polymer template method, which involves the self-assembly of polymer

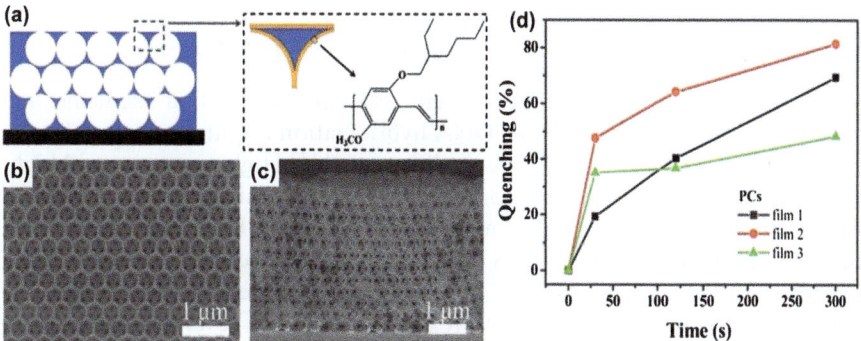

Figure 8.16   (a) Schematic illustration of the silica inverse opal PC coated with MEH-PPV on a porous surface. (b, c) Typical SEM images of top-view and cross section of the PC-based sensor for TNT detection. (d) Time-dependent fluorescence quenching of films 1–3 on inverse opal PC upon exposure to TNT vapor.
Reproduced with permission. Copyright 2011 The Royal Society of Chemistry.

colloidal spheres, infiltration of silica sol and removal of polymer spheres.[142] The PC-based sensors for TNT detection (in Figure 8.16(a)) are prepared by spin coating MEH-PPV solution on an as-prepared inverse opal PC.

As expected, fluorescence intensities of the TNT sensor decrease upon exposure to TNT vapor over elapsing time. In contrast, the quenching efficiency of the optimized MEH-PPV on PC-based sensor reaches ca. 80%, it is 10% higher than that of the control sample (on glass). Higher quenching efficiency of the PC sensor could be attributed to its interconnected porous structure, large specific inner surface areas and more binding sites, which favors the rapid capture of TNT molecules and results in a more effective quenching offluorescence.[139,140] More importantly, the fluorescent signal of the PC-based sensor for TNT detection is greatly amplified compared to that of the control sample. Figure 8.16(d) shows time-dependent fluorescence quenching of films on PC, upon exposure to TNT vapor. There is a distinct quenching effect for the PC-based TNT sensor. Accordingly, the application of PC for a TNT sensor is analogous to the use of a magnifier, which provides us a more distinct detection signal. This fluorescence-amplifying method based on PC-based TNT sensors will present a feasible strategy for improving the sensitivity and resolution, which will open a simple and viable way for the design and development of highly sensitive fluorescent sensors.

The fluorescence enhancement for TNT detection on the optimized PC can be up to 60.6-fold in comparison to that of the control sample. The quenching efficiency achieves 80% after exposure to TNT vapor for 300 s. Furthermore, the fluorescent signal of the PC-based TNT sensor is greatly amplified. This amplification of the fluorescent signal will present a feasible strategy for improving sensitivity and resolution of explosives detection, which will open up a simple and viable way for the design and development of highly sensitive fluorescent sensors.[143]

8.3.2.2 FRET Enhancement Ultrasensitive DNA Detection

Ultrasensitive detection is more widely used and desired in medicine and diseases diagnostics. Detection of DNA hybridization is quite important for the diagnostics of genetic and pathogenic diseases.[144,145] High sensitivity and high-selectivity detection is necessary for early disease diagnosis and treatment.[146,147] Most available methods involve molecular-fluorophore-based assays, in which the target DNA is hybridized with a specific base-sequence probe labeled with a radioisotope or a fluorophore.[148–150] In these DNA detection processes, the amplification steps are important to realize the ultimate in terms of sensitivity. These amplification systems include polymerase chain reaction (PCR) protocols and signal-amplification systems, such as fluorogenic substrate-active enzymes,[151,152] modified liposomes,[153] and nanoparticles.[154–156] However, the relatively complex instrument and specific reagents required restrict their practical application.

Photonic crystals (PCs) are promising materials for the amplification of spontaneous emission as well as the control of the propagation of light with minimal losses.[157,158] Spontaneous radiation from excited molecules can be strongly modified by embedding light sources in the PCs,[159,160] which offer an ideal environment for fluorescence resonance energy transfer (FRET). FRET between energy-donor and -acceptor pairs is a feasible approach to improve the sensitivity and selectivity of the assay.[130,131] In that way, the light-harvesting "antennas" can bring about the amplification of the biosensor signals. Here, we will demonstrate an effective FRET-based DNA hybridization detection method by utilizing a PC to amplify the optical signal.

Figure 8.17 shows the DNA detection system with PCs. The detection system consists of a fluorescein (Fl)-labeled DNA (DNA-Fl) probe and ethidium bromide (EB), which serve as the donor and acceptor, respectively. The DNA-Fl is obtained by introducing an Fl unit at the 5' terminus of a single-stranded DNA (ssDNA) (ssDNA-Fl). The fluorescein group, with an absorption maximum at 488 nm and an emission maximum at 520 nm, is chosen because its emission overlaps with the absorption of EB. FRET from Fl to EB thus becomes energetically feasible.[161–163] EB is a well-known intercalator into double-stranded DNA (dsDNA). If the probe and target ssDNAs are complementary, DNA duplex formation will ensue and EB will intercalate into the dsDNA, resulting in an increase in its fluorescence quantum yield.[164,165] Thus, EB emission upon FRET sensitization from the Fl can be anticipated. If the target ssDNA is not complementary to probe DNA, the formation of the dsDNA structure required for EB intercalation will not take place and FRET to EB is hardly detectable.

It is known that the overlap and suitable orientation between the transition bands and the structures of the donor and acceptor are crucial to the FRET efficiency.[166] The overlap integral provides the analytical expression for how the spectral overlap between the emission of the donor and the absorption of the acceptor influences the rate of transfer. Comparison of the resulting fluorescence reveals a predominant effect on the emission dispersion properties

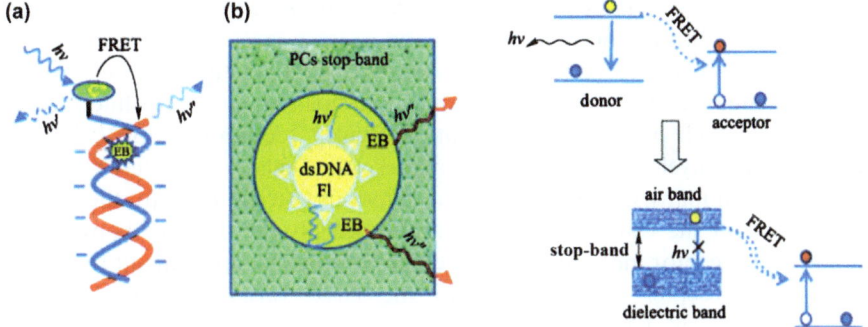

Figure 8.17   (a) DNA sequence detection based on a FRET mechanism. (b) Effect of the PC on FRET.
Reproduced with permission. Copyright 2008 Wiley-VCH Verlag GmbH & Co. KGaA.

of FRET between dsDNA-Fl and EB. The additions of EB cause a decrease in the emission intensities of dsDNA-Fl with a concomitant increase in EB emission intensity. The EB emission intensity of dsDNA-Fl/EB immersed in PCs is about 14.3-fold greater than that of the same system directly excited in solution. The FRET ratio ($I_{EB}/I_{Fl} = 1.71$) in PCs is enhanced by ~ 2.5 times compared with that in solution ($I_{EB}/I_{Fl} = 0.69$). The results indicate that PCs can strongly improve the overall energy-transfer efficiency from Fl to EB. Here, I_{EB}/I_{Fl} gives the proportion of the maximum of the acceptor Fl emission and the donor EB emission to indicate the FRET efficiency. Within the usual FRET system in solution, FRET exhibits poor energy-transfer efficiency from the Fl to the EB emission without being modified. The low efficiency is the result of a certain degree of loss through diffusion and reabsorption in the energy-transfer process.

The prohibited wavelength region matches well with that of the PC stop-band. The results confirm that the stop-band can produce strongly localized states of the donor (dsDNA-Fl) emission. There is a long lifetime for radiation injected into the cavity, and the reaction time between the donor and the acceptor will be lengthened. The donor emission can be readily generated and protected in the energy-transfer process. Thus, the energy loss of the Fl emission can be overcome and most of the energy persists to transfer to EB. As a result, the FRET efficiency can be remarkably enhanced, which leads to great amplification of the detection signal. The method has single-mismatch selectivity and a sensitivity of approximately 13.5 fm, which is hundreds of times greater than those obtained with conventional fluorophore-based methods.[167]

8.3.2.3 Highly Effective Fluoroimmunoassay Protein Detection

The prospect of highly sensitive protein detection offers enormous potential for early clinical diagnosis and understanding of the life.[168,169] Of particular interest is to develop instruments and methods that are capable of achieving a highly sensitive fluoroimmunoassay (FIA). In the following, we will propose a highly effective protein detection method based on FIA through the PC band-edge-induced fluorescence enhancement.

The fluorescence resonance energy transfer (FRET) method has been widely used to enhance the sensitivity of the immunoassay by using fluorescence antenna,[170] such as small fluorescent molecule,[171] quantum dot,[172,173] and conjugated polyelectrolyte.[174,175] However, the energy transfer is highly dependent on many factors, such as the extent of spectral overlap, the relative orientation of the transition dipoles, and especially the distance between the donor and acceptor molecules.[170] For different proteins, it is difficult to control the donor–accepter distance due to the large varieties of the volume, conformation and active sites.[173]

To demonstrate this is an effective and general method, the avidin–biotin system and the colloidal PC are employed in our study (Figure 8.18). As is known, the avidin–biotin system is exploited in a wide range of biochemical assays.[176,177] As Figure 8.18 shows, the PC was biotinylated *via* amide coupling

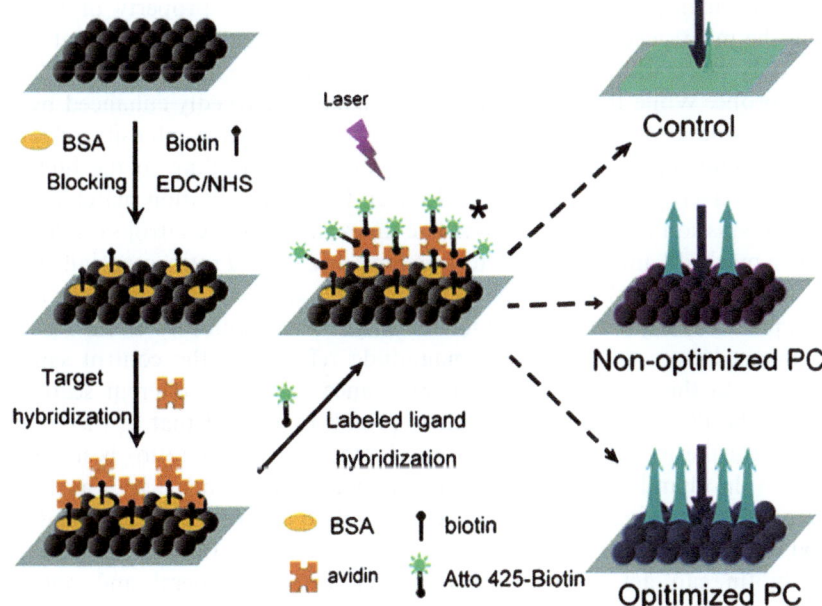

Figure 8.18 Schematic illustration of the enhanced avidin FIA utilizing the colloidal PCs. The colloidal PC is blocked by bovine serum albumin (BSA) and modified with biotin through the amide coupling (A). Then, the target avidin is hybridized on the PC (B). Atto 425-biotin is used to label the avidin (C), and the fluorescence spectra are collected for different samples (D). Compared to the planar control sample, the detection signal can be enhanced due to the large surface area of colloidal PC (the nonoptimized PC) and further markedly enhanced due to the band-edge effect of colloidal PC (the optimized PC).
Reproduced with permission. Copyright 2011 from Elsevier B. V.

between biotin and the bovine serum albumin (BSA) blocked PCs. The target protein (avidin) and the fluorescent-labeled ligand (Atto 425-Biotin) are hybridized on the surface in sequence. Compared to the planar control sample, the PCs have much larger surface area,[178–180] which can enrich the targets. More importantly, the PCs can lead to the band-edge-induced fluorescence enhancement.[181,182] A series of colloidal PCs with different stop-bands were used in the avidin FIA. We tested the detection limit and signal to interference ratio (S/I ratio) of the avidin assay on the PCs.

As noted in our discussions above, PCs can enhance the fluorescent signal for the detection. The large surface area and the band-edge-induced fluorescence enhancement are the two crucial factors for the fluorescence enhancement in the FIA on the PCs. The PC with blue band-edge matching to the excitation wavelength (EX) has the largest enhancement factor as 108.8. The PC with blue band-edge matching to the emission wavelength (EM) has the second largest enhancement factor as 31.9. The PCs with stop-band away from both the excitation and emission wavelength have similar enhancement factors over 16 times. For highly accurate detection, another superior property of the PC should be mentioned, the band-edge-induced fluorescence enhancement takes place only when the band edge overlaps the emission/excitation of the fluorescent probe. While the target fluorescent signal is markedly enhanced by the band-edge-induced fluorescence enhancement, the relative intensity of background fluorescence can be reduced simultaneously, leading to the high S/I ratios. The average S/I ratios of the FIA on PC with excitation matching and PC with emission matching are over five times that of the control sample. The PCs can markedly improve the detection limit and the S/I ratio, especially when the blue band-edge of the colloidal PC matches the laser excitation of fluorescent probe. The fluorescence signal intensity of the avidin FIA on the PCs can be enhanced over two orders of magnitude relative to the control sample, attributed to the large surface area, resonance field and coherent scattering effect of the PCs. The detection limit is shrunk to 1/69 of that of the control sample. Furthermore, the signal to interference ratio (S/I ratio) is increased because the band-edge-induced fluorescence enhancement is wavelength selective. The interference fluorescence does not go up proportionally, while the signal is significantly enhanced by the PCs. It is believed that the PC modified with biotin can act as an effective material for a general and sensitive fluoroimmunoassay.[183]

8.3.3 High-Performance Optical Data Storage

Utilizing bistable photoswitching of the fluorescence emission is one of the most attractive candidate of photon-mode memory systems. Fluorescent contrast (ON/OFF ratio) is an important performance parameter of optical information storage.[184,185] Then, in optical readout processes, amplification of fluorescent contrast is necessary to maximize sensitivity, resolution, and low error rate. To amplify the fluorescent contrast by introducing photonic crystals (PCs) into optical memory system could be an effective and facile approach to solve this issue.

As the scheme in Figure 8.19 shows, a fluorescent photoswitch material 1,2-bis(2'-methyl-5'-phenyl-3'-thienyl)perfluorocyclopentene (BP-BTE), coumarin was added into an amorphous poly(methyl methacrylate) (PMMA) matrix film. The BP-BTE is adopted as a photochromic switch owing to its excellent fatigue resistance, thermally irreversible properties, rapid response time, and reactivity in the solid state.[186,187] The coumarin, as an available fluorescent dye, is selected to characterize the fluorescent signal. When these two components are mixed in a film, the coumarin acts as an energy donor and the closed-form BP-BTE is an energy acceptor. ON/OFF switching of fluorescence can be induced by intermolecular energy transfer from the fluorophore to the photochromic unit under optical stimulation. The significant overlap between the emission band of the coumarin (curve IV) and the absorption band of the closed-form BP-BTE (curve II), can absorb the light emitted by the coumarin. This is an essential criterion for effective energy transfer between the closed-form BP-BTE and the coumarin. In this case, the photoinduced switching of

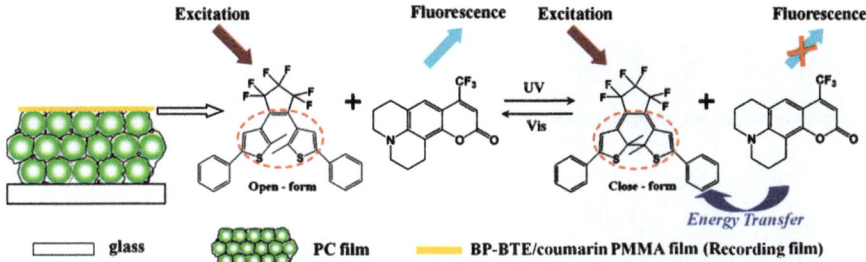

Figure 8.19 Schematic of the BP-BTE/coumarin-loaded PMMA film on the PC surface. When this recoding film is excited by 442-nm light, fluorescence is observed. After irradiation by UV light, the open form of the BP-BTE molecule is transformed to the closed form and fluorescence is quenched due to intermolecular energy transfer from coumarin to the closed-form BP-BTE molecule.
Reproduced with permission. Copyright 2010 Wiley-VCH Verlag GmbH & Co. KGaA.

the closed and open states of the BP-BTE molecule efficiently modulates the ON/OFF switching of fluorescence emission. A 40-fold enhancement of the fluorescence intensity in the ON state on the PC is observed in relation to that on glass, with a 6-fold enhancement obtained in the OFF state. Thus, fluorescent contrast on the PC surface is sevenfold higher than that on glass, which is enhanced from 9:1 to 63:1.

Figure 8.20 displays the optical data record fluorescence readout images on the regular glass and PC substrate. The fluorescence image on the PC exhibits both higher brightness and contrast of the spots compared with that on the glass, indicating that the PC surface does amplify fluorescent contrast effectively to attain high resolution and sensitivity. The fluorescence intensity-associated line profiles from the glass and PC surfaces are shown in Figure 8.20(c). By evaluating the relative intensity, the fluorescent intensity on the PC is evidently enhanced, which corresponds to the high brightness of the PC surface. Additionally, a signal in the background regions is enhanced to a smaller degree than that in the luminescent regions on the PC, which is in accordance with fluorescence emission spectra. Therefore, amplification of fluorescent contrast on the PC surface presents a feasible strategy for improving the ON/OFF ratio in the readout of the stored information.

It should be mentioned that the enhancement value in the OFF state is smaller than that in the ON state on the PC (Figure 8.20(c)), which contributes to the amplification of fluorescence contrast. To explain this phenomenon, fluorescence emission spectra of the BP-BTE/coumarin-loaded PMMA film before and after UV irradiation and reflection spectra of the matched PC are shown in Figure 8.20(d). There is a slight blueshift for the peak position of fluorescence emission after UV irradiation,[188,189] which may be attributed to the partial energy transfer from the coumarin to the BP-BTE resulting from incomplete overlapping between the emission band of coumarin and the

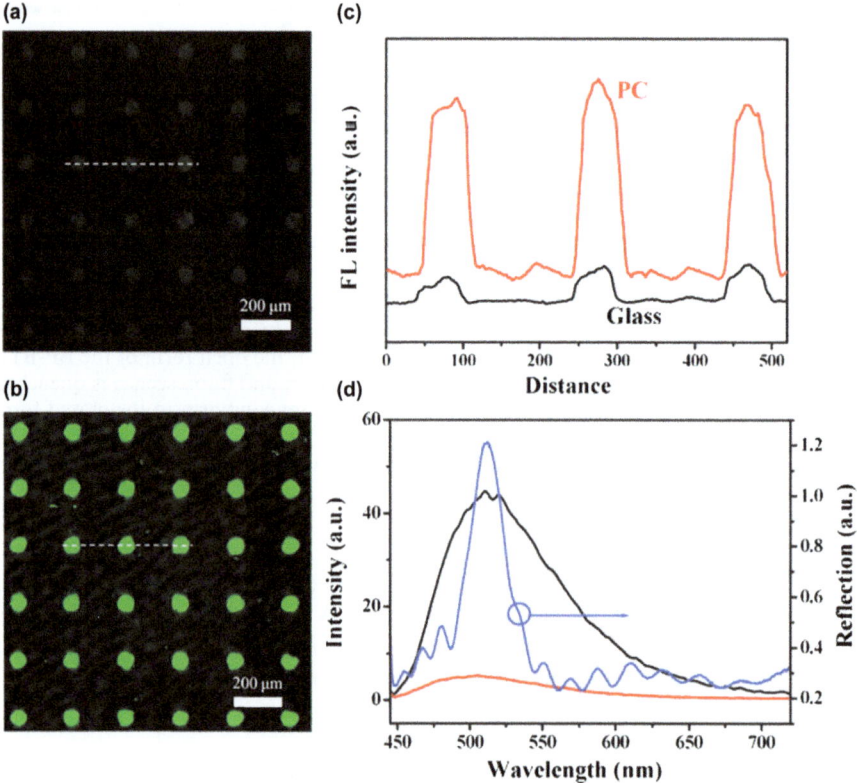

Figure 8.20 (a, b) Fluorescence images generated from irradiation (5 min) at 365 nm of the BP-BTE/ coumarin-loaded PMMA film on the glass and matched PC through a dot-patterned contact mask. The light regions indicate luminescence and the dark regions are weakly luminescent. (c) Fluorescence intensity associated line profiles from the glass and PC surfaces. (d) Fluorescence emission spectra of the BP-BTE/coumarin-loaded PMMA film on the glass surface before (black line) and after (red line) irradiation with UV light and UV-Vis reflection spectra of the matched PC.
Reproduced with permission. Copyright 2010 Wiley-VCH Verlag GmbH & Co. KGaA.

absorption band of closed-form BP-BTE. This blueshift of fluorescence emission reduces the overlapping degree with the stop-band of the PC and induces a smaller enhancement of fluorescence. It can be expected that an even greater amplification of fluorescent contrast will be achieved by either optimizing the PC structure or choosing more efficient energy-transfer systems. PC materials were demonstrated that performed ultrasensitive detection and high-contrast optical data storage. They offer a facile and general approach to fabricate high-performance optical devices and avoids the complicated design and multistep synthesis for the new dye or chemical modification.[190]

8.3.4 High-Efficiency Light Catalysis

Energy and environmental problems are emergent and common issues nowadays. Light catalysis and solar power are expected for the promising projects of the depollution and clean energy.[191] Here, we will discuss several examples of advanced PC materials applications on photochemical hydrogen evolution, photocatalysis organic dye degradation, and efficient output dye-sensitized solar cell.

8.3.4.1 High-Efficiency Photocatalysts Hydrogen Evolution

Hydrogen fuel as a form of clean energy source can provide the ultimate solution for pollution problems. Sunlight-driven water splitting for hydrogen generation provides a promising strategy to transform solar energy into hydrogen fuel.[192] Owing to the cheap, photochemical stability and low toxicity properties, TiO_2 was widely used as a water-splitting photocatalyst. Micro- and nanostructure TiO_2 materials were commonly used to increase the specific surface and improve the catalytic activity.[193–201] PC materials could offer the specific surface nanostructures and special photon-manipulation properties in addition.

Figure 8.21 illustrates that the hierarchical segments of micro- and submicrometer-size peeled off the substrates are preferred for photocatalytic water splitting experiments due to the more efficient interaction with the

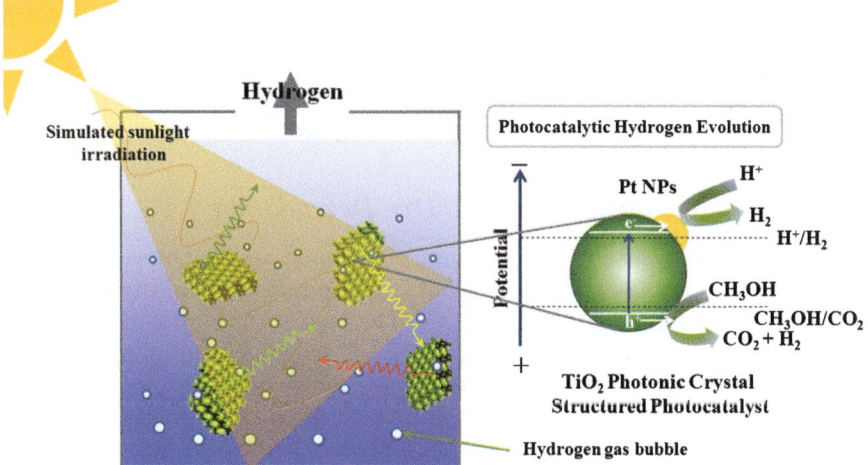

Figure 8.21 Illustration of photocatalytic hydrogen evolution over Pt-loaded TiO_2 PC under xenon lamp irradiation. The left part illustrates the interaction between photons and PC segments dispersed in the methanol aqueous solutions and multiple scatterings among the segments. The right part illustrates the photocatalytic hydrogen evolution on the Pt-loaded TiO_2 PC with methanol as sacrificial reagent.
Reproduced with permission. Copyright 2010 The Royal Society of Chemistry.

photons in the stirring homogeneous solution upon illumination.[202] The PC structure serves to promote light absorption by TiO_2, which may contribute to creating more photogenerated electron–hole pairs and consequently higher hydrogen evolution. When monochromatic light with a certain wavelength is located on the long-wavelength edge of the stop-band of the PC, the slow photon effect is able to promote the absorbance by TiO_2. The hierarchical "PC leaf"-structured TiO_2 can lead to strong light harvesting and provides a promising strategy for efficient photochemical hydrogen evolution.

Pt nanoparticle deposition on TiO_2 can lead to the formation of a Schottky barrier, which serves as an efficient electron trap to avoid electron–hole recombination in the photocatalytic process (as illustrated in Figure 8.21).[203] The obtained TiO_2 displays good anatase crystallinity, which is beneficial for photocatalytic water splitting. In the photocatalytic water-splitting experiment, PC segments dispersed in methanol aqueous solution can act as dielectric mirrors, which contribute to a greater effective light path length and light scattering. When light shines on the Pt-loaded PC segments or Pt-loaded nanocrystalline sample nc-TiO_2 (nc-Pt-TiO_2), hydrogen is continuously obtained. The optimum stop-band position of i-Pt-TiO_2 for highly efficient water splitting upon xenon lamp illumination is 342 nm and the highest amplified ratio is 2.5 for i-Pt-TiO_2-133. The light scattering through the PC segments also plays a key role in the enhancement of the hydrogen evolution compared with nc-Pt-TiO_2. PC segments show great capabilities of strong light harvesting due to the slow-photon enhancement at the stop-band edge and multiple scattering among the segments. It puts forward a facile, low-cost method to improve the photocatalytic efficiency and will throw new light on improving the current artificial photosynthesis systems.[204,205] Here, one study of structuring common TiO_2 materials into hierarchical PC segments with stop-bands and TiO_2 absorption matching is investigated.[206]

8.3.4.2   Photocatalysis Organic Dye Degradation

Focusing on environmental depollution problems, titania photocatalysis has been proven to be a promising method for the purification and treatment of contaminated air and waste water.[207,208] The high surface area of the meso-porous structure and three-dimensionally (3D) interconnected architecture contribute to high photocatalytic efficiency. The application of PC-based ordered macroporous photocatalysts is promising, as photons near photonic stop-band edges can be slowed down (slow photons) and localized in the active material to increase the photocarrier generation efficiency. So, constructing TiO_2 into PC structure can promote light absorbance of TiO_2, which could improve the TiO_2 photocatalytic efficiency by improving light harvesting.[209–213]

Figure 8.22 presents the synthesis diagram of the hierarchically macro-/mesoporous i-Ti-Si PC, which was obtained by combining the colloidal crystal template and the assembly of titanium isopropoxide and tetraethyl ortho-dilicate with amphiphilic triblock copolymer. After calcining to remove the templates, the hierarchically macro/mesoporous i-Ti-Si PC will be obtained.

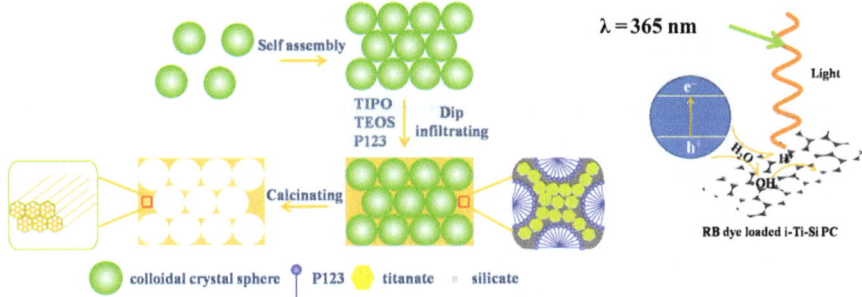

Figure 8.22   (a) Diagram of procedure for preparing the 3D ordered macro-/ mesoporous i-Ti-Si PCs. (b) The schematic photoexcitation process of TiO$_2$ for photodegradation of the RB dye under 365-nm light irradiation. At photon energies near the photonic stop-band from the red side, light can be described as a standing wave with peaks localized in the high-dielectric part of PC, *i.e.* TiO$_2$ part.
Reproduced with permission. Copyright 2009 The American Chemical Society.

It has been demonstrated that the light with photon energy at both sides of the photonic stop-band can be described as a standing wave.[214] The peaks of light waves are localized in different parts of the matrix, depending on the photon energy. For photocatalytic applications, PC serves to promote light absorption of TiO$_2$ through increasing interaction time between photons and the TiO$_2$.[211,213] When the light wavelength locates on the red side of PC, the slow photon effect is able to enhance the absorbance of TiO$_2$ in UV region, thereby producing more electron–hole pairs and leading to enhanced photodegradation efficiency (Figure 8.22 right set).

Hierarchically macro-/mesoporous Ti-Si oxides photonic crystal (i-Ti-Si PC) with highly efficient photocatalytic activity has been synthesized by combining colloidal crystal template and amphiphilic triblock copolymer. It was found that the thermal stability of mesoporous structures in the composite matrix was improved due to the introduction of silica acting as a glue and linking anatase nanoparticles together. The influences of photonic and structural effect of i-Ti-Si PCs on photocatalytic activity were investigated. The photodegradation efficiency of i-Ti-Si PCs was 2.1 times higher than that of TiO$_2$ photonic crystals (i-TiO$_2$ PCs) in the photodegradation of Rhodamine B (RB) dye as a result of higher surface area. When the energy of the slow photons was optimized to the absorption region of TiO$_2$, a maximum enhanced factor of 15.6 was achieved in comparison to nanocrystalline TiO$_2$ films (nc-TiO$_2$), which originated from the synergetic effect of slow-photon enhancement and high surface area.[215]

8.3.4.3   *Photonic Crystal Concentrator: Thermostable Dye-Sensitized Solar Cells*

The solar cell has been proved as one of the most promising clean energy source to solve current energy problems. Dye-sensitized solar cells (DSSCs) based on titania were extensively studied due to their great potential for cost-effective photovoltaic devices.[216] Photonic crystal can selectively modify the propagation of light with a specific wavelength based on its intrinsic periodic structure.[217,218] On coupling PC structures into DSSCs, the light-harvesting enhancement of PCs can increase the light-to-electricity conversion efficiency. However, the large-area DSSCs and PC anode fabrication is a big challenge and has current difficulties. A concentrator is one current and effective method to carry out practical solar energy research. Unfortunately, the nonselective sunlight focus will be due to the photobleaching by UV and thermodegradation of IR. In the following, after light harvesting of the slow-photon effect, we will present one ingenious application of photonic crystal, the PC concentrator.

Figure 8.23(a) shows a schematic illustration of the photovoltaic system containing a PC concentrator. The PCs concentrator was used as a light-harvesting module; it could concentrate light for the DSSC to generate electricity. The light corresponding to the absorbance of dye can be converged when the stop-band of the PCs concentrator just overlaps the absorbance of the dye, and the IR and UV light could be filtered out simultaneously. The PCs concentrators were fabricated by self-assembly of latex spheres on concave watch glasses based on the vertical deposition method. Here, a suitable angle between the normal direction of the concave surface and the gravity direction was necessary to reduce the gravity effect on the assembly and to ensure the well-ordered assembly of the latex spheres upon the concave glass surface. Figure 8.23(b) shows photographs of the as-prepared PCs concentrators assembled from latex spheres with different stop-bands and blue, green and red colors, respectively.

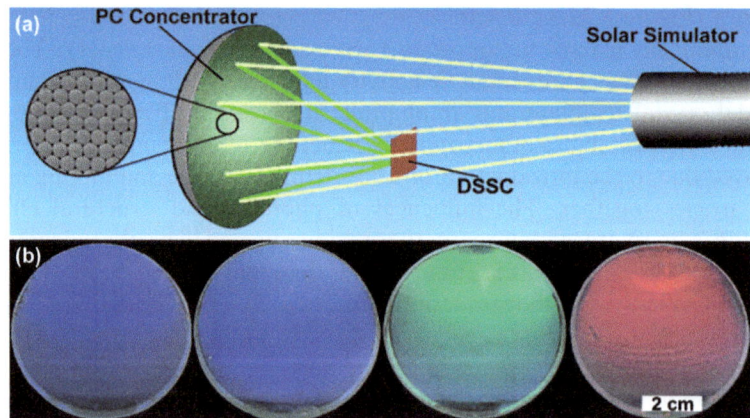

Figure 8.23   (a) Schematic illustration of the photovoltaic system with a PCs concentrator. The insert is a typical SEM image of the PCs concentrator. (b) Typical photographs of the PCs concentrators with different colors, the stop-bands of the PCs concentrators from left to right are 448, 475, 530 and 647 nm, respectively. The scale bar is 2 cm.
Reproduced with permission. Copyright 2008 The Royal Society of Chemistry.

The PCs concentrator DSSCs maximum output power display a 2–5 times increase. The better matching or overlap of PCs stop-band and dye absorbance will offer the higher output power and efficacy. A rapid decrease of the maximum output power for DSSC with increasing irradiation time could be clearly observed when using an Al film concentrator or a regular concentrator without selectivity, while the value with the PCs concentrator produced more stable performance. The light wavelength converged by the PCs concentrator accorded with the stop-band, indicating the corroborative filtration of UV and IR light. The PCs concentrator benefits greatly to the themostable performance of DSSC.[219]

References

1. E. Yablonovitch, *Phys. Rev. Lett.*, 1987, **58**, 2059.
2. S. John, *Phys. Rev. Lett.*, 1987, **58**, 2486.
3. E. Yablonovitch, *J. Phys. Condens. Matter.*, 1993, **5**, 2443.
4. X. H. Wang, R. Z. Wang, B. Y. Gu and G. Z. Yang, *Phy. Rev. Lett.*, 2002, **88**, 0939021.
5. Y. Liu, F. Qin, Z. Y. Wei, Q. B. Meng, D. Z. Zhang and Z. Y. Li, *Appl. Phys. Lett.*, 2009, **95**, 1311161.
6. S. John and J. Wang, *Phys. Rev. Lett.*, 1990, **64**, 2418.
7. D. R. Solli, C. F. McCormick and J. M. Hickmann, *J. Lightwave Technol.*, 2006, **24**, 3864.
8. S. Kinoshita, S. Yoshioka and J. Miyazaki, *Rep. Prog. Phys.*, 2008, **71**, 076401.
9. P. Vukusic and J. R. Sambles, *Nature*, 2003, **424**, 852.
10. Y. Z. Zhang, Z. R. Li, Y. M. Zheng, J. X. Wang, Y. L. Song and L. Jiang, *Acta Polym. Sin.*, 2010, **11**, 1341.
11. H. M. Whitney, M. Kolle, P. Andrew, L. Chittka, U. Steiner and B. J. Glover, *Science*, 2009, **323**, 130.
12. Y. M. Zheng, X. F. Gao and L. Jiang, *Soft Matter*, 2007, **3**, 178.
13. J. Zi, X. D. Yu, Y. Z. Li, X. H. Hu, C. Xu, X. J. Wang, X. H. Liu and R. T. Fu, *Proc. Natl. Acad. Sci.*, 2003, **100**, 12576.
14. X. F. Gao, X. Yan, X. Yao, L. Xu, K. Zhang, J. H. Zhang, B. Yang and L. Jiang, *Adv. Mater.*, 2007, **19**, 2213.
15. F. Liu, B. Q. Dong, X. H. Liu, Y. M. Zheng and J. Zi, *Opt. Exp.*, 2009, **17**, 16183.
16. Z. Z. Gu, H. Uetsuka, K. Takahashi, R. Nakajima, H. Onishi, A. Fujishima and O. Sato, *Angew. Chem. Int. Ed.*, 2003, **8**, 894.
17. J. X. Wang, Y. Z. Zhang, S. T. Wang, Y. L. Song and L. Jiang, *Acc. Chem. Res.*, 2011, **44**, 405.
18. R. N. Wenzel, *Ind. Eng. Chem.*, 1936, **28**, 988.
19. A. Cassie and S. Baxter, *Trans. Faraday Soc.*, 1944, **40**, 546.
20. L. Feng, S. H. Li, Y. S. Li, H. J. Li, L. J. Zhang, J. Zhai, Y. L. Song, B. Q. Liu, L. Jiang and D. B. Zhu, *Adv. Mater.*, 2002, **14**, 1857.
21. O. Sato, S. Kubo and Z. Z. Gu, *Acc. Chem. Res.*, 2009, **42**, 1.

22. Z. Z. Gu, A. Fujishima and O. Sato, *Appl. Phys. Lett.*, 2004, **85**, 5067.
23. J. Liu, M. Z. Li, J. X. Wang, Y. L. Song, L. Jiang, T. Murakami and A. Fujishima, *Environ. Sci. Technol.*, 2009, **43**, 9425.
24. D. Brennan, J. Justice, B. Corbett, T. McCarthy and P. Galvin, *Anal. Bioanal. Chem.*, 2009, **395**, 621.
25. Y. Z. Zhang, J. X. Wang, Y. Huang, Y. L. Song and L. Jiang, *J. Mater. Chem.*, 2011, **21**, 14113.
26. H. L. Ge, G. J. Wang, Y. N. He, X. G. Wang, Y. L. Song, L. Jiang and D. B. Zhua, *ChemPhysChem*, 2006, **7**, 575.
27. J. R. Dorvee, A. M. Derfus, S. N. Bhatia and M. J. Sailor, *Nature Mater.*, 2004, **3**, 896.
28. A. M. Brozell, M. A. Muha, A. Abed-Amoli, D. Bricarello and A. N. Parikh, *Nano Lett.*, 2007, **7**, 3822.
29. T. J. Yao, C. X. Wang, Q. Lin, X. Li, X. L. Chen, J. Wu, J. H. Zhang, K. Yu and B. Yang, *Nanotechnology*, 2009, **20**, 06530.
30. J. X. Wang, Y. Q. Wen, H. L. Ge, Z. W. Sun, Y. M. Zheng, Y. L. Song and L. Jiang, *Macromol. Chem. Phys.*, 2006, **207**, 596.
31. J. X. Wang, Y. Q. Wen, X. J. Feng, Y. L. Song and L. Jiang, *Macromol. Rapid Commun.*, 2006, **27**, 188.
32. J. X. Wang, Y. Q. Wen, J. P Hu, Y. L. Song and L. Jiang, *Adv. Funct. Mater.*, 2007, **17**, 219.
33. J. X. Wang, J. P. Hu, Y. Q. Wen, Y. L. Song and L. Jiang, *Chem. Mater.*, 2006, **18**, 4984.
34. L. Xu, J. X. Wang, Y. L. Song and L. Jiang, *Chem. Mater.*, 2008, **20**, 3554.
35. H. L. Li, L. X. Chang, J. X. Wang, L. M. Yang and Y. L. Song, *J. Mater. Chem.*, 2008, **18**, 5098.
36. H. L. Li, J. X. Wang, L. M. Yang and Y. L. Song, *Adv. Funct. Mater.*, 2008, **18**, 3258.
37. H. Y. Erbil, A. L. Demirel, Y. Avci and O. Mert, *Science*, 2003, **299**, 1377.
38. N. Hadjichristidis, H. Iatrou, M. Pitsikalis, S. Pispas and A. Avgeropoulos, *Prog. Polym. Sci.*, 2005, **30**, 725.
39. A. M. Mathur, B. Drescher, A. B. Scranton and J. Klier, *Nature*, 1998, **392**, 367.
40. C. M. Roland and R. Casalini, *Macromolecules*, 2005, **38**, 8729.
41. Y. K. Lai, X. F. Gao, H. F. Zhuang, J. Y. Huang, C. J. Lin and L. Jiang, *Adv. Mater.*, 2009, **21**, 3799.
42. C. Li, R. W. Guo, X. Jiang, S. X. Hu, L. Li, X. Y. Cao, H. Yang, Y. L. Song, Y. M. Ma and L. Jiang, *Adv. Mater.*, 2009, **21**, 4254.
43. M. J. Liu, Y. M. Zheng, J. Zhai and L. Jiang, *Acc. Chem. Res.*, 2010, **43**, 368.
44. Y. Huang, M. J. Liu, J. X. Wang, J. M. Zhou, L. B. Wang, Y. L. Song and L. Jiang, *Adv. Funct. Mater.*, 2011, **21**, 4436.
45. Y. Huang, J. Wang, J. M. Zhou, L. Xu, Z. R. Li, Y. Z. Zhang, J. J. Wang, Y. L. Song and L. Jiang, *Macromolecules*, 2011, **44**, 2404.

46. L. Xu, H. Li, X. Jiang, J. X. Wang, L. Li, Y. L. Song and L. Jiang, *Macromol. Rapid Commun.*, 2010, **31**, 1422.
47. E. T. Tian, J. X. Wang, Y. M. Zheng, Y. L. Song, L. Jiang and D. B. Zhu, *J. Mater. Chem.*, 2008, **18**, 1116.
48. S. Shinohara, T. Seki, T. Sakai, R. Yoshida and Y. Takeoka, *Angew. Chem. Int. Ed.*, 2008, **47**, 9039.
49. E. T. Tian, Y. Ma, L. Y. Cui, J. X. Wang, Y. L. Song and L. Jiang, *Macromol. Rapid Commun.*, 2009, **30**, 1719.
50. E. T. Tian, L. Y. Cui, J. X. Wang, Y. L. Song and L. Jiang, *Macromol. Rapid Commun.*, 2009, **30**, 509.
51. S. Sacanna, W. T. M. Irvine, P. M. Chaikin and D. J. Pine, *Nature*, 2010, **464**, 575.
52. S. M. Yang, S. H. Kim, J. M. Lim and G. R. Yi, *J. Mater. Chem.*, 2008, **18**, 2177.
53. S. H. Im, U. Jeong and Y. N. Xia, *Nature Mater.*, 2005, **4**, 671.
54. J. P. Ge, Y. Hu, T. Zhang and Y. Yin, *J. Am. Chem. Soc.*, 2007, **129**, 8974.
55. Y. F. Li, J. H. Zhang and B. Yang, *Nano Today*, 2010, **5**, 117.
56. J. W. Haus, *J. Mod. Opt.*, 1994, **41**, 195.
57. R. Biswas, M. M. Sigalas and K. M. Ho, *Phys. Rev. B: Condens. Matter*, 1998, **57**, 3701.
58. T. Ding, K. Song, K. Clays and C. H. Tung, *Adv. Mater.*, 2009, **21**, 1936.
59. T. Ding, K. Song, K. Clays and C. H. Tung, *Langmuir*, 2010, **26**, 11544.
60. H. Lee, Y. Song, I. D. Hoseina and C. M. Liddell, *J. Mater. Chem.*, 2009, **19**, 350.
61. S. H. Lee and C. M. Liddell, *Small*, 2009, **5**, 1957.
62. S. Wong, V. Kitaev and G. A. Ozin, *J. Am. Chem. Soc.*, 2003, **125**, 15589.
63. J. F. Bertone, P. Jiang, K. S. Hwang, D. M. Mittleman and V. L. Colvin, *Phys. Rev. Lett.*, 1999, **83**, 300.
64. H. Mi'guez, N. Tetreault, B. Hatton, S. M. Yang, D. Perovic and G. A. Ozin, *Chem. Commun.*, 2002, **38**, 2736.
65. T. Ruhl and G. P. Hellmann, *Macromol. Chem. Phys.*, 2001, **202**, 3502.
66. T. Ruhl, P. Spahn and G. P. Hellmann, *Polymer*, 2003, **44**, 7625.
67. J. Wang, Y. Wen, H. Ge, Z. Sun, Y. Zheng, Y. Song and L. Jiang, *Macromol. Chem. Phys.*, 2006, **207**, 596.
68. B. Lange, N. Metz, M. N. Tahir, F. Fleischhaker, P. Theato, H.-C. Schröder, W. E. G. Müller, W. Tremel and R. Zentel, *Macromol. Rapid Commun.*, 2007, **28**, 1987.
69. J. G. McGrath, R. D. Bock, J. M. Cathcart and L. A. Lyon, *Chem. Mater.*, 2007, **19**, 1584.
70. L. A. Lyon, J. D. Debord, S. B. Debord, C. D. Jones, J. G. McGrath and M. J. Serpe, *J. Phys. Chem. B*, 2004, **108**, 19099.
71. X. Chen, L. Wang, Y. Wen, Y. Zhang, J. Wang, Y. Song, L. Jiang and D. Zhu, *J. Mater. Chem.*, 2008, **18**, 2262.
72. B. You, N. Wen, L. Shi, L. Wu and J. Zi, *J. Mater. Chem.*, 2009, **19**, 3594.
73. E. Tian, L. Cui, J. Wang, Y. Song and L. Jiang, *Macromol. Rapid Commun.*, 2009, **30**, 509.

74. E. Tian, Y. Ma, J. X. Wang, L. Cui, Y. L. Song and L. Jiang, *Macromol. Rapid Commun.*, 2009, **30**, 1719.

75. Y. Q. Zhang, X. Hao, J. M. Zhou, J. X. Wang, Y. L. Song and L. Jiang, *Macromol. Rapid Commun.*, 2010, **31**, 2115.

76. Y. F. Shi, F. Zhang, Y. S. Hu, X. H. Sun, Y. C. Zhang, H. I. Lee, L. Q. Chen and G. D. Stucky, *J. Am. Chem. Soc.*, 2010, **132**, 5552.

77. J. M. Zhou, H. L. Li, L. Ye, J. Liu, J. X. Wang, T. Zhao, L. Jiang and Y. L. Song, *J. Phys. Chem. C*, 2010, **114**, 22303.

78. K. B. Singh and M. S. Tirumkudulu, *Phys. Rev. Lett.*, 2007, **98**, 218302.

79. P. Xu, A. S. Mujumdar and B. Yu, *Drying Technol.*, 2009, **27**, 636.

80. T. Kanai, T. Sawada, A. Toyotama and K. Kitamura, *Adv. Funct. Mater.*, 2005, **15**, 25.

81. T. Kanai and T. Sawada, *Langmuir*, 2009, **25**, 13315.

82. A. A. Chabanov, Y. Jun and D. J. Norris, *Appl. Phys. Lett.*, 2004, **84**, 3573.

83. L. K. Wang and X. S. Zhao, *J. Phys. Chem. C*, 2007, **111**, 8538.

84. B. Griesebock, M. Egen and R. Zentel, *Chem. Mater.*, 2002, **14**, 4023.

85. C. J. Jin, M. A. McLachlan, D. W. McComb, R. M. D. L. Rue and N. P. Johnson, *Nano Lett.*, 2005, **5**, 2646.

86. B. Hatton, L. Mishchenko, S. Davis, K. H. Sandhage, and J. Aizenberg, *Proc. Natl. Acad. Sci. U. S. A.*, 2010, **107**, 10354.

87. J. M. Zhou, J. X. Wang, Y. Huang, G. M. Liu, L. B. Wang, S. R. Chen, X. H. Li, D. J. Wang, Y. L. Song and L. Jiang, *NPG Asia Mater.*, 2012, **4**, e21.

88. Y. Huang, J. Zhou, B. Su, L. Shi, J. Wang, S. Chen, L. Wang, J. Zi, Y. Song and L. Jiang, *J. Am. Chem. Soc.*, 2012, **134**, 17053.

89. L. Cui, Y. Zhang, J. Wang, Y. Ren, Y. Song and L. Jiang, *Macromol. Rapid Commun.*, 2009, **30**, 598.

90. H. Yang and P. Jiang, *Langmuir*, 2010, **26**, 13173.

91. H.-Y. Ko, J. Park, H. Shin and J. Moon, *Chem. Mater.*, 2004, **16**, 4212.

92. D. Wang, M. Park, J. Park and J. Moon, *Appl. Phys. Lett.*, 2005, **86**, 241114.

93. J. Park, J. Moon, H. Shin, D. Wang and M. Park, *J. Colloid Interface Sci.*, 2006, **298**, 713.

94. L. Cui, Y. Li, J. Wang, E. Tian, X. Zhang, Y. Zhang, Y. Song and L. Jiang, *J. Mater. Chem.*, 2009, **19**, 5499.

95. H. Kim, J. P. Ge, J. Kim, S. Choi, H. Lee, W. Park, Y. D. Yin and S. H. Kwon, *Nature Photonics*, 2009, **3**, 534.

96. L. B. Wang, J. X. Wang, Y. Huang, M. J. Liu, M. X. Kuang, Y. F. Li, L. Jiang and Y. L. Song, *J. Mater. Chem.*, 2012, **40**, 21405.

97. S. Y. Lin, E. Chow, V. Hietala, P. R. Villeneuve and J. D. Joannopoulos, *Science*, 1998, **282**, 274.

98. S. Fan, P. R. Villeneuve, J. D. Joannopoulos and E. F. Schubert, *Phys. Rev. Lett.*, 1997, **78**, 3294.

99. E. P. Petrov, V. N. Bogomolov, I. I. Kalosha and S. V. Gaponenko, *Phys. Rev. Lett.*, 1998, **81**, 77.

100. S. V. Gaponenko, V. N. Bogomolov, E. P. Petrov, A. M. Kapitonov, D. A. Yarotsky, I. I. Kalosha, A. A. Eychmueller, A. L. Rogach, J. McGilp, U. Woggon and F. Gindele, *J. Lightwave Technol.*, 1999, **17**, 2128.

101. L. Bechger, P. Lodahl and W. L. Vos, *J. Phys. Chem. B*, 2005, **109**, 9980.

102. M. Megens, J. E. G. J. Wijnhoven, A. Lagendijk and W. L. Vos, *J. Opt. Soc. Am. B*, 1999, **16**, 1403.

103. (a) P. Lodahl, A. F. van Driel, I. S. Nikolaev, A. Irman, X. Overgaag, D. Vanmaekelbergh and W. L. Vos, *Nature*, 2004, **430**, 654; (b) A. F. Koenderink and W. L. Vos, *Phys. Rev. Lett.*, 2003, **91**, 213902.

104. R. C. Schroden, M. Al-Daous and A. Stein, *Chem. Mater.*, 2001, **13**, 2945.

105. F. Fleischhaker and R. Zentel, *Chem. Mater.*, 2005, **17**, 1346.

106. M. Müller, R. Zentel, T. Maka, S. G. Romanov and C. M. Sotomayor Torres, *Chem. Mater.*, 2000, **12**, 2508.

107. M. Boroditsky, T. F. Krauss, R. Coccioli, R. Vrijen, R. Bhat and E. Yablonovitch, *Appl. Phys. Lett.*, 1999, **75**, 1036.

108. Y. R. Do, Y. C. Kim, Y. W. Song, C. Cho, H. Jeon, Y. J. Lee, S. H. Kim and Y. H. Lee, *Adv. Mater.*, 2003, **15**, 1214.

109. V. Lousse and J. P. Vigneron, *Phys. Rev. B*, 2001, **64**, 201104.

110. R. Wang, X. H. Wang, B. Y. Gu and G. Z. Yang, *Phys. Rev. B*, 2003, **67**, 155114.

111. Y. Q. Zhang, J. X. Wang, Z. Y. Ji, W. P. Hu, L. Jiang, Y. L. Song and D. B. Zhu, *J. Mater. Chem.*, 2007, **17**, 90.

112. J. R. Sheats, H. Antoniadis, M. Hueschen, W. Leonard, J. Miller, R. Moon, D. Roitman and A. Stocking, *Science*, 1996, **273**, 884.

113. S. L. Jones, D. Kumar, K. G. Cho, R. Singh and P. H. Holloway, *Displays*, 1999, **19**, 151.

114. M. Y. Gao, B. Richter and S. Kirstein, *Adv. Mater.*, 1997, **9**, 802.

115. F. Li, G. Cheng, Y. Zhao, J. Feng, S. Y. Liu, M. Zhang, Y. G. Ma and J.C. Shen, *Appl. Phys. Lett.*, 2003, **83**, 4716.

116. H. Kanno, R. J. Holmes, Y. Sun, S. Kena-Cohen and S. R. Forrest, *Adv. Mater.*, 2006, **18**, 339.

117. M. Z. Li, Q. Liao, Y. Liu, Z. Y. Li, J. X. Wang, L. Jiang and Y. L. Song, *Appl. Phys. A*, 2010, **98**, 85.

118. H. J. Kim, S. Kim, H. Jeon, J. Ma, S. H. Choi, S. Lee, C. Ko and W. Park, *Sens. Actuators B-Chem.*, 2007, **124**, 147.

119. J. Hu, X. W. Zhao, Y. J. Zhao, J. Li, W.Y. Xu, Z. Y. Wen, M. Xu and Z. Z. Gu, *J. Mater. Chem.*, 2009, **19**, 5730.

120. P. Lodahl, A. F. van Driel, I. S. Nikolaev, A. Irman, K. Overgaag, D. L. Vanmaekelbergh and W. L. Vos, *Nature*, 2004, **430**, 654.

121. A. C. Arsenault, T. J. Clark, G. Von Freymann, L. Cademartiri, R. Sapienza, J. Bertolotti, E. Vekris, S. Wong, V. Kitaev, I. Manners, R. Z. Wang, S. John, D. Wiersma and G. A. Ozin, *Nature Mater.*, 2006, **5**, 179.

122. M. N. Shkunov, Z. V. Vardeny, M. C. DeLong, R. C. Polson, A. A. Zakhidov and R. H. Baughman, *Adv. Funct. Mater.*, 2002, **12**, 21.

123. J. I. L. Chen, G. von Freymann, S. Y. Choi, V. Kitaev and G. A. Ozin, *J. Mater. Chem.*, 2008, **18**, 369.

124. L. Bechger, P. Lodahl and W. L. Vos, *J. Phys. Chem. B*, 2005, **109**, 9980.

125. F. Jin, Y. Song, X. Z. Dong, W. Q. Chen and X. M. Duan, *Appl. Phys. Lett.*, 2007, **91**, 031109.

126. G. R. Maskaly, M. A. Petruska, J. Nanda, I. V. Bezel, R. D. Schaller, H. Htoon, J. M. Pietryga and V. I. Klimov, *Adv. Mater.*, 2006, **18**, 343.

127. S. H. Fan, P. R. Villeneuve, J. D. Joannopoulos and E. F. Schubert, *Phys. Rev. Lett.*, 1997, **78**, 3294.

128. M. Zelsmann, E. Picard, T. Charvolin, E. Hadji, M. Heitzmann, B. Dal'zotto, M. E. Nier, C. Seassal, P. Rojo-Romeo and X. Letartre, *Appl. Phys. Lett.*, 2003, **83**, 2542.

129. H. Li, J. X. Wang, F. Liu, Y. L. Song and R. M. Wang, *J. Colloid Interface Sci.*, 2011, **356**, 63.

130. E. M. Talavera, R. Bermejo, L. Crovetto, A. Orte and J. M. Alvarez-Pez, *Appl. Spectrosc.*, 2003, **57**, 208.

131. W. R. Algar, M. Massey and U. J. Krull, *J. Fluoresc.*, 2006, **16**, 555.

132. J. H. Wosnick, J. H. Liao and T. M. Swager, *Macromolecules*, 2005, **38**, 9287.

133. S. J. Zhang, F. T. Lu, L. N. Gao, L. P. Ding and Y. Fang, *Langmuir*, 2007, **23**, 1584.

134. C. M. Deng, P. Gong, Q. G. He, J. G. Cheng, C. He, L. Q. Shi, D. F. Zhu and T. Lin, *Chem. Phys. Lett.*, 2009, **483**, 219.

135. D. M. Gao, Z. Y. Wang, B. H. Liu, L. Ni, M. H. Wu and Z. P. Zhang, *Anal. Chem.*, 2008, **80**, 8545.

136. T. Naddo, Y. K. Che, W. Zhang, K. Balakrishnan, X. M. Yang, M. Yen, J. C. Zhao, J. S. Moore and L. Zang, *J. Am. Chem. Soc.*, 2007, **129**, 6978.

137. S. Y. Tao, G. T. Li and J. X. Yin, *J. Mater. Chem.*, 2007, **17**, 2730.

138. D. F. Zhu, Q. G. He, H. M. Cao, J. G. Cheng, S. L. Feng, Y. S. Xu and T. Lin, *Appl. Phys. Lett.*, 2008, **93**, 261909.

139. S. Y. Tao, J. X. Yin and G. T. Li, *J. Mater. Chem.*, 2008, **18**, 4872.

140. Q. L. Fang, J. L. Geng, B. H. Liu, D. M. Gao, F. Li, Z. Y. Wang, G. J. Guan and Z. P. Zhang, *Chem.-Eur. J.*, 2009, **15**, 11507.

141. C. P. Chang, C. Y. Chao, J. H. Huang, A. K. Li, C. S. Hsu, M. S. Lin, B. R. Hsieh and A. C. Su, *Synth. Met.*, 2004, **144**, 297.

142. Y. Q. Zhang, J. X. Wang, X. Chen, L. Jiang, Y. L. Song and D. B. Zhu, *Appl. Phys. A: Mater. Sci. Process.*, 2007, **87**, 271.

143. H. Li, J. X. Wang, Z. L. Pan, L. Y. Cui, L. Xu, R. M. Wang, Y. L. Song and L. Jiang, *J. Mater. Chem.*, 2011, **21**, 1730.

144. A. P. Shuber, L. A. Michalowsky, G. S. Nass, J. Skoletsky, L. M. Hire, S. K. Kotsopoulos, M. F. Phipps, D. M. Barberio and K.W. Klinger, *Hum. Mol. Genet.*, 1997, **6**, 337.

145. R. K. Saiki, S. Scharf, F. Faloona, K. Mullis, G. T. Horn, H. A. Erlich and N. Arnheim, *Science*, 1985, **230**, 1350.

146. F. F. Chehab and Y. W. Kan, *Proc. Natl. Acad. Sci.*, 1989, **86**, 9178.

147. T. R. Golub, D. K. Slonim, P. Tamayo, C. Huard, M. Gaasenbeek, J. P. Mesirov, H. Coller, M. L. Loh, J. R. Downing, M. A. Caligiuri, C. D. Bloomfield and E. S. Lander, *Science*, 1999, **286**, 531.

148. O. Seitz, *Angew. Chem. Int. Ed.*, 2000, **39**, 3249.

149. D. M. Hammond, A. Manetto, J. Gierlich, V. A. Azov, P. M. E. Gramlich, G. A. Burley, M. Maul and T. Carell, *Angew. Chem. Int. Ed.*, 2007, **46**, 4184.

150. W. C. W. Chan and S. M. Nie, *Science*, 1998, **281**, 2016.

151. F. Patolsky, E. Katz and I. Willner, *Angew. Chem. Int. Ed.*, 2002, **41**, 3398.

152. A. Saghatelian, K. M. Guckian, D. A. Thayer and M. R. Ghadiri, *J. Am. Chem. Soc.*, 2003, **125**, 344.

153. F. Patolsky, A. Lichtenstein and I. Willner, *J. Am. Chem. Soc.*, 2001, **123**, 5194.

154. T. A. Taton, C. A. Mirkin and R. L. Letsinger, *Science*, 2000, **289**, 1757.

155. Y. W. C. Cao, R. C. Jin and C. A. Mirkin, *Science*, 2002, **297**, 1536.

156. S. Brakmann, *Angew. Chem. Int. Ed.*, 2004, **43**, 5730.

157. S. John and T. Quang, *Phys. Rev. A*, 1994, **50**, 1764.

158. M. S. Thijssen, R. Sprik, J. Wijnhoven, M. Megens, T. Narayanan, A. Lagendijk and W. L. Vos, *Phys. Rev. Lett.*, 1999, **83**, 2730.

159. E. Yablonovitch, *Phys. Rev. Lett.*, 1987, **58**, 2059.

160. S. John, *Phys. Rev. Lett.*, 1987, **58**, 2486.

161. S. C. Benson, P. Singh and A. Glazer, *Nucleic Acids Res.*, 1993, **21**, 5727.

162. B. Q. Ferguson and D. C. H. Yang, *Biochemistry*, 1986, **25**, 5298.

163. S. Wang, B. S. Gaylord and G. C. Bazan, *J. Am. Chem. Soc.*, 2004, **126**, 5446.

164. J. B. LePecq and C. Paoletti, *J. Mol. Biol.*, 1967, **27**, 87.

165. A. R. Morgan and D. E. Pulleyblank, *Biochem. Biophys. Res. Commun.*, 1974, **61**, 346.

166. T. Förster, *Ann. Phys.*, 1948, **2**, 55.

167. M. Z. Li, F. He, Q. Liao, J. Liu, L. Xu, L. Jiang, Y. L. Song, S. Wang and D. B. Zhu, *Angew. Chem., Int. Ed.*, 2008, **47**, 7258.

168. K.-B. Lee, E.-Y. Kim, C. A. Mirkin and S. M. Wolinsky, *Nano Lett.*, 2004, **4**, 1869.

169. Y. Xie, T. Yin, W. Wiegraebe, X. C. He, D. Miller, D. Stark, K. Perko, R. Alexander, J. Schwartz, J. C. Grindley, J. Park, J. S. Haug, J. P. Wunderlich, H. Li, S. Zhang, T. Johnson, R. A. Feldman and L. Li, *Nature*, 2009, **457**, 97.

170. K. E. Sapsford, L. Berti and I. L. Medintz, *Angew. Chem., Int. Ed.*, 2006, **45**, 4562.

171. E. F. Ullman, M. Schwarzberg and K. E. Rubenstein, *J. Biol. Chem.*, 1976, **251**, 4172.

172. G.-X. Liang, H.-C. Pan, Y. Li, L.-P. Jiang, J.-R. Zhang and J.-J. Zhu, *Biosens. Bioelectron.*, 2009, **24**, 3693.

173. I. L. Medintz, A. R. Clapp, H. Mattoussi, E. R. Goldman, B. Fisher and J. M. Mauro, *Nature Mater.*, 2003, **2**, 630.

174. P. S. Heeger and A. J. Heeger, *Proc. Natl. Acad. Sci.*, 1999, **96**, 12219.

175. Y. Wang and B. Liu, *Biosens. Bioelectron.*, 2009, **24**, 3293.

176. P. Rigler and W. Meier, *J. Am. Chem. Soc.*, 2006, **128**, 367.

177. M. Wilchek and E. A. Bayer, *Anal. Biochem.*, 1988, **171**, 1.

178. H. J. Kim, S. Kim, H. Jeon, J. Ma, S. H. Choi, S. Lee, C. Ko and W. Park, *Sens. Actuators B*, 2007, **124**, 147.

179. Z. Yang, Z. Xie, H. Liu, F. Yan and H. Ju, *Adv. Funct. Mater.*, 2008, **18**, 3991.

180. S. Zheng, H. Zhang, E. Ross, T. V. Le and M. J. Wirth, *Anal. Chem.*, 2007, **79**, 3867.

181. N. Ganesh, W. Zhang, P. C. Mathias, E. Chow, J. A. N. T. Soares, V. Malyarchuk, A. D. Smith and B. T. Cunningham, *Nature Nanotechnol.*, 2007, **2**, 515.

182. P. C. Mathias, N. Ganesh and B. T. Cunningham, *Anal. Chem.*, 2008, **80**, 9013.

183. W. Z. Shen, M. Z. Li, L. Xu, S. T. Wang, L. Jiang, Y. L. Song and D. B. Zhu, *Biosens. Bioelectron.*, 2011, **26**, 2165.

184. A. J. Myles, B. Gorodetsky and N. R. Branda, *Adv. Mater.*, 2004, **16**, 922.

185. M. Tomasulo, S. Giordani and F. M. Raymo, *Adv. Funct. Mater.*, 2005, **15**, 787.

186. M. Irie, *Chem. Rev.*, 2000, **100**, 1685.

187. H. Tian and Y. L. Feng, *J. Mater. Chem.*, 2008, **18**, 1617.

188. J. K. Lee, H. J. Kim, T. H. Kim, C. H. Lee, W. H. Park, J. S. Kim and T. S. Lee, *Macromolecules*, 2005, **38**, 9427.

189. K. K. Haldar, T. Sen and A. Patra, *J. Phys. Chem. C*, 2008, **112**, 11650.

190. H Li, J. X. Wang, L. Xu, W. Xu, R. M. Wang, Y. L. Song and D. B. Zhu, *Adv. Mater.*, 2010, **22**, 1237.

191. N. S. Lewis and D. G. Nocera, *Proc. Natl. Acad. Sci.*, 2006, **103**, 15729.

192. K. Maeda, K. Teramura, D. L. Lu, T. Takata, N. Saito, Y. Inoue and K. Domen, *Nature*, 2006, **440**, 295.

193. G. K. Mor, K. Shankar, M. Paulose, O. K. Varghese and C. A. Grimes, *Nano Lett.*, 2005, **5**, 191.

194. J. H. Park, S. Kim and A. J. Bard, *Nano Lett.*, 2006, **6**, 24.

195. Y. X. Yin, Z. G. Jin and F. Hou, *Nanotechnology*, 2007, **18**, 495608.

196. J. F. Zhu and M. Zach, *Curr. Opin. Colloid Interface Sci.*, 2009, **14**, 260.

197. H. B. Yi, T. Y. Peng, D. N. Ke, D. Ke, L. Zan and C. H. Yan, *Int. J. Hydrogen Energy*, 2008, **33**, 672.

198. N. Lakshminarasimhan, E. Bae and W. Choi, *J. Phys. Chem. C*, 2007, **111**, 15244.

199. Z. Y. Liu, B. Pesic, K. S. Raja, R. R. Rangaraju and M. Misra, *Int. J. Hydrogen Energy*, 2009, **34**, 3250.

200. M. Matsuoka, M. Kitano, M. Takeuchi, M. Anpo and J. M. Thomas, *Top. Catal.*, 2005, **35**, 305.

201. R. Asahi, T. Morikawa, T. Ohwaki and Y. Taga, *Science*, 2001, **293**, 269.

202. A. Kudo and Y. Miseki, *Chem. Soc. Rev.*, 2009, **38**, 253.

203. Z. C. Shan, J. J. Wu, F. F. Xu, F. Q. Huang and H. M. Ding, *J. Phys. Chem. C*, 2008, **112**, 15423.
204. X. J. Feng, T. J. LaTempa, J. I. Basham, G. K. Mor, O. K. Varghese and C. A. Grimes, *Nano Lett.*, 2010, **10**, 948.
205. A. J. Bard and M. A. Fox, *Acc. Chem. Res.*, 1995, **28**, 141.
206. J. Liu, G. Liu, M. Li, W. Shen, Z. Liu, J. Wang, J. Zhao, L. Jiang and Y. L. Song, *Energ. Environ. Sci.*, 2010, **3**, 1503.
207. A. Fujishima and K. Honda, *Nature*, 1972, **238**, 37.
208. A. Fujishima, X. Zhang and D. A. Tryk, *Surf. Sci. Rep.*, 2008, **63**, 515.
209. J. E. G. J. Wijnhoven and W. L. Vos, *Science*, 1998, **281**, 802.
210. M. Ren, R. Ravikrishna and K. T. Valsaraj, *Environ. Sci. Technol.*, 2006, **40**, 7029.
211. J. I. L. Chen, G. Freymann, S. Y. Choi, V. Kitaev and G. A. Ozin, *Adv. Mater.*, 2006, **18**, 1915.
212. Y. Li, T. Kunitake and S. Fujikawa, *J. Phys. Chem. B*, 2006, **110**, 13000.
213. J. I. L. Chen, E. Loso, N. Ebrahim and G. A. Ozin, *J. Am. Chem. Soc.*, 2008, **130**, 5420.
214. S. Nishimura, N. Abrams, B. A. Lewis, L. I. Halaoui, T. E. Mallouk, K. D. Benkstein, J. van de Lagemaat and A. J. Frank, *J. Am. Chem. Soc.*, 2003, **125**, 6306.
215. J. Liu, M. Z. Li, J. X. Wang, Y. L. Song, L. Jiang, T. Murakami and A. Fujishima, *Environ. Sci. Technol.*, 2009, **43**, 9425.
216. B. O'Regan and M. Grätzel, *Nature*, 1991, **353**, 737.
217. S. John, *Phys. Rev. Lett.*, 1987, **58**, 2486.
218. E. Yablonovitch, *Phys. Rev. Lett.*, 1987, **58**, 2059.
219. Y. Z. Zhang, J. X. Wang, Y. Zhao, J. Zhai, L. Jiang, Y. L. Song and D. B. Zhu, *J. Mater. Chem.*, 2008, **18**, 2650.

CHAPTER 9

Magnetic Assembly and Tuning of Colloidal Responsive Photonic Nanostructures

LE HE, MINGSHENG WANG AND YADONG YIN*

Department of Chemistry, University of California, Riverside,
California 92521, USA
*Email: yadongy@ucr.edu

9.1 Introduction

Self-assembly of colloidal particles represents a powerful bottom-up method for the fabrication of functional materials, particularly colloidal photonic crystals in which the dielectric contrast between the building blocks and the surrounding medium in the periodic arrays creates a photonic bandgap that inhibits the propagation of light within certain wavelength ranges.[1,2] Structural-colored photonic materials operating in the visible regime are of special interest as important chromatic materials with wide applications in color painting and printing, information storage, displays, and sensors.[3–6] Unlike typical top-down approaches, colloidal assembly strategies for photonic crystal fabrication are technologically favorable due to mild processing conditions, low cost, and potential for scale-up.[2] Although a variety of colloidal strategies have been developed, the precision of the structural and orientational control, the efficiency, and the scalability of the colloidal assembly routes still need to be greatly improved before they can be used for a wide range of practical applications.

RSC Smart Materials No. 5
Responsive Photonic Nanostructures: Smart Nanoscale Optical Materials
Edited by Yadong Yin
© The Royal Society of Chemistry 2013
Published by the Royal Society of Chemistry, www.rsc.org

In many cases, it is also highly desirable to have the capability to tune the optical properties of the photonic materials by chemical or physical means.[6,7] Colloidal assembly approaches are particularly suitable for this purpose because stimuli-responsive components can be easily incorporated into colloidal crystals by modifying the building blocks or their surroundings to realize tuning of their diffraction colors in response to the application of the external stimuli.[8] Although various responsive mechanisms have been developed, such as mechanical stretching,[9] solvent swelling,[5] and temperature-dependent phase change,[10] many challenges still exist, including limited tunability of the photonic properties, a slow response to the external stimuli, incomplete reversibility, and difficulty of integration into existing photonic devices. To broaden the tuning range of diffraction color, the external stimulus must be able to induce large changes in either the refractive index of the components, or the symmetries, lattice parameters or orientations of the ordered arrays. New mechanisms need to be established to also significantly enhance the response rate of the active components to the external stimuli in order to offer dynamic optical modulation that can meet the demands of practical applications.

To this end, a magnetic field has been regarded as an effective stimulus to guide the assembly of magnetic particles into periodic colloidal arrays and tune the diffraction of photonic structures.[11] Strong magnetic interactions can be initiated instantly by the application of an external magnetic field, providing enough driving force for rapid assembly of colloidal particles even within one second.[12–14] Magnetic forces, acting at a large distance, drive the formation of ordered colloidal arrays and provide convenient control over the photonic properties by changing the interparticle distance. Magnetic fields can also induce phase transitions in preassembled 3D colloidal crystals, in which the static or dynamic structural changes in these systems are always accompanied by the switching of photonic properties.[15] Magnetic interactions, directional in nature, not only guide the formation of anisotropic photonic structures,[16,17] but also enable additional control over the orientation of the magnetic assemblies so that one can conveniently tune their optical properties through rotational manipulation.[18] The complexity achievable in the spatial distribution of magnetic fields also makes it possible to define patterns of photonic structures by controlling the local assembly behavior of dipolar particles.

In this chapter, we present recent efforts towards magnetic assembly of colloidal particles into different photonic nanostructures and magnetically manipulating the optical properties of colloidal photonic structures. We first briefly introduce the magnetic interactions exerted on colloidal particles and then discuss magnetic assembly of colloidal particles into different ordered structures. By using several examples, we then show the exploitation of magnetic fields as convenient stimuli to tune the photonic properties of the assemblies. By taking advantage of the unique features of our magnetically responsive photonic system, we also highlight several unique practical and fundamental applications in structural color printing, anticounterfeiting devices, humidity sensors, rewritable paper and colloidal force measurement.

9.2   Magnetic Interactions

There are two types of magnetic interactions experienced by dipolar colloidal particles in external magnetic fields, which originate from their permanent or induced dipole moments.[19,20] The interparticle dipole–dipole force describes the interaction of a dipole with the magnetic field induced by another dipole, while the packing force results from the gradient of the external magnetic field. Among the many types of magnetic materials, superparamagnetic colloidal particles are presumably the most suitable building blocks for reversible assembly and manipulation as their strong and widely tunable dipolar interactions can be fully initiated and controlled by the external magnetic fields.[21] Ferromagnetic particles have net magnetic dipole moments in the absence of a magnetic field so they tend to aggregate in colloidal dispersions due to the magnetic dipole–dipole attraction. Conversely, there is no effective magnetic attraction between paramagnetic particles in the absence of magnetic fields as their dipole moments are only initiated in the external magnetic fields. Compared with typical paramagnetic materials, superparamagnetic particles have much higher magnetic susceptibility, rendering them significantly more responsive to an external magnetic field. As an example, we first discuss the magnetic interactions experienced by superparamagnetic particles in magnetic fields.

An external magnetic field induces a magnetic dipole moment in a super-paramagnetic particle $m = \chi H V$, where χ is the volume susceptibility of the particle, H the local magnetic field, and V the volume of the particle. When the external magnetic field is adequately strong, the magnetic moment of the particle will reach a saturated value. For a spherical particle (particle 1) with a magnetic moment of m, its induced magnetic field H_1 felt by another particle (particle 2) can be described as $H_1 = [3(m \cdot r)r - m]/d^3$, where r is the unit vector parallel to the line pointed from the center of particle 1 to that of particle 2 and d is the center–center distance. The dipole–dipole interaction energy of particle 2 with the same magnetic moment m can be thus written as $U_2 = m \cdot H_1 = (3\cos^2\alpha - 1)m^2/d^3$, where α, ranging from $0°$ to $90°$, is the angle between the external magnetic field and the line connecting the center of the two particles. The dipole force exerted on particle 2 induced by particle 1 can be expressed as:

$$F_2 = \nabla(m \cdot H_1) = \frac{3(1 - 3\cos^2\alpha)m^2}{d^4} \cdot r$$

The above equation clearly shows the dependence of the dipole–dipole force on the configuration of the two dipoles. At the critical angle of $54.09°$, the interaction approaches zero. The dipole–dipole interaction is attractive when $0° \le \alpha < 54.09°$ and repulsive in cases where $54.09° < \alpha \le 90°$ (Figure 9.1). When the interaction energy is large enough to overcome thermal fluctuations, the magnetic dipole–dipole force drives the self-assembly of particles into 1D chain-like structures along the dipole moment (Figure 9.1(d)).

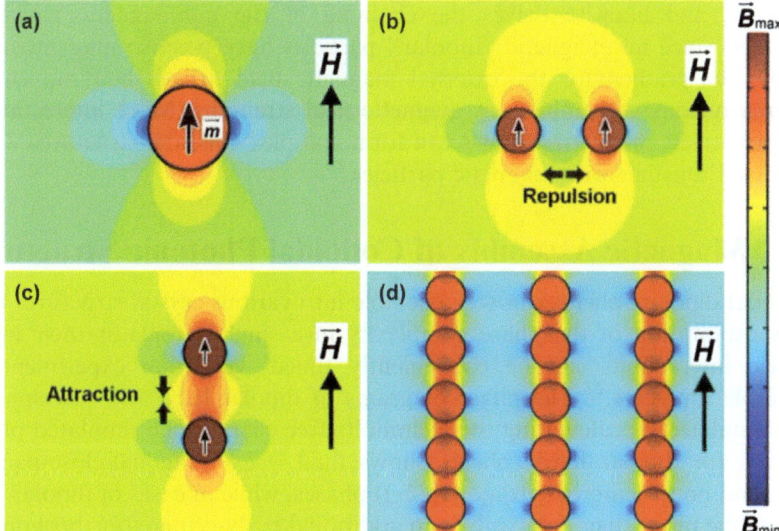

Figure 9.1 (a) Magnetic-field distribution around a superparamagnetic particle with a dipole moment in the same direction as the external magnetic field. The repulsive (b) and attractive (c) dipole–dipole forces in different particle configurations drive the formation of particle chains along the magnetic field (d). The color bar on the right shows the relative strength of the local magnetic field.
Reproduced with permission from Ref. 11 © 2012 American Chemical Society.

Similarly, the packing force can be understood as the interaction of a magnetic dipole with the external magnetic field. The interaction energy of a magnetic dipole m in an external magnetic field H is expressed by $U_m = m \cdot H$ so the packing force can be written as $F_p = \nabla(m \cdot H)$. The packing force drives the movement of magnetic particles towards regions with the maximum magnetic field strength and induces a particle concentration gradient[22] or crystallization.[23] As discussed in the following sections, the particle concentration plays a critical role in their assembly behavior. More importantly, it allows state-of-the-art design of spatial magnetic field gradients to control the local assembly behavior that may eventually lead to high-resolution, *e.g.* pixel-level, manipulation in parallel.[24]

Nonmagnetic particles typically have no effective response to normal external magnetic fields due to their negligible induced magnetic moments. However, when dispersed in a magnetized ferrofluid, a nonmagnetic particle behaves as a magnetic "hole" with an imaginary magnetic moment that can be approximately considered as equal to the total moment of the displaced ferrofluid but in the opposite direction of the magnetic field, $m = -V\chi_{eff}H$, where V is the volume of the particles, χ_{eff} is the effective volume susceptibility of the ferrofluid and H is the local magnetic field strength.[25] The dipole–dipole interaction between magnetic holes has the same directional nature as the real moments, and likewise drives the chaining of the nonmagnetic particles. However, since nonmagnetic "dipolar" particles have positive magnetostatic energy, the gradient of the external magnetic field drives their movement towards regions with minimum magnetic field strengths. More interestingly, such negatively magnetized holes can form complex colloidal structures when assembled together with magnetic particles.[26]

9.3   Magnetic Assembly of Colloidal Photonic Structures

Colloidal dipolar spheres can self-assemble into various crystal structures that allow exploration of the phase complexity in a single sample as their inter-particle interactions can be conveniently tuned within an experimentally accessible timescale.[27] The phase diagrams of dipolar hard and soft spheres were simulated by calculating the Helmholtz free energy. The simulated phase diagram for dipolar hard spheres shows fluid, *fcc*, hexagonal-close-packed (*hcp*), and body-centered-tetragonal (*bct*) phases, while in cases of dipolar soft spheres the phase diagram exhibits an additional body-centered-orthorhombic (*bco*) phase.[27] The key factors that determine the thermodynamic equilibrium state of dipolar colloidal suspensions are the local particle concentration and the interparticle interaction potentials from both repulsive and attractive forces. Magnetic fields can affect both factors through magnetic forces, which not only induce the self-assembly of one-dimensional (1D), two-dimensional (2D) and three-dimensional (3D) structures, but also cause phase transitions between the different structures. For example, the magnetic-field gradient can drive the redistribution of colloidal particles to change local particle concentrations through the magnetic packing force. On the other hand, the strength of the external magnetic field can determine the induced moment in super-paramagnetic particles, which controls the interparticle dipole-dipole potentials. The mechanism of magnetic assembly can be simply understood as magnetic control of local particle concentration and interparticle potential in colloidal suspensions.

The transition between Brownian fluids and 1D particle chains represents the simplest switch between random and ordered states. When the interparticle dipole-dipole potential is strong enough to overcome thermal fluctuations, the alignment of dipolar particles along the direction of their magnetic moments is the direct result of the directional dipole–dipole force. For "hard" spheres, the particles inside the chains are in contact with each other due to the absence of long-range repulsive forces. Therefore, the periodicity is defined by the size of the particles, which also determines the photonic properties of the particle chains. Recently, we successfully characterized the particle chain structures in different ways, providing direct evidence of the 1D structure (Figure 9.2).[16,17,28] In cases of colloidal "soft" particles, where long-range interparticle repulsions exit, such as electrostatic force, the magnetic dipole–dipole attraction can be balanced by the repulsions. Therefore, a force equilibrium can be established, leading to a defined interparticle separation. If the force balance is altered, the interparticle separation will change accordingly, which allows dynamic control

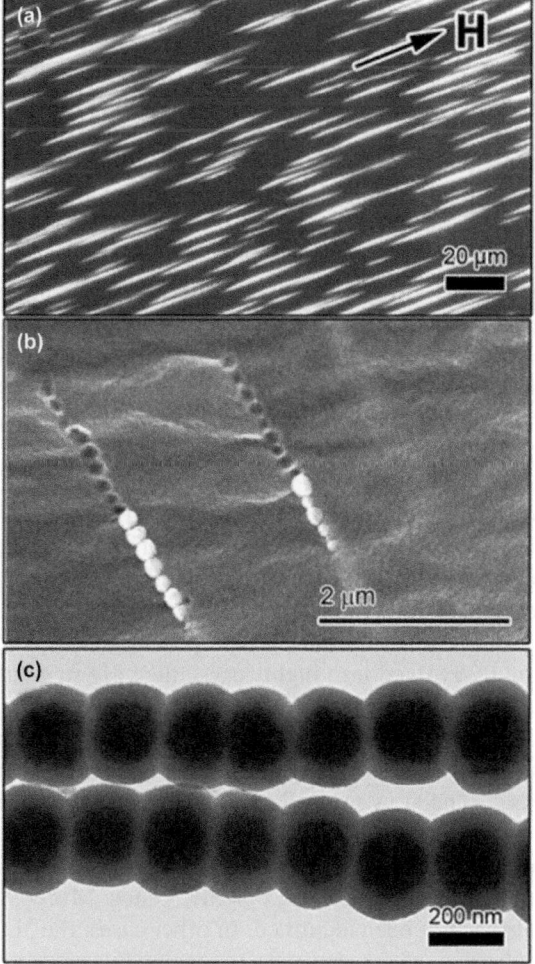

Figure 9.2 (a) Dark-field optical image showing 1D chain-like structures of CNCs along a magnetic field. (b) SEM image of $Fe_3O_4@SiO_2$ particle chains embedded in a PEGDA matrix showing the periodic arrangement of the particles inside each chain. (c) TEM image of nanochains of Fe_3O_4 CNCs fixed by wrapping in a layer of SiO_2 through a sol-gel process. Reproduced with permission from Ref. 11 © 2012 American Chemical Society.

over the lattice constant of the particle chains. Most studies on 1D dipolar photonic chains utilize soft superparamagnetic particles so an external magnetic field can induce the formation of dynamic particle chains with long-range 1D order at a relatively large interparticle separation. The large inter-particle separation also permits a broader choice of materials due to higher tolerance of the monodispersity of the building blocks.

In 1D dipolar particle chains, the interchain forces, such as magnetic dipole-dipole repulsion and other types of repulsive forces, keep the chains away from each other. However, when the concentration of the particle chains, as well as the interparticle dipole–dipole interactions, reaches an extent such that the interchain repulsive forces are sufficiently strong that they become thermodynamically unstable, the chains start to aggregate into zigzag multiple chains or 2D labyrinth structures to minimize the repulsive potentials.[29,30] As discussed above, the dipole–dipole force reaches its minimum at the critical angle of 54.09°, which drives the formation of zigzag structures to minimize the interparticle magnetic dipole-dipole repulsion. However, the dynamic 2D structures are difficult to permanently fix for further characterization. We have found few reports on the detailed study of the photonic properties of these 2D structures.[31] Nevertheless, it is expected that the long-range periodicity of the 2D assemblies would lead to interesting optical anisotropy. For example, magnetically rotating the 2D structures may cause a dramatic change in their photonic properties.

In some studies, 2D structures were considered as the intermediate state between 1D dipolar chains and 3D structures. For example, we previously reported different phase transitions of submicrometer nonmagnetic colloidal particles in a ferrofluid that was exposed to external magnetic fields.[31] As the particle concentration and the interparticle potential were increased, the 2D structures eventually evolved into highly crystalline 3D structures (Figure 9.3). Sacanna and Philipse reported a magnetic-field-induced self-assembly of magnetic poly(methyl methacrylate) latex spheres, which were synthesized by a standard emulsion polymerization method using magnetite-stabilized emulsion droplets as seeds.[23] The slow crystallization process induced by the magnetic field gradients was realized through the local concentrating of particles driven by the magnetic packing forces. However, the assembly efficiency was very low due to weak magnetic response of the polymer beads. Moreover, the structure of the 3D crystals has not been identified. In other cases, the 3D structures were formed in the absence of interparticle magnetic interactions. For example, Asher and coworkers reported the formation of crystalline colloidal arrays with *fcc* structures from highly charged monodisperse superparamagnetic colloidal spheres due to the minimization of interparticle electrostatic repulsive potentials.[23,32,33] The introduction of magnetic interactions could then induce the compression of the preformed arrays along the magnetic field gradient. Besides *fcc* structures, theoretical studies predict a number of other potential 3D structures in the phase diagram of dipolar colloidal suspensions.[27] Surprisingly, the phase diagrams of both dipolar soft and hard spheres predict a large portion of the *hcp* phase, which is unstable in normal bulk suspensions. Unfortunately, advancement in phase transitions of dipolar spheres has been limited by slow progress in the experimental observation of rich varieties of the computed phases. For example, the predicted transition between *fcc* and *hcp* for dipolar spheres has not yet been identified experimentally. Recently, we successfully observed the magnetically induced phase transition from polycrystalline *fcc* structures to single-crystalline-like *hcp*

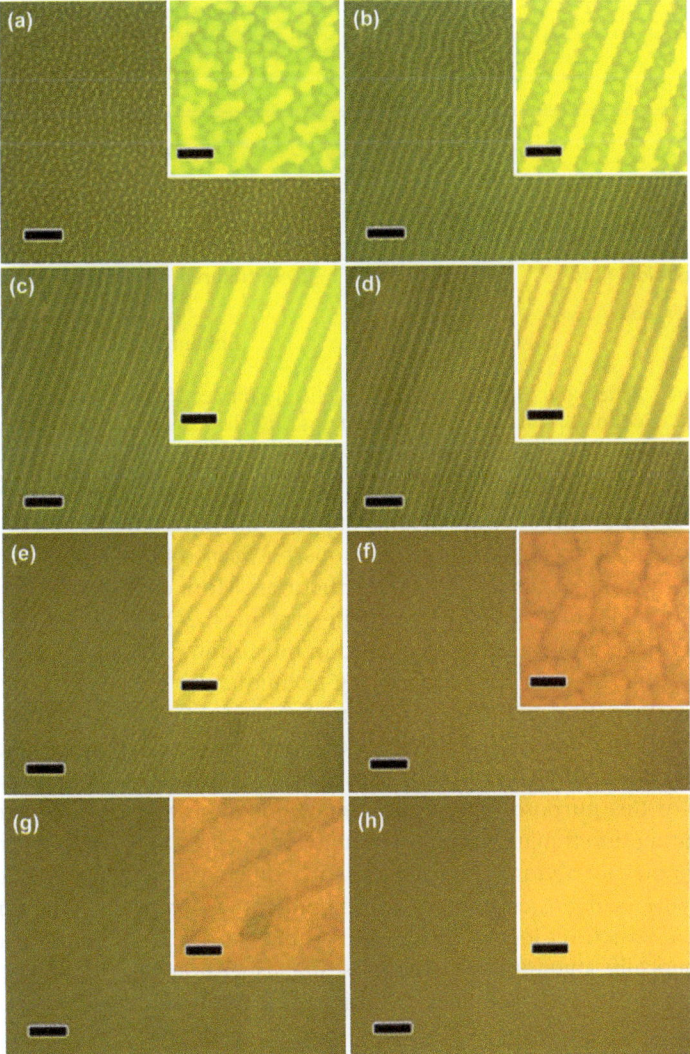

Figure 9.3   Optical microscope images showing the structure evolution of assembled 185-nm PS beads dispersed in ferrofluid in increasingly strong magnetic fields: (a) 1050 G and 1900 G/cm, (b) 1200 G and 2160 G/cm, (c) 1220 G and 2190 G/cm, (d) 1240 G and 2220 G/cm, (e) 1260 G and 2255 G/cm, (f) 1300 G and 2320 G/cm, (g) 1380 G and 2460 G/cm, (h) 1460 G and 2600 G/cm. The volume fractions are 4% for both PS and Fe_3O_4. The mixed solution is sealed in a glass cell with a thickness of 1 mm. The direction of the magnetic field is parallel to the viewing angle. The scale bars are 20 μm. Insets show the corresponding enlarged images. The lightness and contrast in the insets were adjusted to show the assembled patterns. The scale bars are 5 μm for all insets.
Reproduced with permission from Ref. 31 © 2010 American Chemical Society.

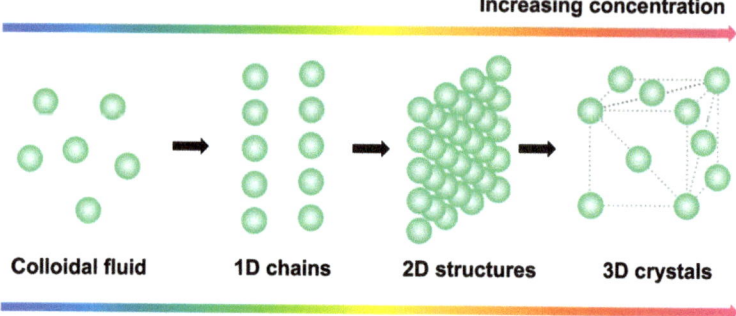

Increasing concentration

Colloidal fluid   **1D chains**   **2D structures**   **3D crystals**

Enhancing magnetic field

Figure 9.4   Schematic illustration of magnetically induced phase transitions between colloidal fluids, 1D particle chains, 2D sheets and 3D structures.

structures in suspensions of highly charged superparamagnetic colloidal spheres (Figure 9.4).[15] Our results suggest the interparticle dipole–dipole interaction not only drives the transitions between different phases but also enhances the long-range order in the structures. More complicated transitions from colloidal fluids to 1D dipolar chain structures, 2D hexagonal sheets, and finally 3D structures was also confirmed in our follow-up studies of the system by using a microradian X-ray scattering technique.

It is worth noting that the transitions between different structures usually require great changes of local particle concentrations and particle interaction potentials. Magnetic assembly of superparamagnetic particles is particularly suitable for this purpose. The induced magnetic moment can be controlled by manipulating the strength of the external magnetic field to tune the interparticle dipole–dipole interactions. The packing force due to the magnetic field gradient can induce a local particle concentration gradient, which benefits the formation of different structures locally in a single sample. As discussed below, the ability to magnetically manipulate structures in suspensions of dipolar colloidal particles is of great importance for tuning their photonic properties.

9.4 Magnetic Tuning of Photonic Properties

When the periodicity of and the dielectric contrast in the assemblies of dipolar particles match visible diffraction conditions, they may display iridescent structural colors. The diffraction wavelength (or bandgap position) of the colloidal arrays can be described by Bragg's law, $m\lambda = 2nd\sin\theta$, where m is the diffraction order, λ is the wavelength of incident light, n is the effective refractive index, d is the lattice spacing, and θ is the glancing angle between the incident light and diffraction crystal plane.[34] To magnetically tune the optical diffractions, the external magnetic field must be able to induce changes in either the refractive index of the components, or the symmetries, lattice parameters or orientations of the ordered arrays.[8] Monodisperse superparamagnetic colloidal spheres are presumably the most suitable building blocks for magnetically responsive photonic nanostructures since their magnetic responses are much stronger than normal paramagnetic materials and their dipole–dipole interactions can be fully initiated and controlled by external magnetic fields. Tuning the photonic properties by the application of external magnetic fields is realized through the manipulation of the magnetic interaction potentials and the local particle concentrations, which eventually results in changes in the lattice constant d, the orientation θ, or the crystal structure (which changes the effective refractive index and/or the lattice constant). In the following, we discuss these three types of magnetic tuning. As there are limited reports on 2D photonic structures, we limit the discussion to 1D and 3D structures.

Magnetically tuning the lattice constant (interparticle spacing) of colloidal photonic structures without causing significant disturbance to the ordering seems to be the simplest and most predictable way of controlling the optical properties. The interparticle spacing can be tuned by the manipulation of either the interparticle dipole–dipole interaction or the packing force. For example, Asher and coworkers developed magnetically controllable 3D crystalline colloidal arrays made of monodisperse polymer beads that exhibited a broad shift of the diffraction wavelength in response to nonuniform external magnetic fields.[32,33] Owing to the gradient of the external magnetic field, the packing force drives the movement of particles towards the field gradient maximum, resulting in the compression of the colloidal arrays and a blueshift of the diffraction wavelength. The effect of the magnetic dipole–dipole force is minor and only causes a minor distortion of the crystal structure in this case. As the magnetic content in the polymer beads is quite low, the response rate of the 3D arrays is relatively slow.

On the other hand, the interparticle distance can be more rapidly and conveniently tuned in 1D particle chains by controlling the interparticle dipole–dipole forces by using an external magnetic field. Bibette and coworkers pioneered the magnetic assembly of uniform emulsion droplets containing concentrated ferrofluids into 1D chains with optical diffractions tunable in a varying magnetic field.[34,35] However, the thermodynamic instability of the emulsion droplets, their incompatibility with nonaqueous solvents, and the complicated steps necessary for obtaining uniform droplets greatly limit the practical use of their system. We have recently developed 1D magnetically tunable photonic chain-like structures by assembling superparamagnetic Fe_3O_4 colloidal nanocrystal clusters (CNCs) with overall diameters in the range of 100–200 nm.[12–14,36] Utilizing clusters of ~10 nm superparamagnetic Fe_3O_4 nanocrystals not only increases their magnetic responses and thereby magnetic interactions, but also avoids the superparamagnetic–ferromagnetic transition (at a domain size of ~30 nm for Fe_3O_4). As a result, we are able to instantly assemble them into ordered structures (less than 1 s) and rapidly tune the photonic properties across the whole visible region through the application of a relatively weak (typically 50–500 Oe) external magnetic field (Figure 9.5). The rapid tuning of the diffraction of dynamic photonic chains is realized by controlling the interparticle separation d in response to an external magnetic field with varying strengths. For example, enhancing the magnetic-field

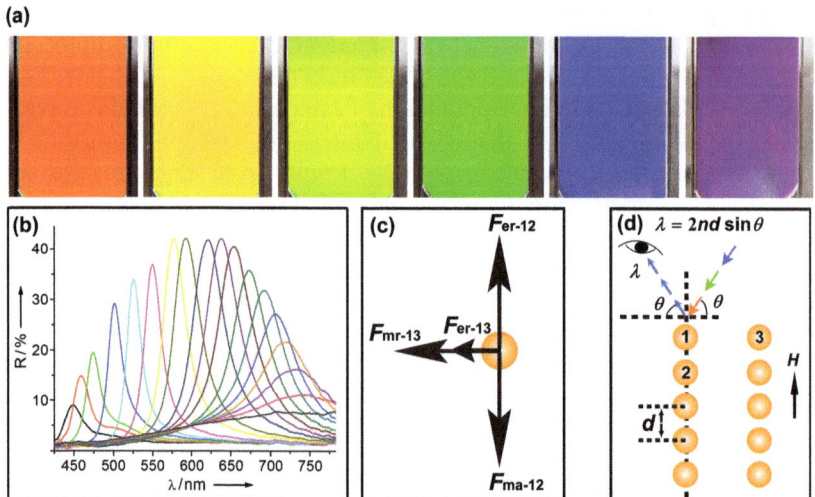

Figure 9.5   (a) Digital photos showing the diffraction color change in a typical CNC dispersion encapsulated in a capillary tube with a width of 1 cm in response to a magnetic field with increasing strengths from left to right. (b) Reflectance spectra of the same sample in different magnetic fields. (c) Illustration of interparticle force inside the chain and between different chains. (d) Scheme of Bragg diffraction from the chains of CNCs.
Reproduced with permission from Ref. 11 © 2012 American Chemical Society.

strength induces a stronger magnetic attraction along the chain, which brings the particles closer and consequently produces a blueshift of the diffraction. The key to the successful assembly of CNCs and tuning of the optical diffraction is the coexistence of a highly tunable magnetic dipole–dipole force and comparable long-range electrostatic force, both of which are separation dependent. Along the chain, the magnetic dipole–dipole attraction is balanced by the electrostatic force, while the interchain magnetic repulsive force, as well as the electrostatic force, keep the chains away from each other. The surface of the original CNCs is grafted with a layer of polyelectrolyte, polyacrylate, during the synthesis, which provides a strong interparticle electrostatic force in aqueous solution. Engineering the surface of Fe_3O_4 CNCs with a layer of silica through the sol-gel process increases their compatibility with nonaqueous solvents; further modifying the silica surface with a layer of hydrophobic silane enables the dispersion of the particles in nonpolar solvents.[36,37] The charges carried by the silica-coated Fe_3O_4 produce electrostatic interactions that can work with the magnetically induced attraction and enable successful assembly.[36] However, it is difficult to establish electrostatic repulsive interactions in nonpolar solvents due to the high energy barrier to forming surface charges. We have addressed this challenge by introducing reverse micelles to reduce the energy barrier of charge separation in nonpolar solvents, thus creating electrostatic repulsive interactions that can counteract the magnetic attraction to allow ordering of the superparamagnetic colloids.[37] An important feature of the 1D tunable system is that one can achieve a high diffraction intensity at a very low particle concentration ($\sim 0.1\%$ volume fraction *vs.* $\sim 74\%$ for close-packed colloidal crystals). The low particle density, as well as the dynamic chaining structure with only one translational order, allows instant and reversible switching between the highly ordered and completely disordered states. Our recent measurement suggested a switching rate of ~ 30 Hz for the aqueous system.

Besides controlling the interparticle spacing d, magnetic manipulation of the orientation (θ) of colloidal assemblies provides an alternative means of constructing magnetically responsive photonic structures.[18] For example, Asher and coworkers developed a magnetically controllable photonic system using ferromagnetic 3D polymerized crystalline colloidal arrays (PCCAs).[38] The preformed 3D arrays were magnetized in a strong magnetic field to induce a permanent magnetic dipole inside each particle along the $<111>$ direction. As the position and the arrangement of the particles relative to each other were locked inside the polymerized arrays, one was able to observe a net magnetization of the PCCAs. A weaker magnetic field can then induce the reorientation of the photonic crystals as the magnetic dipoles always tend to align along the external magnetic field to minimize the magnetic potential. Later, we demonstrated new types of 1D magnetically rotatable photonic units with reduced sizes by fixing photonic chains directly through a silica-coating process or inside an emulsion droplet of UV-curable resins.[16,17] Due to the dipole–dipole interaction between neighboring particles, the superparamagnetic particle chains tend to align along the magnetic field to minimize their interaction energy so that the optical properties of the nanochains can be controlled by magnetically tuning their orientations relative to the incident light.

In the case of the 1D photonic chains, we also demonstrated that the lattice constant and the orientation can be controlled synergetically to realize magnetic control of the photonic properties. As can be seen from the Bragg equation, the diffraction color of the 1D dynamic photonic chains can be tuned through controlling the interparticle spacing d and/or their orientation θ, while d can be easily varied by adjusting the magnetic field strength, the orientation of the photonic chains follows the direction of the magnetic field. As an example, we have recently studied the assembly behavior of silica-modified superparamagnetic CNCs ($Fe_3O_4@SiO_2$) in response to a complex magnetic field produced by a nonideal linear Halbach array and found that a horizontal magnetic field sandwiched between two vertical fields could allow one to change the orientation of the particle chains, producing high contrast color patterns (Figures 9.6(a) and (b)).[24] When subjected to a spatial magnetic field with large variance in field strength and direction, both the interparticle spacing and orientation of the photonic chains can be modulated to display multiple colors from different areas in a single sample. A rainbow-like color effect can be successfully created in the dispersion of magnetic particles near the edge of a cubic magnet (Figures 9.6(c) and (d)). The blueshift of the diffraction color

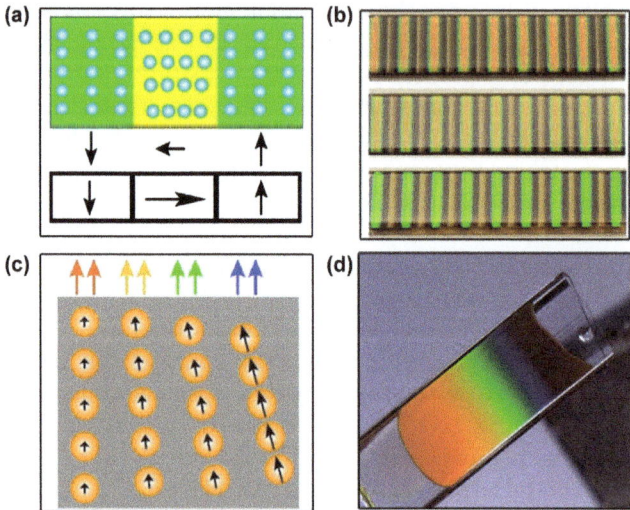

Figure 9.6 (a) Schematic illustration and (b) digital photo showing the assembly of
Fe$_3$O$_4$@SiO$_2$ particles in a patterned magnetic field with alternating field
orientation. (c) Schematic illustration and (d) digital photo showing a
suspension of CNCs displaying rainbow-like colors due to the variance in
the interparticle spacing d and the orientation θ in different regions
controlled by the magnetic field.
Reproduced with permission from Ref. 11 © 2012 American Chemical
Society.

from left to right is due to both compression of the chains and the tilting of the
chaining away from the viewing angle.

Magnetic field can also induce phase transitions in suspensions of colloidal
dipolar spheres along with a shifting of their photonic properties. For example,
we demonstrated the magnetic assembly of monodisperse nonmagnetic
polymer beads from 1D chains to 2D plate-like structures and, eventually,
high-quality 3D photonic crystals.[31] The assembly process was found to be
driven by the interplay of the magnetic forces and electrostatic forces
experienced by the nonmagnetic particles. For example, in a strong magnetic
field with high gradient, the strong packing force results in a substantial
concentration gradient of nonmagnetic particles driven by the minimization of
their magnetostatic potentials. When the local concentration of nonmagnetic
particles reaches a critical value, the interparticle magnetic dipole-dipole
potential and electrostatic repulsive potential become effective and drive the
formation of 3D photonic crystals. However, the structural transition in
response to the external magnetic fields during the assembly process is still not
clearly characterized. Nevertheless, we were able to observe the shift of the
optical diffraction towards shorter wavelength and higher intensity. Very
recently, we demonstrated a new type of 3D magnetically tunable photonic
system and experimentally observed the predicted reversible transition from
polycrystalline *fcc* to single-crystalline-like hexagonal-packing structures

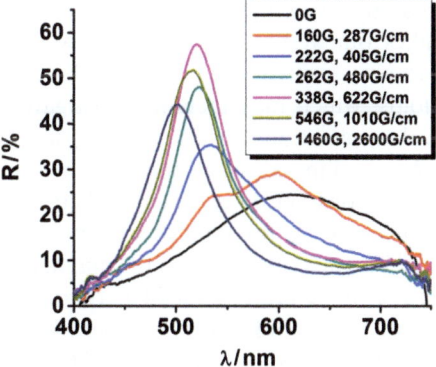

Figure 9.7   Reflectance spectra of a 180-nm (120-nm core @ 30-nm shell) Fe_3O_4@SiO_2 aqueous sample (volume fraction $\varphi \sim 0.15$) in response to external magnetic fields.
Reproduced with permission from Ref. 15 © 2012 Royal Society of Chemistry.

controlled by the strength of the external magnetic field in suspensions of superparamagnetic soft spheres.[15] Above a critical concentration, which depends on the solvents and particle surface charge, CNCs can self-assemble into nonclose-packed polycrystalline *fcc* structures driven by the interparticle electrostatic repulsive force. Applying an increasing magnetic field with a gradient to the CNC suspension can then induce the gradual transition towards single-crystalline-like hexagonal-packing structures with higher packing density. During the transition, we observed the blueshift of the diffraction color from the suspension (Figure 9.7). The diffraction intensity first increases and then slightly decreases owing to the change in the degree of ordering. Similar to the 1D photonic chains, the assembly and structural color tuning process was found to be fast and reversible due to the strong response of the CNCs to the magnetic fields. Different from the 1D magnetically responsive photonic structures we previously reported, the 3D photonic crystals can reach a higher diffraction intensity in the presence of an external magnetic field due to the much higher density of diffraction units.

9.5   Applications of Magnetically Responsive Photonic Nanostructures

Magnetic assembly provides an effective route towards the fabrication of a variety of colloidal photonic nanostructures with unique optical properties. A magnetic field can also be used as a convenient stimulus to manipulate the optical properties of the colloidal photonic assemblies. Besides the direct magnetochromatic effect, magnetically responsive colloidal photonic nanostructures have promising practical applications in many fields, such as sensing, color printing, and reflective displays. In this last section, we highlight several applications of our magnetically responsive photonic system. The instant formation of ordered arrays with widely, rapidly, and reversibly tunable structural colors makes our system not only a new platform for chromatic applications, but also a straightforward and convenient tool for the fundamental study of colloids, which may include investigations of colloidal interactions. For practical applications, while direct utilization of the colloidal particle suspension may be limited, encapsulation of this liquid suspension in the form of microdroplets or fixation of assembled structures inside a polymer matrix provides a convenient approach for easier manufacturing and more reliable performance.[39] The instant nature of the magnetic assembly also makes this method suitable for rapid fabrication of permanent structural colors by combining assembly with a rapid photopolymerization process to fix the assembled structure. For example, a dispersion of CNC particles in a photocurable resin can be directly used as a magnetic ink for producing multiple colors by controlling the strength of the applied magnetic field in each "magnetic tuning and photopolymerization" cycle.[40] When the orientation of photonic chains is programmed in each "magnetic tuning and photopolymerization" cycle, patterns with angular-dependent color contrast can be produced, which show switchable color distribution when the angle of incident light changes or the samples are tilted.[41] In addition, by taking advantage of the swelling property of the polymer matrix in response to solvents, these photonic-structures-embedded polymer films can be engineered as rewritable photonic papers or humidity sensors.[42,43] Orientation-dependent photonic structures can also be fabricated by fixing the periodic chains individually or inside polymer microspheres.[16,17] For fundamental research, since the separation of colloidal particles in the photonic structures can be easily calculated from the diffraction wavelength, our system becomes a unique tool for probing colloidal interactions, which may help improve the understanding of how microscopic objects interact with each other.[44]

9.5.1 Flexible Magnetically Responsive Photonic Film

Encapsulating a dispersion of magnetic particles in a solid flexible film makes it convenient to integrate the magnetically responsive photonic structures into complex devices for practical applications. This concept was demonstrated by dispersing an ethylene glycol solution of $Fe_3O_4@SiO_2$ particles in a liquid prepolymer of polydimethylsiloxane (PDMS) in the form of emulsion droplets with typical sizes of several micrometers, followed by solidification of the polymer matrix.[36,39] Applying a uniform magnetic field to the film then induces chaining of the magnetic particles inside each droplet with the same diffraction wavelength so that the film displays a homogeneous color that can be recognized by the naked eye. The composite film maintains the excellent flexibility of the PDMS matrix and is still able to show a rapid response to external magnetic fields even if folded into various shapes. Color-gradient effects can be easily created by controlling the assembly of each droplet in parallel. By incorporation of different types of CNC particle solutions that have similar background colors but show different visible responses to the magnetic

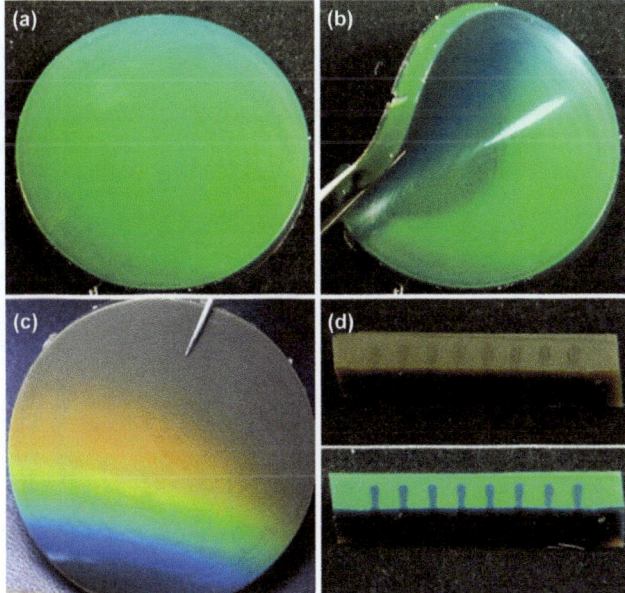

Figure 9.8 Photograph showing a flexible PDMS film displaying structural color in (a, b) a uniform magnetic field and (c) a nonuniform magnetic field. (d) Digital photos of a PDMS film that can display color patterns upon the application of an external magnetic field. The diameter of the disk in (a–c) is ~47 mm, and the length of the bar in (d) is ~25 mm.
Reproduced with permission from Ref. 11 © 2012 American Chemical Society.

field, for example, by choosing CNCs with different sizes, a flexible film with a patterned display is created. Without an external field, the film shows the native brown color of iron oxide with essentially no contrast. Upon the application of the magnetic field, however, a color pattern can be clearly observed due to the diffraction contrast resulting from the different sizes of CNC particles. This type of graphic display may find use in applications such as anticounterfeiting devices or switchable signage where prestored information can be hidden unless activated by external stimuli (Figure 9.8).

9.5.2 Rewritable Photonic Paper by Solvent Swelling

Thanks to the compatibility of $Fe_3O_4@SiO_2$ particles with prepolymers, such as poly(ethylene glycol) diacrylate (PEGDA), the assembled chain-like photonic structures can be permanently fixed inside a water-swellable polymer film through a magnetic tuning and photopolymerization process.[42] A paper/ink system can then be realized by using water as an ink to write on these photonic-particle-chain-embedded polymer composite films. We have developed a rewritable photonic paper using a $Fe_3O_4@SiO_2$-particle-chain-embedded PEGDA film as "paper" and an aqueous solution of hygroscopic

Figure 9.9 Left: Schematic illustrations of the mechanism of writing and erasing realized by infiltrating or removing the hygroscopic salt. Right: Digital photographs of flexible photonic paper (3 × 4 cm) on the plastic substrate with and without letters printed. The blank paper is uniform in color and the color gradient in the photos is caused by the flash lamp illuminating as a point light source.
Reproduced with permission from Ref. 42 © 2009 Wiley.

salt such as LiCl, MgCl$_2$, CaCl$_2$, ZnCl$_2$, or AlCl$_3$ as an "ink". Figure 9.9(a) schematically illustrates the mechanism of writing or erasing by using this method. Upon application, the ink swelled the polymer matrix, thus increasing the interparticle spacing within the chains and redshifting the diffraction. As shown in Figures 9.9(b)–(d), the color contrast led to a visible ink mark, which may last indefinitely, as the solution can hardly evaporate at ambient temperature and humidity owing to the hygroscopic nature of the salts. The ink marks can be easily erased by rinsing the paper in distilled water to dissolve the residual salts when necessary. The interparticle spacing decreases after drying, thus blueshifting the diffraction to its original state and making the photonic paper fully rewritable for many cycles. The entire writing and erasing process is safe for daily usage, as the photonic paper, salt ink, and rinsing water are nontoxic and environmentally benign.

9.5.3 Humidity Sensor

The same photonic-particle-chain-embedded polymer composite film was also demonstrated to be an indicator of air humidity.[43] The polymer matrix, which is composed of hygroscopic poly(ethylene glycol) methacrylate (PEGMA) and crosslinker PEGDA, strongly absorbs water and therefore swells upon contact with humid air, leading to the increase of the lattice constant and a color

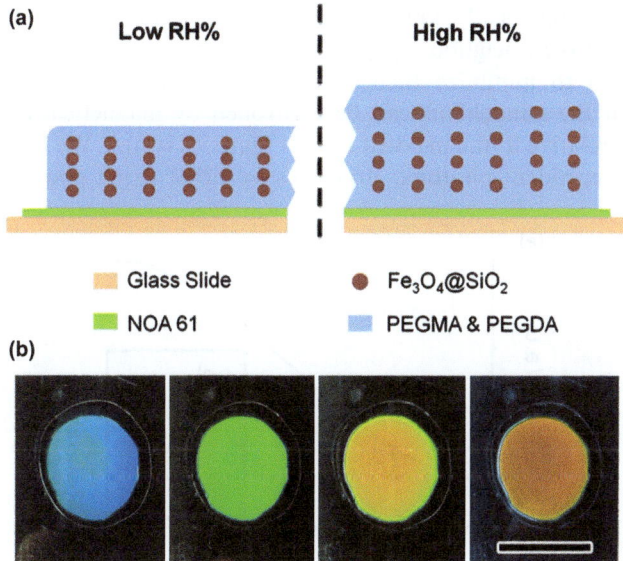

Figure 9.10   (a) Schematic illustration of the structure of the humidity sensing film and the mechanism of color switching between low and high relative humidity (RH) environments. (b) Digital photos of a typical film with a 50% crosslink level in different humidity environments, showing bluish-green, green, yellow and red, accordingly. The scale bar is 1 cm. Reproduced with permission from Ref. 43 © 2011 Royal Society of Chemistry.

change from bluish-green to red (Figure 9.10). The colors are even brighter when the film is directly in contact with water. PEG-based acrylates are known as optically transparent polymers with low toxicity compared to acrylamide-based polymers; making them promising for many optical applications. The diffraction peak redshifts *ca.* 220 nm when the humidity changes from 22% to 100%, meaning that the color change can cover the whole visible range if an appropriate starting lattice constant is utilized. The response sensitivity and speed can be optimized by tuning the crosslinking level and the thickness of the polymer matrix. Such a self-display sensor has intrinsic advantages for daily and household usage, as the colors are visually accessible and reproducible. It also represents a unique humidity sensing system that can be conveniently made into large-area colorimetric indicators that require no power for long-term operation.

9.5.4   Magnetically Responsive Orientation-Dependent Photonic Structures

Another strategy for tuning the diffraction of photonic units is changing their orientations since structural colors strongly rely on the glancing angle. As discussed above, the photonic chains assembled from superparamagnetic particles are magnetically anisotropic in nature. Due to the dipole–dipole interaction between neighboring particles, the chains tend to align along the magnetic field to minimize their magnetostatic energy so that the optical properties of the nanochains can be controlled by magnetically tuning their orientations relative to the incident light (Figure 9.11(a)).[11] We have fabricated orientation-dependent photonic structures by either fixing photonic chains in

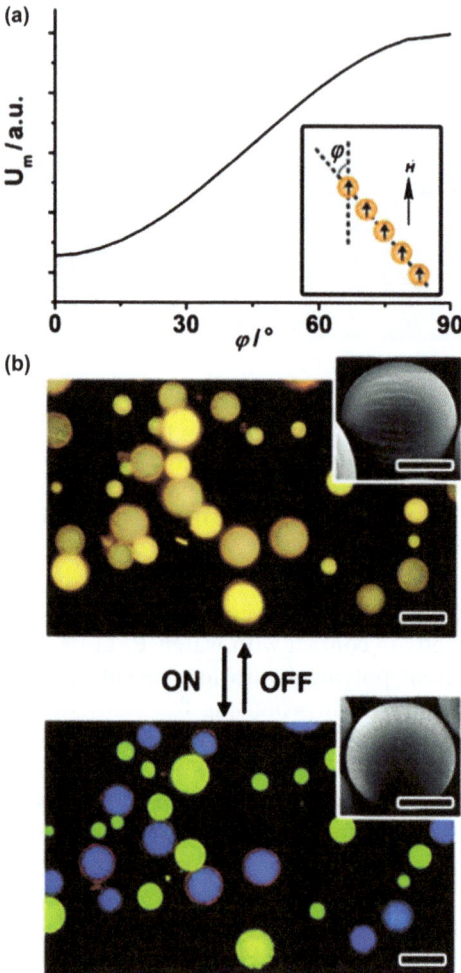

Figure 9.11   (a) Plot of magnetic dipole-dipole energy of a nanochain versus the orientation with respect to the external magnetic field. (b) "ON/OFF" switching of the color of a mixture of two types of microspheres, imaged by dark-field optical microscopy, from the native light brown of iron oxide to blue and green by tuning the magnetic field orientation from horizontal to vertical. Insets are SEM images of the microspheres in horizontal (off) and vertical (on) orientations. All scale bars are 20 μm. Reproduced with permission from Ref. 11 © 2012 American Chemical Society.

polymer microspheres through the magnetically induced assembly of particles and UV-induced polymerization, or wrapping photonic chains with another layer of silica *via* the sol-gel method.[16,17] The photonic properties of the resulting photonic units can be conveniently tuned by magnetically controlling their orientations. Since the magnetic assembly and tuning processes are separated into individual steps, color mixing can be easily realized by mixing photonic units with different diffraction colors. Figure 9.11(b) demonstrates the switching of the diffraction of photonic-chain-blended microspheres between "on" and "off" states by rotating the external magnetic field parallel or perpendicular to the viewing direction. Separation of the assembly and tuning into two steps has several advantages including long-term stability of optical response, improved tolerance to environmental variances such as ionic strength and solvent hydrophobicity, and greater convenience for incorporation into many liquid or solid matrices.[45]

9.5.5 Structural Color Printing and Encoding

Benefiting from the instantaneous nature of the magnetic creation and tuning of structural colors as well as photopolymerization, structural colors can be fixed locally in a polymer film with a high spatial resolution. By using a spatially modulated focused UV beam as the printing tool, and a mixture of CNC particles and UV curable resin as the ink, we have developed a fast and high-resolution color printing technique for producing various multicolored patterns with a single ink by repeating the "magnetic tuning" and "maskless lithographical fixing" processes, which can be programmed to print different color regions sequentially and eventually yield a large complex multicolored pattern (Figures 9.12(a) and (b)).[39,40] Since the ultimate resolution can reach 2 μm, it is easy to realize grayscale modulation and spatial color mixing by varying the density and color of primary color units in a pixel that is smaller than the resolution of a human eye. Graphic coding can also be realized by a similar technique to generate complicated functional microparticles with complex nanostructured compartments.[46] Due to the variation in structures, each compartment possesses different photonic properties. It is demonstrated that a single particle with up to eight different color encodings can be fabricated within approximately one second, superior to conventional techniques that usually involve binary encoding and require many cycles of consecutive patterning of colorful materials at high resolution. With advantages including multicolor production from a single ink, no need to move the substrate or change any physical photomask during printing, and the intrinsic characteristics of structural colors such as iridescence and metallic appearance and high color durability without bleaching, this technique represents a new platform for high-resolution color printing or encoding with immediate applications ranging from forgery protection and structurally colored graphic design to biomedical applications.

In addition to tuning the interparticle separation, the orientation of the photonic chains can be programmed in each "magnetic tuning and

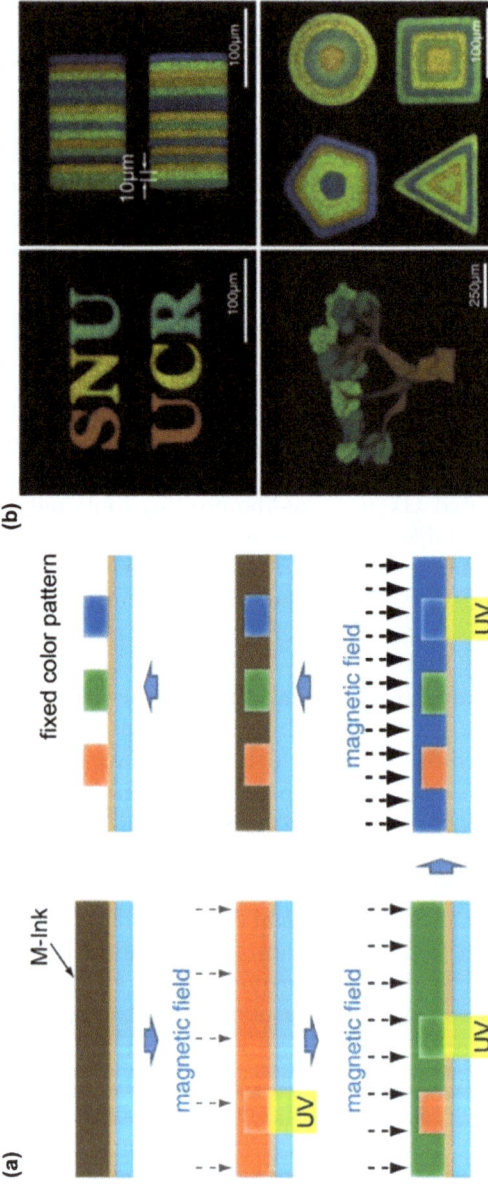

Figure 9.12   High-resolution patterning of multiple structural colors with a single magnetic ink. (a) Schematic illustration of multicolor patterning with a single ink by the sequential steps of "tuning and fixing". (b) High-resolution multicolor patterns produced using magnetic ink.
Reproduced with permission from Ref. 39 © 2010 Royal Society of Chemistry.

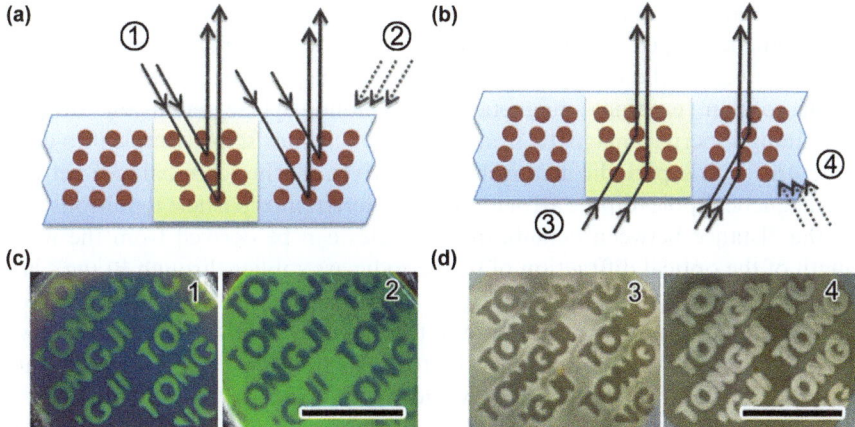

Figure 9.13 (a, b) Scheme and (c, d) digital photographs of photonic labels in (a, c) reflection and (b, d) transmission modes. The yellow and blue sections in the scheme correspond to characters and the background, respectively. Photographs 1 and 2 (or 3 and 4) show the same sample as the incident light is projected along opposite directions. The scale bars in (c, d) are 1 cm.
Reproduced with permission from Ref. 41 © 2011 American Chemical Society.

lithographical photopolymerization" cycle to print patterns with angular-dependent color contrast, which shows switchable color distribution when the angle of incident light changes or the samples are tilted.[41] Again, only a single magnetic ink is needed for printing a photonic crystal film with different crystal orientation distributions. As shown in Figure 9.13, upon incidence of light, different structural colors appear in areas with different chain orientations. When the incident angle is close to the longitudinal direction of the particle chains, the corresponding areas diffract longer-wavelength light (green, characters in case 1), while the other areas diffract shorter-wavelength light (blue, background in case 1). On the contrary, as the incident light is projected close to the chain direction in the background area (case 2), the characters and background instantly switch their colors. With multiple masks, it is expected that more complicated and unique patterns can be fabricated, which may be particularly interesting for anticounterfeiting applications.

9.5.6 Colloidal Force Measurement

Magnetic assembly of colloidal particles into a thermodynamic equilibrium state usually requires the force balance between different types of interparticle interactions. As discussed above, magnetic manipulation of the interparticle force balance provides an effective and convenient way of tuning the photonic properties through controlling the lattice constant or crystal structures. On the other hand, the study of colloidal assembly in turn provides insight into the fundamental understanding of colloidal interactions. For example, Bibette and coworkers developed a magnetic chaining technique to directly probe repulsive forces between colloidal particles in aqueous suspension.[34] Upon the application of an external magnetic field, monodisperse emulsion droplets containing superparamagnetic iron-oxide nanocrystals self-assemble into chain structures with a periodic arrangement as driven by the force balance between the magnetic dipole–dipole attraction and different interparticle repulsions.[47–50] As the distance between neighboring particles can be derived from the wavelength of the optical diffraction of the periodic assemblies through Bragg's law, different long-range and short-range repulsive forces have been studied. Compared to conventional techniques, this method allows simple control over the interparticle distance and thus the force–distance law can be conveniently derived. However, the apparent drawbacks of the approach include the lack of long-term stability of the structures from emulsion droplets, and the complicated and time-consuming steps needed for the fabrication of uniform emulsion droplets due to the involvement of multiple size-selection processes. More importantly, the surface property of the droplet particles is determined by the surfactants, which complicates the application of the technique to the study of many other short-range interactions such as the solvation force.

The widely used DLVO theory fails in explaining the superior stability of silica colloids in aqueous suspensions under conditions of high ionic strength where the electrostatic forces are effectively screened.[51,52] Recent studies have shown that another repulsive force, the solvation force, may become dominant when the distance between silica surfaces drops to the range of several nanometers.[53–55] While the physical origin of this strong short-range force is still under debate, it has been mostly attributed to the formation of a thin rigid layer of solvent molecules in the vicinity of silica surfaces through hydrogen bonding.[55–57] However, the effective determination of the thickness of this solvation layer remains a challenge. Recently, we extended the magnetic assembly technique to the measurement of solvation-layer thickness in many aqueous solvents by taking advantage of the magneto-optical response of $Fe_3O_4@SiO_2$ colloids.[44] The key to the successful measurement is the effective screening of the electrostatic force so that the neighboring particles interact with each other through the repulsion resulting from the overlap of solvation layers, which balances the magnetically induced attraction to form ordered assemblies. The thickness of the solvation layer can then be estimated by using Bragg's law (Figure 9.14(a)). $Fe_3O_4@SiO_2$ particles can be well dispersed in many polar solvents or their mixtures, thus enabling the measurement of the solvation-layer thickness from different solvents. We first studied how the solvation-layer thickness changes upon the addition of ethanol. As shown in Figure 9.14(b), the solvation layer on the surface of $Fe_3O_4@SiO_2$ particles was found to be thickest in pure water (~ 4.4 nm) and the addition of ethanol gradually contracted the solvation layer to a minimum of ~ 2 nm. It is believed that the replacement of water with ethanol, which has relatively weaker hydrogen-bonding ability, interferes with the hydrogen-bonding network so that the overall rigid solvation layer becomes thinner. Accordingly, increased

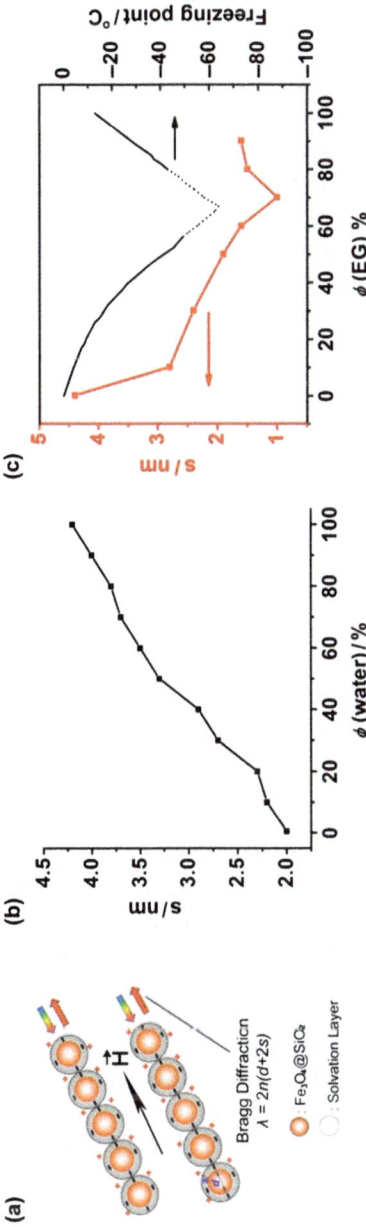

Figure 9.14 (a) Schematic illustration of the magnetic assembly strategy to determine the thickness of the solvation layer s on the surface of $Fe_3O_4@SiO_2$ colloidal particles. In an external magnetic field, the $Fe_3O_4@SiO_2$ particles self-assemble into photonic chains along the magnetic field and Bragg diffract visible light. When the electrostatic force is effectively screened, the solvation layers contact and produce a repulsive force to balance the magnetic attraction between the colloids. The thickness of the solvation layer (s) can be calculated by using the Bragg Equation with the effective refractive index (n), the diameter of colloids (d), and the measured wavelength (λ) of diffraction of the chains. (b) The dependence of the solvation-layer thickness (s) on the volume fraction of water in an ethanol/water mixture, determined by the magnetophotonic method using $Fe_3O_4@SiO_2$ particles. (c) The dependence of the solvation-layer thickness on the volume fraction of EG in water/EG mixtures (red curve). The dependence of the freezing point of water/EG mixtures on the volume fraction EG is also plotted using literature data (black curve). The dotted curve represents the metastable freezing temperatures.
Reproduced with permission from Ref. 44 © 2012 American Chemical Society.

replacement happens as the volume fraction of ethanol increases, further decreasing the solvation layer thickness. We also studied the change in solvation layer thickness in different aqueous ethylene glycol (EG) solutions and found, interestingly, that they exhibit a different profile from the case of ethanol/water mixtures, first shrinking and then expanding with increasing EG concentration (Figure 9.14(c)). The initial drop at 70% volume fraction can be explained by the disruption of the solvation layer by replacing water with EG, which has much weaker hydrogen-bonding ability similar to the case of ethanol. However, unlike the case of the ethanol/water mixture, the thickness of the solvation layer increases when the EG concentration increases to above 70%. This concentration dependence is in fact consistent with the solid–liquid phase equilibrium of the EG/water mixture, as plotted in the same graph by using data from literature, suggesting the most severe disruption of the hydrogen-bonding network occurs at $\sim 70\%$ EG concentration.[37–39] A relationship between the hydrogen-bonding ability of the solvents and the thickness of the solvation layer on $Fe_3O_4@SiO_2$ colloidal surface has been identified, which is consistent with the prior understanding of a hydrogen-bonding origin of the solvation force. Compared to conventional techniques, our magneto-photonic strategy represents an easily accessible method with a simple and inexpensive setup that produces consistent results, and is a promising tool for the measurement of other short-range forces involved in colloidal particle interactions.

9.6   Conclusion

In this chapter, we introduce the utilization of magnetic fields to organize colloidal particles into photonic structures and tune their optical properties through controlling the periodicity, orientation or the crystal structure. The study of magnetic assembly of dipolar colloidal particles is not only of fundamental importance as a model system for mimicking the behavior of simple atomic liquids and solids, but also of technological significance for fabricating colloidal responsive photonic structures. Magnetic fields can rapidly induce dramatic changes of local particle concentration and interparticle interaction potentials in suspensions of dipolar colloids. A magnetic field is regarded as a convenient stimulus to tune the optical properties of many photonic structures through controlling the lattice constant d, the orientation θ, or the crystal structures (which changes the effective refractive index and/or the lattice constant). The instant formation of ordered arrays with widely, rapidly, and reversibly tunable structural colors makes this system not only a new platform for chromatic applications, such as color displays, anticountfeiting devices, information storage and sensors, but also a straightforward and convenient tool for the fundamental study of colloidal interactions. Future work will be focused on the optimization of the assembly efficiency, response rate, optical tuning range and the quality of the photonic nanostructures. The study of 2D responsive structures remains a challenge, but may lead to many interesting results. Another important direction is the miniaturization of the responsive system to the micrometer scale, which would allow pixel-level control of photonic properties in parallel towards real display applications.

Acknowledgements

We are thankful for the financial support from the National Science Foundation (DMR-0956081). Yin also thanks the Research Corporation for Science Advancement for the Cottrell Scholar Award and DuPont for the Young Professor Grant.

Biographical Information

Le He received his BS in Chemistry from Nanjing University in 2008. He is now a PhD candidate under the supervision of Prof. Yadong Yin at the University of California, Riverside. His research interests include the synthesis and self-assembly of nanostructured materials.

Mingsheng Wang received his BS and MS in Materials Science from the University of Science and Technology of China in 2007 and 2010. He is now a PhD candidate under the supervision of Prof. Yadong Yin at the University of California, Riverside. His research interests include the fabrication and application of photonic nanostructures.

Yadong Yin received his PhD in Materials Science and Engineering from the University of Washington in 2002. He then worked as a postdoctoral fellow at the University of California, Berkeley and the Lawrence Berkeley National Laboratory. In 2006 he joined the faculty in the Department of Chemistry at the University of California, Riverside. His research interests include the synthesis and application of nanostructured materials, self-assembly, surface functionalization, and colloidal chemistry.

References

1. Y. Xia, B. Gates, Y. Yin and Y. Lu, *Adv. Mater.*, 2000, **12**, 693–713.
2. F. Li, D. P. Josephson and A. Stein, *Angew. Chem. Int. Ed.*, 2011, **50**, 360–388.
3. R. C. Schroden, M. Al-Daous, C. F. Blanford and A. Stein, *Chem. Mater.*, 2002, **14**, 3305–3315.
4. S. H. Foulger, P. Jiang, A. Lattam, D. W. Smith, J. Ballato, D. E. Dausch, S. Grego and B. R. Stoner, *Adv. Mater.*, 2003, **15**, 685–689.
5. H. Fudouzi and Y. Xia, *Langmuir*, 2003, **19**, 9653–9660.
6. A. C. Arsenault, D. P. Puzzo, I. Manners and G. A. Ozin, *Nature Photon.*, 2007, **1**, 468–472.
7. J. H. Holtz and S. A. Asher, *Nature*, 1997, **389**, 829–832.
8. J. Ge and Y. Yin, *Angew. Chem. Int. Ed.*, 2011, **50**, 1492–1522.
9. H. Fudouzi and T. Sawada, *Langmuir*, 2005, **22**, 1365–1368.
10. U. Jeong and Y. Xia, *Angew. Chem. Int. Ed.*, 2005, **44**, 3099–3103.
11. L. He, M. Wang, J. Ge and Y. Yin, *Acc. Chem. Res.*, 2012, **45**, 1431–1440.

12. J. Ge, Y. Hu and Y. Yin, *Angew. Chem. Int. Ed.*, 2007, **119**, 7572–7575.
13. J. Ge, Y. Hu, T. Zhang, T. Huynh and Y. Yin, *Langmuir*, 2008, **24**, 3671–3680.
14. J. Ge and Y. Yin, *J. Mater. Chem.*, 2008, **18**, 5041–5045.
15. L. He, V. Malik, M. Wang, Y. Hu, F. E. Anson and Y. Yin, *Nanoscale*, 2012, **4**, 4438–4442.
16. J. Ge, H. Lee, L. He, J. Kim, Z. Lu, H. Kim, J. Goebl, S. Kwon and Y. Yin, *J. Am. Chem. Soc.*, 2009, **131**, 15687–15694.
17. Y. Hu, L. He and Y. Yin, *Angew. Chem. Int. Ed.*, 2011, **50**, 3747–3750.
18. B. Gates and Y. Xia, *Adv. Mater.*, 2001, **13**, 1605–1608.
19. Y. Kraftmakher, *Eur. J. Phys.*, 2007, **28**, 409–414.
20. J. Ge, L. He, Y. Hu and Y. Yin, *Nanoscale*, 2010, **3**, 177–183.
21. U. Jeong, X. W. Teng, Y. Wang, H. Yang and Y. Xia, *Adv. Mater.*, 2007, **19**, 33–60.
22. R. M. Erb, D. S. Sebba, A. A. Lazarides and B. B. Yellen, *J. Appl. Phys.*, 2008, **103**, 063916.
23. S. Sacanna and A. P. Philipse, *Langmuir*, 2006, **22**, 10209–10216.
24. L. He, Y. Hu, X. Han, Y. Lu, Z. Lu and Y. Yin, *Langmuir*, 2011, **27**, 13444–13450.
25. A. T. Skjeltorp, *Phys. Rev. Lett.*, 1983, **51**, 2306–2309.
26. R. M. Erb, H. S. Son, B. Samanta, V. M. Rotello and B. B. Yellen, *Nature*, 2009, **457**, 999–1002.
27. A.-P. Hynninen and M. Dijkstra, *Phys. Rev. Lett.*, 2005, **94**, 138303.
28. J. Ge, Y. Hu, M. Biasini, W. P. Beyermann and Y. Yin, *Angew. Chem.*, 2007, **119**, 4420–4423.
29. M. F. Islam, K. H. Lin, D. Lacoste, T. C. Lubensky and A. G. Yodh, *Phys. Rev. E*, 2003, **67**, 021402.
30. H. Singh, P. E. Laibinis and T. A. Hatton, *Langmuir*, 2005, **21**, 11500–11509.
31. L. He, Y. Hu, H. Kim, J. Ge, S. Kwon and Y. Yin, *Nano Lett.*, 2010, **10**, 4708–4714.
32. X. L. Xu, G. Friedman, K. D. Humfeld, S. A. Majetich and S. A. Asher, *Adv. Mater.*, 2001, **13**, 1681–1684.
33. X. L. Xu, G. Friedman, K. D. Humfeld, S. A. Majetich and S. A. Asher, *Chem. Mater.*, 2002, **14**, 1249–1256.
34. F. L. Calderon, T. Stora, O. Mondain Monval, P. Poulin and J. Bibette, *Phys. Rev. Lett.*, 1994, **72**, 2959.
35. A. Koenig, P. Hebraud, C. Gosse, R. Dreyfus, J. Baudry, E. Bertrand and J. Bibette, *Phys. Rev. Lett.*, 2005, **95**, 128301.
36. J. Ge and Y. Yin, *Adv. Mater.*, 2008, **20**, 3485–3491.
37. J. Ge, L. He, J. Goebl and Y. Yin, *J. Am. Chem. Soc.*, 2009, **131**, 3484–3486.
38. X. Xu, S. A. Majetich and S. A. Asher, *J. Am. Chem. Soc.*, 2002, **124**, 13864–13868.
39. J. Ge, S. Kwon and Y. Yin, *J. Mater. Chem.*, 2010, **20**, 5777–5784.

40. H. Kim, J. Ge, J. Kim, S. Choi, H. Lee, H. Lee, W. Park, Y. Yin and S. Kwon, *Nature Photon.*, 2009, **3**, 534–540.
41. R. Xuan and J. Ge, *Langmuir*, 2011, **27**, 5694–5699.
42. J. Ge, J. Goebl, L. He, Z. Lu and Y. Yin, *Adv. Mater.*, 2009, **21**, 4259–4264.
43. R. Xuan, Q. Wu, Y. Yin and J. Ge, *J. Mater. Chem.*, 2011, **21**, 3672–3676.
44. L. He, Y. Hu, M. Wang and Y. Yin, *ACS Nano*, 2012, **6**, 4196–4202.
45. J. Kim, Y. Song, L. He, H. Kim, H. Lee, W. Park, Y. Yin and S. Kwon, *Small*, 2011, **7**, 1163–1168.
46. J. Kim, L. He, Y. Song, Y. Yin and S. Kwon, *Chem. Commun.*, 2012, **48**, 6091–6093.
47. J. Bibette, *J. Magn. Magn. Mater.*, 1993, **122**, 37–41.
48. P. John, O. Mondain-Monval, F. L. Calderon and J. Bibette, *J. Phys. D: Appl. Phys.*, 1997, **30**, 2798.
49. T. D. Dimitrova and F. Leal-Calderon, *Langmuir*, 1999, **15**, 8813–8821.
50. R. Dreyfus, D. Lacoste, J. Bibette and J. Baudry, *Eur. Phys. J. E*, 2009, **28**, 113–123.
51. J. Depasse and A. Watillon, *J. Colloid Interface Sci.*, 1970, **33**, 430–438.
52. H. Yotsumoto and R.-H. Yoon, *J. Colloid Interface Sci.*, 1993, **157**, 426–433.
53. N. V. Churaev and B. V. Derjaguin, *J. Colloid Interface Sci.*, 1985, **103**, 542–553.
54. R. G. Horn, D. T. Smith and W. Haller, *Chem. Phys. Lett.*, 1989, **162**, 404–408.
55. N. A. M. Besseling, *Langmuir*, 1997, **13**, 2113–2122.
56. G. Vigil, Z. Xu, S. Steinberg and J. Israelachvili, *J. Colloid Interface Sci.*, 1994, **165**, 367–385.
57. C. Eun and M. L. Berkowitz, *J. Phys. Chem. B*, 2009, **113**, 13222–13228.

CHAPTER 10

Chemical Routes to Fabricate Three-Dimensional Magnetophotonic Crystals

OANA PASCU, GERVASI HERRANZ* AND ANNA ROIG*

Institut de Ciència de Materials de Barcelona (ICMAB-CSIC), Campus de la UAB, E08193 Bellaterra, Spain
*Email: gherranz@icmab.es; roig@icmab.es

10.1 Introduction

10.1.1 Photonics in Our Future

Today's society and our lifestyle progressively demand high-speed technologies in telecommunications, data processing, entertainment, consumer electronics or microelectronics. This steady progress puts stringent demands on the semiconductor technology in terms of portability, power consumption, speed, dissipative heat and data transfer. It is expected that solutions for some of these challenges will be achieved by replacing electrons with photons, which are faster information carriers (due to the absence of charge and mass), less costly and ensure negligible heat dissipation.[1] Indeed, a recent report of the European Commission[2] emphasizes the important role of photonics in providing innovative products and services for relevant societal challenges, such as sustainable economy, climate change or ageing society. Thus, applications in photonics are foreseen in many different fields including telecommunications, green energy production, photonic sensors, biophotonics and health care. In particular, recent progresses have led to spectacular advances in the control of light flow

RSC Smart Materials No. 5
Responsive Photonic Nanostructures: Smart Nanoscale Optical Materials
Edited by Yadong Yin
© The Royal Society of Chemistry 2013
Published by the Royal Society of Chemistry, www.rsc.org

and light–matter interaction through plasmonics[3] and optical metamaterials,[4] as well as in photonic crystal (PC) technology.[5,6] Photonic devices based on PCs are designed so that light cannot propagate in the crystals within a given range of frequencies; this feature can be harnessed to confine the electromagnetic waves along designed paths or cavities.[3] Additionally, photonic band effects can be exploited to generate enhanced optical responses. Beyond these applications, responsive PCs – sensitive to external stimuli such as electric, magnetic or strain fields – are being considered for novel applications in which the photonic responses can be tuned externally.

10.1.2   Photonic Crystals: Photonic Bandgap Effects

PCs were designated as "semiconductors of light" by Eli Yablonovitch (1987) because they are materials presenting a periodic modulation of the permittivity affecting the light propagation in an analogy to the semiconductor bandgap affecting the motion of electrons. Due to the refraction-index modulation, the interfaces between the dielectric media behave as light-scattering centers and, as a consequence, light undergoes multiple reflections and refractions as it is transmitted through the photonic material. As a result, a photonic band structure emerges in which the propagation of electromagnetic waves is forbidden within some spectral regions – the *photonic bandgaps* (PBG).[5] Figure 10.1 illustrates the band structure of a three-dimensional PC, in which the light wavevector k – confined to the first Brillouin zone – is plotted as a function of energy (ω). It can be seen that a PBG opens at the edge of the first Brillouin zone (Figure 10.1) that, depending on the wavevector direction, occurs at different energies. At the photonic bandgap edges, the energy bands flatten[7] (see Figure 10.1 – black square) and the group velocity of light $v_g = d\omega/dk$ is much reduced. Consequently, the effective optical path – the time taken by an electromagnetic wave to traverse the sample – is dramatically increased, enhancing the light–matter interaction and the optical responses at specific wavelengths. In particular, these PBG features can be used to increase

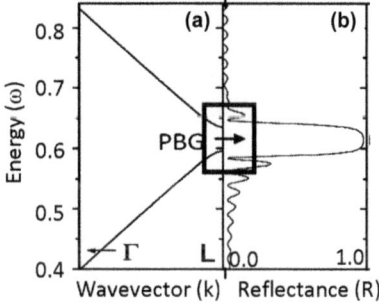

Figure 10.1   The band structure (a) and optical spectra (b) of a photonic crystal. The black square highlights the edges of the photonic bandgap (PBG) where the propagating electromagnetic waves flatten.
Adapted from Ref. 7.

substantially the magneto-optical activity at selected wavelengths. Strategies to increase the magneto-optical activity of engineered magnetophotonic crystals are thus the focus of this study.

10.1.3   Magnetophotonic Crystals: Magneto-Optical Effects

Magneto-optic phenomena result from the interaction between light and magnetized media and can be observed as an induced rotation and ellipticity of linearly polarized light when it is transmitted through the magnetized material (Faraday effects).[8] Magneto-optical rotation (birefringence) arises from the different propagation speeds – different refractive indices – of the left-hand (LCP) and right-hand (RCP) components of a linearly polarized light, whereas ellipticity (dichroism) results from the differential LCP and RCP absorption (see Figure 10.2).[9,10] The difference in absorption between LCP light and RCP light, induced in a sample by a strong magnetic field oriented parallel to the direction of light propagation, can be measured by magnetic circular dichroism (MCD) spectroscopy.[11]

Because of the nonreciprocal character of magneto-optic effects, the polarization state is not invariant to the reversal of the propagation direction of light; from this feature a number of useful nonreciprocal polarization devices can be designed. For instance, optical isolators can be built on magneto-optic devices that only allow the light traveling in one direction, thus preventing harmful backreflections.[12,13] Faraday rotators are usually made from magnetic garnet crystals of large dimensions,[13] achieving rotation angles over 45°. These isolators, however, are still not small enough to be considered for the next generation of integrated magneto-optic devices. Furthermore, most garnet

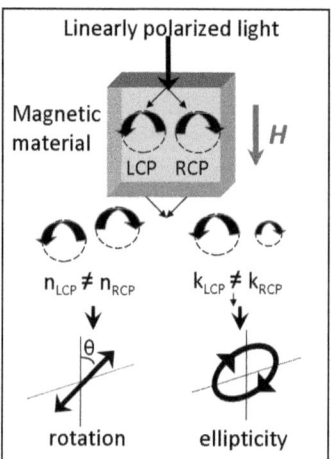

Figure 10.2   Schematic illustration of magneto-optical effects: Faraday rotation and Faraday ellipticity.

Faraday rotators and modulators work at infrared wavelengths where their magneto-optic activity is reasonably high and the transmittance is also high enough. However, in spite of exhibiting a significantly larger Faraday rotation, they do not operate at shorter (visible) wavelengths because the transmittance becomes very low in this spectral region[14] making them impractical for commercial applications at these frequencies. An alternative to achieve high magneto-optic activity with low optical losses is based on the addition of a magnetic element into a photonic crystal, the so-called *magnetophotonic crystals (MPCs)*, which provides an interesting strategy to be explored with the aim at designing materials with large enhanced magneto-optical responses at the desired operating frequencies in the visible with reasonably high optical transmittances. MPCs could then be used for the development of novel, fast and compact optical isolators[15] with much reduced thickness and optical losses and operating at visible wavelengths. Other applications of MPCs that can be considered are high-speed sensors, optical circulators, or ultrahigh-speed spatial light modulators.[16–19]

10.1.4   Three-Dimensional Magnetophotonic Crystals Fabrication Approaches

A number of one- (1D) and two-dimensional (2D) MPCs have been fabricated to achieve enhanced magneto-optical activity at band-edge frequencies.[10,20–26] Confirming theoretical predictions[9,27] about huge Faraday rotations, enhancements by factors up to $\approx 10^2$ were reported experimentally in both Faraday rotation angle and intensity of second-harmonic generation signals.[21,28] These remarkable results have prompted the investigation of three-dimensional (3D)-MPCs in which the periodicity in the three directions should lead to a complete photonic bandgap in the optical region providing a better integration in devices with an enhanced functionality.[10,29,30] Nevertheless, the achievement of high-quality 3D-MPCs is much more complex and, therefore, the attainment of an optimal (magneto) optical response—comparable at least to that of 1D-MPCs—remains a challenging issue.[31,32] The main difficulties are related to the preservation of long-range three-dimensional structural order and the achievement of high enough magnetic content in the structures. So far, a few methods have been used to produce 3D-MPCs with a defect-free structural order of hundreds of μm^2 in the surface and up to $100 \, \mu m$ in thickness.[33] However, it is further demanded to both increase the area (*i.e.* for high-quality materials) ensuring at the same time relatively high transmissivity and reduce fabrication costs to guarantee the feasibility of the future applications of magnetophotonic integrated devices. There are a few types of 3D-PCs that have been used for fabrication of 3D-MPCs: artificial direct (spheres of polystyrene or silica)[25,34–38] and inverse opals (SiO_2, Al_2O_3),[39,40] functionalized or core–shell colloidal crystals[32,41–45] and lately natural (bio) templates (e.g. butterfly wings).[46] The magnetic component can be introduced either by physical or chemical methods.

In *physical methods* such as sputtering,[10] molecular beam epitaxy[25] or electrodepositon[35,37,47] the voids of a direct opal (spheres of polystyrene or silica) are coated with the magnetic component in the form of a thin film followed by the removal of the opal spheres. The result is a structure with air holes in a magnetic matrix. The main advantage of these methods is the precise control of the infiltration degree, infiltration uniformity and desired magnetic composition. Even so, they are costly, time consuming and the magnetic component is in the form of a film and not as monodispersed magnetic nanoparticles.

In *chemical methods* the voids of a direct/inverse opal or even biotemplates, are infiltrated either with liquid dispersions of presynthesized magnetic nanoparticles (e.g. iron oxide, Ni,) – infiltration with *ex-situ* formed nanoparticles[38,39,45,48,49] – or with a solution containing the metal precursors and its subsequent chemical reactions (*e.g.* thermal decomposition, coprecipitation) resulting in the *in-situ* nanoparticles formation.[25,34,46,50,51] Adding the magnetic component in the form of *ex-situ* nanoparticles has the advantage that nanoparticles sizes can be controlled so as to present superparamagnetic behavior, thus ensuring large responses at small magnetic fields and lack of remnant magnetization, which are prerequisites for optoelectronic devices. Additionally, a better control over size, shape, polydispersity and magnetic properties of the nanoparticles can be achieved. Some drawbacks associated with this method are related to eventual nonuniform infiltrations within the structures, such as accumulation of nanoparticles at the opal surface, causing diffuse light scattering and decrease of material transmissivity, or structural damage during infiltration. These drawbacks, however, can be overcome by *in-situ* nanoparticles formation using either an artificial or natural 3D photonic structure as templates.

An alternative to solid MPCs are liquid MPCs produced by self-assembly in a liquid media of i) magnetic colloidal particles – core–shell type – such as SiO_2,[32,43,52,53] PS[41] spheres embeded with superparamagnetic NPs or SiO_2,[45,54,55] PS[56,57] sheres coated with superparamagnetic NPs, ii) colloidal superparamagnetic clusters[45,58–60] or iii) magnetic (ferro)fluids[61–64] under the influence of an external magnetic field. The structural lattice parameters and thus the optical properties can then be tuned by varying the strength of the magnetic field. Besides the advantages of reversibility and relatively fast tuning, liquid MPCs also present some drawbacks such as the need to work in liquids. This can be turned into an advantage by fabricating responsive polymer composites[58] liquid displays[56,65] or sensors.[59]

10.2   3D Magnetophotonic Crystals by Infiltration with *ex-situ* Synthesized Magnetic Nanoparticles to the Opal Structure

In this section, we present the magnetophotonic response of 3D opals infiltrated with *ex-situ* synthesized nickel nanoparticles. By disentangling their intrinsic magneto-optical response from other sources of optical activity we could demonstrate a strong modification of the magneto-optical spectral response of the nickel nanoparticles, especially prominent near the photonic band edges.

10.2.1   Nanoparticles Synthesis, Structural and Functional (Magnetic, Magneto-Optical) Characterization

10.2.1.1   Nickel Nanoparticles Synthesis

Nickel nanoparticles were synthesized by a high-temperature organometallic decomposition route. It is a heat-up method[66] involving a slow heat-up of the reaction mixture (organometallic precursor with surfactants and the solvent) from room to high temperature. Then the reaction is held at the aging temperature for a specific time before cooling to room temperature. An adapted procedure from those reported by Chen *et al.*[67] and Murray *et al.*[66] was used. A detailed description of the experimental method is given elsewhere.[40] Briefly, the synthesis consisted in mixing the organometallic precursor $Ni(acac)_2$ with the oleylamine solvent, containing two types of surfactants, oleic acid (OAc) and tri-*n*-octylphosphine (TOP). The mixture was magnetically stirred and was first heated in an inert atmosphere up to 130 °C and kept for 20 min at this temperature. During this step, the solution changed its color from blue turquoise (specific for the dissolved Ni precursor) to very clear green (formation and accumulation of an intermediary compound). In a second step, the reaction solution was further heated up to the reflux point (250 °C) and maintained at this temperature for an additional 30 min. During heating, nanoparticles nucleation occurred slightly before the reflux point, being accompanied by an almost instantaneous color change (from green to black). During the aging time (30 min), nickel nanoparticles growth took place. After cooling the reaction solution to room temperature, the nanoparticles were separated by adding ethanol, followed by two-times centrifugation. The precipitate was dried over night in an oven (at 70 °C), redispersed in an organic nonpolar solvent (hexane) and finally kept in hexane as a concentrated dispersion that was used for further characterizations and opal infiltration.

Because of the interest in fabricating magnetophotonic materials, the aim was to produce nanoparticles with the following characteristics: spherical shape, small size (< 20 nm), superparamagnetic behavior at room temperature, high magnetic susceptibility and high saturation magnetization. Thus, some experimental parameters were varied to study their influence on the particle size and size distribution and to find the optimal conditions for the production of the targeted nanoparticles. It turned out that small nanoparticles were the most suitable for the opal infiltration, so that we will henceforth focus the discussion on two Ni nanoparticles systems, with average sizes of 8 and 15 nm, respectively.

10.2.1.2   Structural and Morphological Analysis

The presence and type of surfactants are crucial in colloidal chemistry. In the particular case of magnetic nanoparticles (metallic or ferrites), due to the additional magnetic interactions, the presence of surfactant acting as a protective shell against agglomeration is a must. To produce monodispersed metallic magnetic nanoparticles, intensive research has been done in recent years, worth noting are the works reported by Murray's group.[66] They demonstrated that using a pair of surfactants – one strongly bound (oleic acid) and one weakly bound tri-*n*-octylphosphine to the nanoparticles surface – the growth, stability and oxidation of the nanoparticles can be controlled. By varying the molar ratio between these two surfactants, we synthesized nanoparticles with different sizes. When a molar ratio of OAc to TOP of 2:1 was used, 15 nm Ni NPs were prepared. By inverting the surfactants molar ratio, (OAc:TOP of 1:2) 8-nm Ni NPs were obtained. TEM images (see Figures 10.3(a)–(d)) revealed the production of spherical nanoparticles with an excellent size distribution histogram with polydispersions of 12.5% and 8% for 8- and 15-nm NPs, respectively. Moreover, the crystalline phase for both systems, detected either by X-ray or electron diffractions (see Figure 10.3(e)) corresponded to fcc nickel, without any detectable peaks for NiO.

10.2.1.3 Magnetic and Magneto-Optical Analysis

Magnetometry measurements using a superconducting quantum interference device (SQUID) of temperature- and field-dependent magnetization, $M(T)$ and $M(H)$, were acquired to assess the superparamagnetic blocking temperature

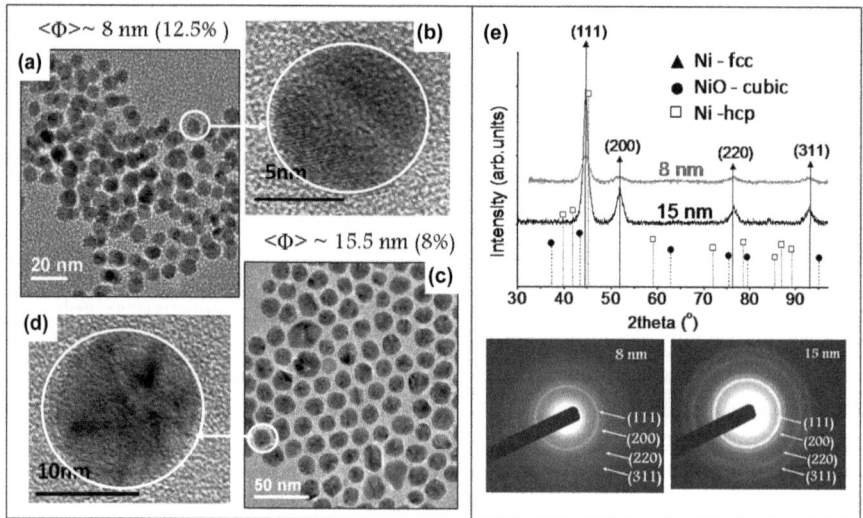

Figure 10.3 Structural and morphological characterization of 8- and 15-nm Ni NPs. (a, c) TEM images of two systems; (b, d) HRTEM of a single NP; (e) X-ray diffraction patterns and comparison with the standards of several Ni phases (SAED of the two NPs sizes are included as insets). Adapted with permission from Ref. 40 © 2010 American Chemical Society.

(T_B, K), saturation magnetization value (M_s, emu/g Ni) and the coercivity (H_c, Oe) of the synthesized nanoparticles. The magnetic units of emu/gNi were based on the mass of metallic Ni (excluding the surfactant mass). Figure 10.4(a) presents the $M(H)$ magnetization loops measured at room temperature with saturation magnetization of 18 and 44 emu/g for 8 and 15 nm particles. Additionally, the superparamgnetism was confirmed by the zero-field-cool–field-cool magnetization curve (Figure 10.4(b)), with blocking temperatures (T_B) of 59 K and 231 K, respectively.

Magnetic circular dichroism (MCD) spectroscopy can additionally be applied to analyze the magnetic properties of the magnetic colloids in liquid media; a scheme of the setup and the raw data of the measured materials are included in Figure 10.4(c). This methodology has numerous advantages over the commonly used SQUID magnetometry, in the sense that a much lower magnetic content can be analyzed at much faster rates[40] and, that the method is suitable for the analysis of magnetic liquids, even for extremely diluted ones, which is of interest for many applications and, in particular, for nanomedicine. MCD spectroscopy provides additional information that is not accessible by SQUID magnetometry: the spectral MCD response probes the electronic structure of the system with photons of different energies. We thus have the possibility to directly monitor changes of the electronic structure such as for instance those resulting from a reduction of the particle size or from the capping agent. Furthermore, MCD spectroscopy can be exploited to ascertain the concentration of magnetic nanoparticles in colloidal liquids.[40] For that

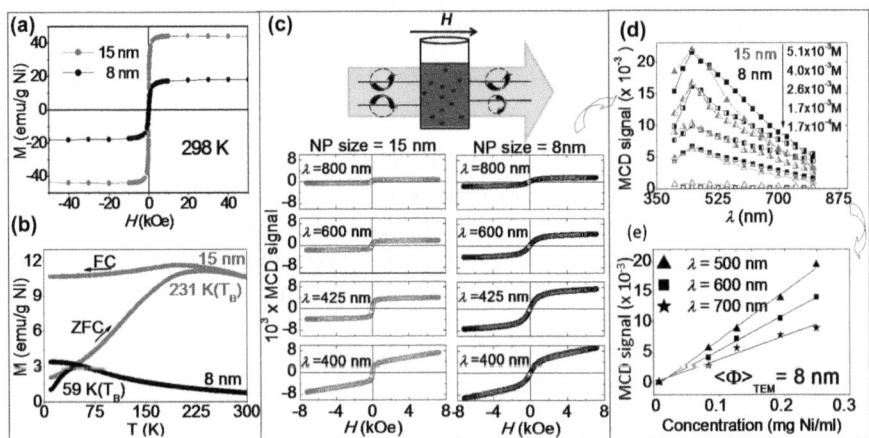

Figure 10.4 Magnetic and magneto-optical analysis of the two sets of Ni nanoparticles: (a) $M(H)$ at 300 K and (b) ZFC-FC curves at 50 Oe. (c) Schematic illustration of MCD measurement principle and raw data of measured MCD signal for two Ni sizes at several wavelengths (d) magneto-optical spectral behavior and (e) linear magneto-optical response of Ni colloids at different concentrations.
Adapted with permission from Ref. 40 © 2010 American Chemical Society.

purpose we carried out MCD experiments for different concentrations in the range 1.7×10^{-4}–5.1×10^{-3} M. At a given concentration, superparamagnetic loops were measured with a magnitude that was a strong function of the probe light wavelength (Figure 10.4(c)). From the saturated value, we obtained the MCD spectral dependence within the range 400–800 nm (Figure 10.4(d)). We note that the observed MCD spectral shape is in agreement with the reported optical dielectric response for Ni bulk, in which discernible features appear close to the energies at which the MCD response is peaked ($\lambda \approx 450$ nm, photon energy of about 2.75 eV).[68,69] In addition, for a given λ, plotting the MCD values at the different concentrations, a linear relationship is observed (Figure 10.4e) with a slope that depends on the wavelength of light (Figure 10.4(e)), with an error bar below 5% in the range of the analyzed concentrations. These linear correlations proof that MCD spectroscopy can be exploited to determine, with high accuracy, the concentration of magnetic nanoparticles dispersed in liquid media. Indeed, for the specific experimental conditions used in our experiments, NP concentrations as small as 1 µg/ml could be detected. We note, finally, that the magneto-optical response of Ni nanoparticles randomly distributed in magnetic colloids were used as reference MCD spectra to be further compared with the magneto-optical spectral response of the same nanoparticles infiltrated within an ordered structure of a photonic crystal.

10.2.2 Inverse Magnetophotonic Crystal System. Structural and Optical/Magneto-Optical Properties. Enhanced Magneto-Optical Effects

10.2.2.1 *Inverse Magnetic Opals by ex-situ Produced Nickel Nanoparticles*

Infiltration was carried out through a step-motor driven vertical dip coating (see Figure 10.5) of an inverse opal into stable colloidal dispersions.[70]

The degree and homogeneity of the infiltration was controlled by the lifting speed, the number of infiltration cycles and the size and concentration of the colloidal dispersion. In the materials discussed here, the infiltration was done in one or two cycles by dipping Al_2O_3 inverse opals into stable hexane colloidal dispersions of 8 and 15 nm Ni NPs (5×10^{-4} M), obtaining opals without nanoparticles aggregations at the surfaces and uniform NP distributions across the opal depth. The resulting materials are presented in Figure 10.6. On the one hand, the images noticeably reveal that the tiny bright spots on the spherical Al_2O_3 surfaces and within the voids – which are better visible in the zoomed image of the inset of Figures 10.6(a) and (b) – are the Ni nanoparticles. A uniform infiltration can be seen for the 8-nm NPs (Figures 10.6(a) and (b)) with no surface accumulation (Figure 10.6(a)). The infiltration efficiency of 15-nm NPs was significantly lower than that of the 8-nm NPs (Figure 10.6(e)), probably due to both the smaller diffusivity and lower total NPs volume/opal surface ratio in the case of the bigger NPs.

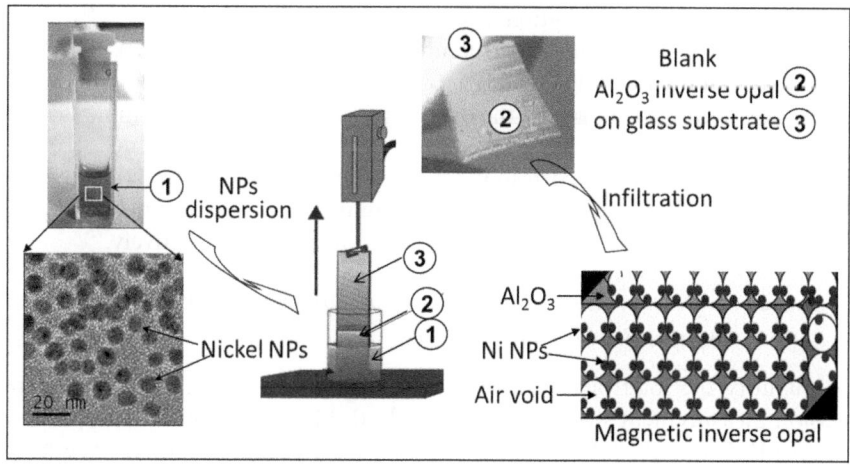

Figure 10.5   Schematic illustration of step-motor-assisted dip-coating setup.
Adapted with permission from Ref. 73 © 2011 American Chemical Society.

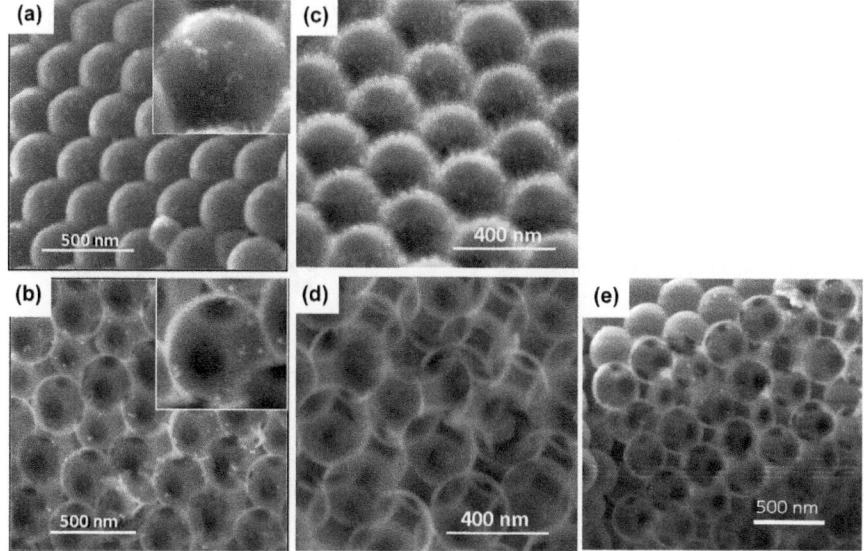

Figure 10.6   SEM images of magnetic opals fabricated by step-motor-assisted dip coating: (a,c) surface and (b,d) cross section of inverse opal infiltrated in one-cycle and two cycles, respectively, with 8-nm Ni NPs; (e) inverse opal with 15-nm Ni NPs for one-cycle infiltration.
Adapted with permission from Ref. 73 © 2011 American Chemical Society and from Ref. 79 © 2011 Royal Society of Chemistry.

To increase the amount of infiltrated nanoparticles, further tests were run with a higher concentration of colloidal dispersion (3×10^{-3} M) of 8-nm Ni and with more infiltration cycles at the same lifting speed. Unfortunately, the infiltration could not be enhanced in this way. For instance, two cycles of infiltration at higher dispersion concentration resulted in the accumulation of NPs at the surface rather than within the structure. (Figures 10.6(c) and (d)). The SEM cross-sectional image of the 3D-MPCs (Figure 10.6(b)) was used to estimate the infiltration homogeneity across the depth of the opals and to determine the void and connecting windows diameters, which were found ~ 288 nm and ~ 100 nm, respectively. The values corresponding to the void and connecting window diameters were then used to calculate the volume filling fractions of air (f_{air}) and alumina (f_{Al2O3}), which were found to be 0.87 and 0.13, respectively.[73]

10.2.2.2   Optical, Magneto-Optical Properties and Enhanced Magneto-Optical Effects

To quantify the load of magnetic nanoparticles infiltration, an optical transmission measurement was performed for both blank and infiltrated opals (see Figures 10.7(a) and 10.7(b)). In the case considered here, a blank inverse Al_2O_3 opal with sphere diameter of 316.8 nm exhibited a stop-band centered at $\lambda_B \approx 537$ nm, whereas after the infiltration with the Ni nanoparticles the stop-band position was shifted to higher wavelength (redshift) $\lambda_B \approx 552$ nm (see Figure 10.7(d) – spectrum 1). For a uniform infiltration, without nanoparticles

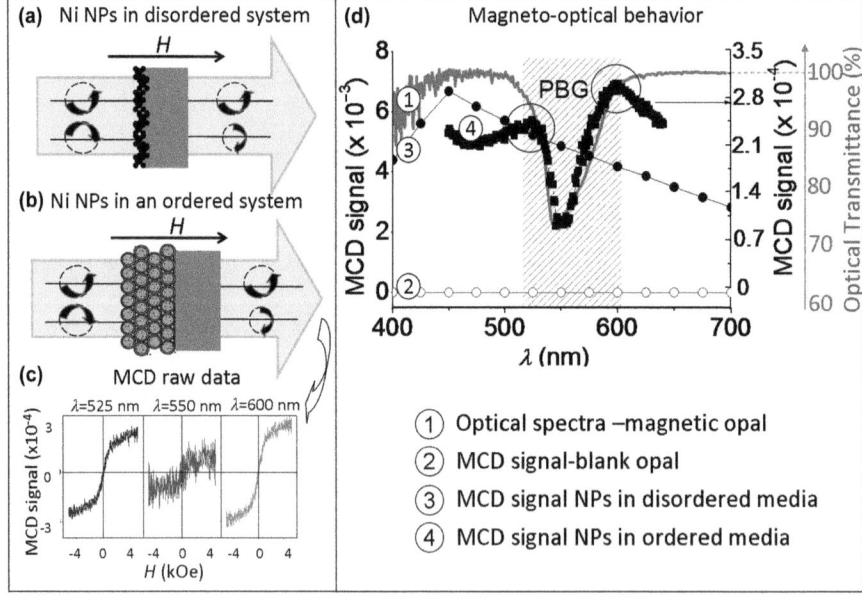

Figure 10.7   Optical and magneto-optical analysis of 8-nm Ni NPs in an ordered/ disordered media. (a,b) schematic illustration of MCD measurements; (c) raw data of MCD signal at three wavelengths; λ: 525, 550 and 600 nm. (d) comparison of magneto-optical response of Ni NPs in disordered (left axis) and ordered media (right axis), compared with the optical spectra of the Al_2O_3 opal after infiltration having the minimum at $\sim 75\%$ transmission and maximum at 100% (second right axis).
Adapted with permission from Ref. 73 © 2011 American Chemical Society.

surface accumulations, the magnitude of the stop-band redshift is directly related to the degree of the opal infiltration. To quantify this parameter, we calculated the effective refractive index (n_{eff}) of the material before and after infiltration from the experimental λ_B using the Bragg law for a *fcc* closed packed structure (10.1). [71]

$$\lambda_B = \left(\frac{8}{3}\right)^{\frac{1}{2}} D n_{eff} \tag{10.1}$$

where D is the sphere diameter of the photonic crystal ($D = 316.8$ nm in our case, which is obtained from the addition of the void diameter – 288 nm – plus twice the Al_2O_3 shell thickness). From eqn (10.1), $n_{eff} = 1.039$ for the original opal and $n_{eff} = 1.073$ for the opals infiltrated with Ni nanoparticles of 8 nm were obtained. Thus, the volume fractions f_{Ni} could be extracted by using effective medium approximations, *i.e.* Maxwell–Garnett equation (10.2).

$$\frac{n_{eff}^2 - 1}{n_{eff}^2 + 2} = f_{Ni} \frac{n_{Ni}^2 - 1}{n_{Ni}^2 + 2} + f_{Al_2O_3} \frac{n_{Al_2O_3}^2 - 1}{n_{Al_2O_3}^2 + 2} + f_{air} \frac{n_{air}^2 - 1}{n_{air}^2 + 2} \tag{10.2}$$

or the Bruggeman equation (10.3).

$$0 = f_{Ni} \frac{n_{Ni}^2 - n_{eff}^2}{n_{Ni}^2 + 2 \cdot n_{eff}^2} + f_{Al_2O_3} \frac{n_{Al_2O_3}^2 - n_{eff}^2}{n_{Al_2O_3}^2 + 2 \cdot n_{eff}^2} + f_{air} \frac{n_{air}^2 - n_{eff}^2}{n_{air}^2 + 2 \cdot n_{eff}^2} \tag{10.3}$$

where the refractive indices of air ($n_{air} = 1$) and nickel ($n_{Ni} = 1.78$) were used. [72] Using the f_{air} and the f_{Al2O3} values calculated above and the observed redshift of the gap, the effective filling fractions of Ni could be determined. These estimates were virtually the same for both approaches (eqns (10.2) and (10.3)), and yielded $f_{Ni-8\,nm} \approx 9\%$ for the 3D-MPCs infiltrated with 8-nm nanoparticles. Smaller filling fractions were obtained when using 8-nm Ni nanoparticle dispersions of high concentrations ($>10^{-3}$ M $= 0.1$ mg Ni NPs/ml solvent). Thus, optimal infiltration of 3D-MPCs requires small enough nanoparticle size and low concentrations.

Once the structural quality of the photonic crystals and the efficiency of the infiltration method were accomplished, the investigation of the functional optical and magneto-optical properties of the 3D-MPCs was addressed. Due to their particular compact *fcc* periodic structure, the nonmagnetic 3D photonic crystals can exhibit a significant optical activity near the stop-band frequencies and, thus, a significant fraction of the optical activity of the 3D-MPCs may have a nonmagnetic origin resulting from the geometry of the opal structure. To obtain the intrinsic magneto-optical activity of the 3D-MPCs we recorded the whole series of full-hysteresis MCD loops with 1-nm step resolution, from $\lambda = 400$ nm to $\lambda = 800$ nm, [73] and the maximum amplitude of each hysteresis loop was then used to compute the MCD spectral response. Confirmation that the intrinsic magneto-optical activity of the materials is being measured is given by the lack of signal obtained when measuring the noninfiltrated opal with the same setup, Figure 10.7(d) – spectrum 2. As an illustration, we show in

Figure 10.7(d) – spectrum 4) the measured MCD spectrum, while Figure 10.7(c) displays the full hysteresis loops at three selected wavelengths.

We discuss now the magneto-optical spectral responses of the photonic crystals and their relation to their optical properties measured in inverse Al_2O_3 opals infiltrated with 8 nm Ni nanoparticles. The optical transmission (Figure 10.7(d) – spectrum1) and MCD response (Figure 10.7(d) – spectrum 4) were obtained at the same location of the MPC structure, being the spot diameter ~ 2 mm. Additionally, Figure 10.7(d) shows the MCD spectrum of a sample consisting of 8-nm nanoparticles distributed randomly on a glass substrate (full circles – spectrum 3). This spectrum serves as the reference of the MCD response of the Ni nanoparticles free of any crystal-induced photonic effect. The figure immediately reveals a deep modification of the MCD spectra of the 3D-MPCs with respect to that of the randomly distributed Ni nano-particles. More specifically, a clear signature is observed of a magneto-optic enhancement in the form of two prominent shoulders in the MCD spectra (Figure 10.7(d) – spectrum 4) around the two stop-band edge frequencies (which are indicated by the empty circles). The intensive modification of the magneto-optical response in close proximity of the band-edge frequencies is in agreement with the theoretical calculations[20] that predict a magneto-optic enhancement as a direct consequence of the strong increase of light–matter interaction due to photonic band flattening and the resulting reduction of the light group velocity at band-edge wavelengths.[7] The observation of such large photonic effects in the magneto-optic response is indicative of an optimal infiltration and of the preservation of the crystal quality. This approach is then suitable to generate tailored magneto-optical responses operative at least of the order of a few millimeters (the probe light spot size in our experiments). The structural order is preserved in the same length scale, which makes still more appealing the approach here described for applications using lasers as light sources. We note that the superparamagnetic properties of the nanoparticles ensure an agile response of those systems with the magnetic field, in contra-position with a continuous magnetic layer that would lead to hysteretic ferromagnetic behavior and a more limited agility.

10.3   3D Magnetophotonic Crystals Using *in-situ* Deposition of Magnetic Nanoparticles

An alternative strategy to achieve high magnetic load in photonic crystals while preserving the crystal quality was also designed. For that purpose, a metho-dology for *in-situ* deposition of magnetic nanoparticles based on microwave-assisted sol-gel chemistry was undertaken. The traditional liquid phase sol-gel chemistry is a commonly used route for the fabrication of 3D inverse opals using the direct opals as sacrificial templates. Compared to dry processes (ALD, CVD, melting) it has the advantages to achieve maximal filling of the inorganic materials into the 3D templates voids, lower cost, and availability for a wide range of inorganic materials (e.g. magnetic ferrites, MFe_2O_4, M=Ni,

Co, Mn). Extremely good quality magnetophotonic materials with enhanced functionalities could be fabricated by combining a microwave-assisted sol-gel synthesis of ferrite nanoparticles with photonic materials.

Microwave-assisted chemistry. Microwave heating[74–76] is based on the ability of some compounds to absorb electromagnetic energy and to transform it into heat. While a domestic microwave has a multimode cavity producing nonhomogeneous heating, the one for chemical synthesis is specially designed for a single mode, resulting in uniform heating of the cavity. Compared to other conventional heating methods (e.g., heating plates, oil bath), microwave radiation affects only the solvent and reactants, passing through the reaction vessel without heating it and avoiding temperature gradients. Thus, by internal heating, the temperature is increased in the whole volume simultaneously and uniformly. Besides drastically decreasing the required time for the synthesis, the technique offers a combination of kinetics and selectivity characteristics and shows several advantageous features. The most important are (i) the acceleration of the chemical reaction rate; (ii) heating selectivity (e.g., polar substances absorb electromagnetic radiation and are intensively heated, while the nonpolar ones do not absorb); and (iii) greater movement of the molecules, higher diffusivity and collision probability due to the induced molecular vibration and the subsequent temperature increase, facilitating the conformal deposition onto intricate frames.

10.3.1 Magnetic Ferrite Nanoparticles

10.3.1.1 Structure and Morphology of Synthesized Ferrite Nanoparticles

A nonaqueous benzyl alcohol route for ferrite nanoparticles, adapted from the one first reported by Niederberger and coworkers,[77] was used.[78,79] The microwave experiments were carried out using a CEM Discover reactor (Explorer 12-Hybrid) operating at a frequency of 2.45 GHz and a power input of 200 W. For a typical run the parameters that can be controlled are the reaction temperature (T_{set}), the time at the constant T_{set} (t_{set}) and the maximum power input (200 or 300 W). The microwave reactor was connected to a computer that regulated the temperature during the chemical reaction (monitored by a volume-independent infrared sensor) by adjusting the power input. After completion of the reaction, the mixture was rapidly cooled (up to 3 min) by compressed N_2 (being a built-in cooling feature of the apparatus) (see Figure 10.8). Using this setup, a T_{set} of 160 °C and t_{set} of 5 min (with an overall process time of ~20 min) was enough to produce crystalline monodispersed ferrite NPs using organometallic precursors in benzyl alcohol solvent in the absence of any surfactant. The optimized experimental procedure was used to deposit ferrite NPs over photonic structures with the only difference that the photonic crystals (direct SiO_2 and inverse Al_2O_3 opals) of about 5 mm × 5 mm in size – grown on a glass slide substrate – were vertically immersed in the

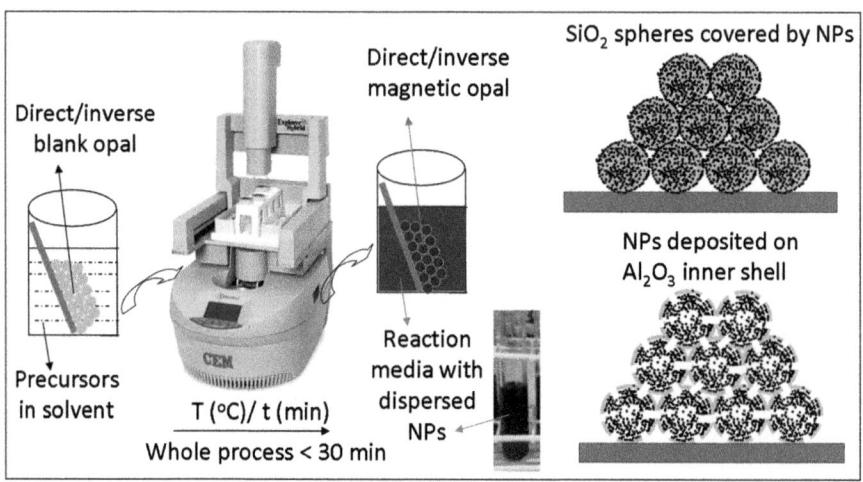

Figure 10.8 Schematic illustration of the microwave-assisted *in-situ* deposition process.

reaction solution using a 10-ml closed pressurized vessel. A detailed description of experimental methodology is reported elsewhere.[78]

10.3.1.2 Structure and Morphology of Synthesized Ferrite Nanoparticles

Size, polydispersity, shape and crystallinity of the as-prepared nanoparticles have been further investigated by TEM and HRTEM. Figures 10.9(a)–(d) show TEM micrographs for iron oxide (a,b) and $MnFe_2O_4$ (c,d) NPs synthesized in the absence of any surfactant. The images show that the two sets of nanoparticles are fairly rounded in shape with mean sizes of ~ 7 nm (23% polydispersity) for iron oxide and 5 nm (16% polydispersity) for $MnFe_2O_4$. The cristallinity of the nanoparticles is evidenced in the HRTEM images (Figures 10.9(b) and (d)). Powder X-ray diffraction (XRD) and selected-area electron diffraction (SAED) patterns of $MnFe_2O_4$ and iron oxide are presented in Figure 10.9(e). Concerning $MnFe_2O_4$, all reflections (pattern a) could be indexed with those of the cubic inverse spinel manganese ferrite phase (ICDD PDF075-0035). For iron oxide (pattern b), the distinction between magnetite and maghemite was not possible due to the very close reflection peaks of the two standards (ICDD PDF019-0629 and PDF039-1346) and the broad reflection peaks observed for the as-prepared NPs (pattern b). Thus, we concluded that the synthesized nanoparticles are a mixture of the two phases and for that reason were labeled as iron oxide.

10.3.1.3 Magnetic Properties

The magnetization loops displayed in Figure 10.10(a) demonstrate superparamagnetic behavior at room temperature, with a high saturation

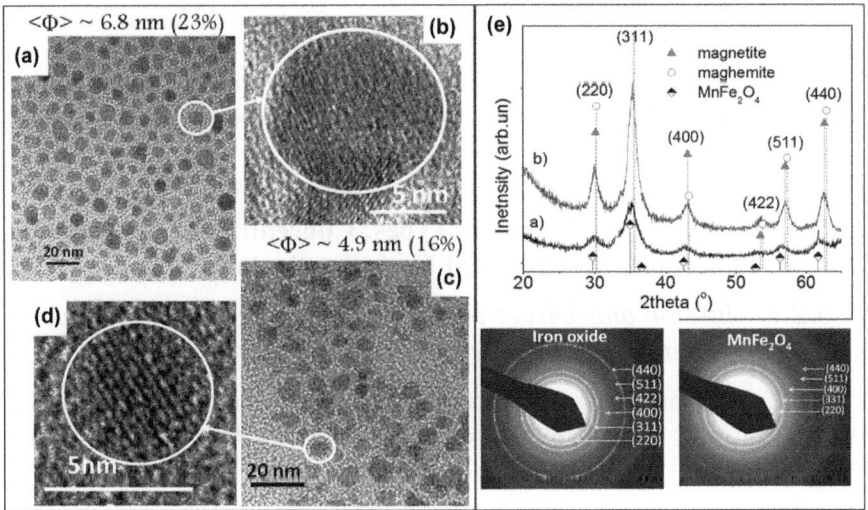

Figure 10.9 Structural and morphological characterization of iron oxide and MnFe₂O₄NPs. (a, c) TEM images of two systems; (b, d) HRTEM of a single NPs of ∼ 7 (iron oxide) and 5 nm (MnFe₂O₄); (e) XRD diffraction patterns compared with the corresponding bulk material, SAED of the two NPs sizes as insets.
Adapted with permission from Ref. 78 © 2012 American Chemical Society.

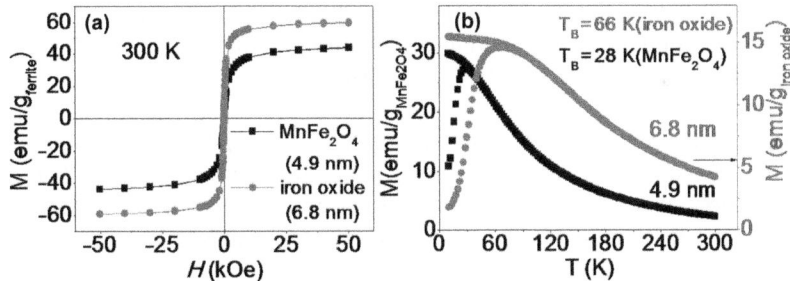

Figure 10.10 Magnetic analysis of the two sets of ferrite nanoparticles: (a) M(H) at 300 K and (b) ZFCFC curves at 50 Oe with blocking temperature (T_B).
Adapted with permission from Ref. 78 © 2012 American Chemical Society.

magnetization for both systems that, depending on the nanoparticles size could reach a value approx. 58 emu/g for iron oxide and ≈42 emu/g for MnFe₂O₄ (M_S of bulk Fe₃O₄ and MnFe₂O₄ being 92 emu/g and 80 emu/g respectively).[80] These optimal magnetic properties point to a high degree of cristallinity of the nanoparticles for both systems. Superparamagnetism at room temperature is also confirmed by the ZFC-FC magnetization curves (Figure 10.10(b)). The ZFC magnetization increases with temperature until reaching a maximum value corresponding to the blocking temperature (T_B) at 28 K for the $MnFe_2O_4$ and 55 K for the iron oxide. Above this temperature, the thermal energy (k_BT) becomes larger than the magnetic energy barrier and the nanoparticles enter in the superparamagnetic regime.

10.3.2 Magnetophotonic Crystals – Structural Characterization (Influence of Precursor Types, Concentration, Deposition Time, Surface Chemical Functionalization)

10.3.2.1 *Inverse and Direct Magnetic Opals by* in-situ *Produced Ferrite Nanoparticles*

For the fabrication of magnetic opals, several experimental parameters were varied to investigate their influence on the final material quality and functionalities (optical and magneto-optical):

- Two metal precursor concentrations (0.15 and 0.3 M) were used while keeping the same microwave set parameters.
- The reaction time (at 170 °C) was 5 or 10 min while the overall irradiation time was in the range 8–25 min.
- Two types of blank opals with two different spheres or void diameters were used, namely direct opals of SiO_2 spheres of 326 and 260 nm and Al_2O_3 inverse opals with void diameter of 270 and 234 nm. The final material obtained from the SiO_2 direct opal has been labeled henceforth as *MD.x*, where MD correspond to magnetic direct opal and x to the sample number, while the magnetic material obtained from Al_2O_3 inverse opals will be labeled as *MI.x* (MI corresponding to the magnetic inverse opal).
- Triethylamine is a catalyst for the condensation reaction.[81] Assuming that the growth of ferrite nanoparticles is based on the surface condensation reaction, the pristine opal (both direct and inverse) was functionalized with triethyalmine to study the influence of amine terminations on the magnetic coverage degree.

10.3.2.2 *Influence of Precursor Type on Magnetic Deposition*

Both iron oxide (magnetite/maghemite) and manganese ferrite ($MnFe_2O_4$) nanoparticles were deposited over silica direct opals (*MD.12* and *MD.7*, respectively). For similar microwave experimental parameters (0.15 M precursor concentration, 160/170 °C reaction time and ~8 min irradiation time) we evaluate first the final material quality in terms of the preservation of the structural order, the amount, crystalline phase and uniformity of the *in-situ* formed magnetic component. SEM images at lower resolution displayed in Figures 10.11(a) and (d) show a very good material quality, indicating that microwave deposition is a nondestructive method preserving the structural order of the original photonic material. The surface cracks that are visible in

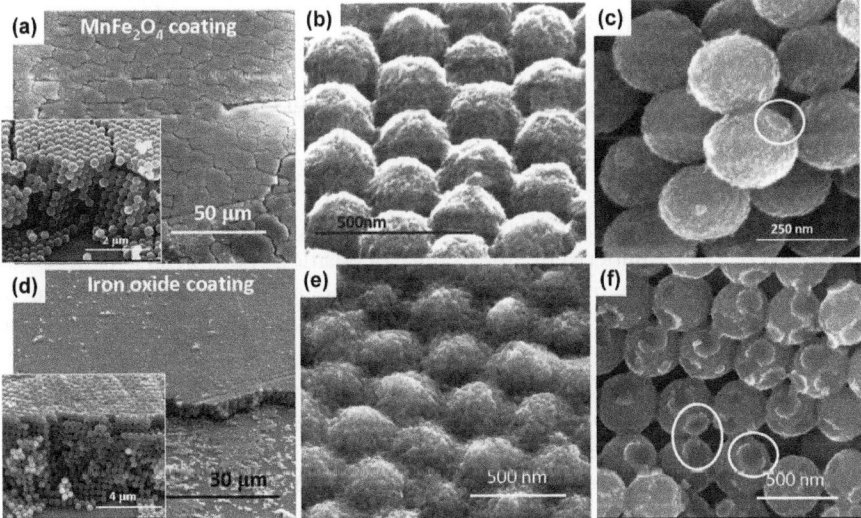

Figure 10.11   SEM images of direct opals coated with a layer of MnFe₂O₄ NPs (MD.7) (a, b, c) and iron oxide (MD.12) (d, e, f). Both opals are presented at low magnification (a, d) with magnified images in the insets. (b, e) are the surface and (c, f) cross-sectional images of the opals. The contact surfaces between two spheres without NPs coverage are highlighted by white circles.

Figure 10.11(a) after deposition were already present in the pristine opal. Moreover, the size of these cracks will have an insignificant effect on the optical and magneto-optical properties. A closer look at the cross section of the magnetophotonic material – by further increasing the magnification (Figures 10.11(a) and (d) insets) reveals the *fcc* structural order across the whole depth of the opal, certifying the suitability of the used approach for the fabrication of magnetophotonic materials. Pristine opals present a thickness gradient, due to the used fabrication approach, which in turn influences the thickness of the magnetic coating. It is worth mentioning that the quality of the blank opal plays a fundamental role in the morphology and functionality of the final magnetophotonic structures.

The morphologies of direct opals covered with a conformal layer of iron oxide and MnFe₂O₄ NPs are presented through their surface and cross-sectional SEM images. The images revealed a conformal continuous manganese ferrite (*MD.7*) (Figures 10.11(b) and (c)) and iron oxide (*MD.12*) (Figures 10.11(e) and (f)) layers confirming the homogeneous and uniform coverage of the opal structures, over the surface as well across the depth. The absence of coating in some particular places (emphasized by white circles in Figures 10.11(c) and (f)), especially at the contact surface between two silica spheres, highlights the uniformity of the magnetic layer. Moreover, the coating also reaches the inferior part of the crystals (bottom layer – images not shown).

10.3.2.3   Influence of Precursors Concentration on Magnetic Deposition

While in the case of microwave-assisted nanoparticles synthesis higher precursor concentrations lead to larger nanoparticles, in the microwave-assisted deposition of opals a higher coverage degree has been observed as a function of the precursor concentration. Figure 10.12 presents the surfaces and cross sections of direct (*MD.6* and *MD.7*) and inverse opals (*MI.2* and *MI.3*), respectively, that underwent the same deposition parameters (170 °C as reaction temperature, 5 min reaction time with a total irradiation time of 9 ± 1 min). While in the case of the direct opal the thickness of the conformal magnetic layer increases with the concentration (Figures 10.12(a) and (b)), for the inverse opal the voids are filled with more nanoparticles (Figures 10.12(c) and (d)). The surface chemistry is an important aspect in microwave-assisted deposition. In both direct and inverse opals, the hydroxyl-terminated oxide surfaces of the SiO_2 and Al_2O_3 scaffolds are more susceptible to absorb microwave energy than the bulk of the material, activating in this way the opal surfaces by locally increasing the temperature[82,83] (hot spots) and promoting the nucleation and the growth of the magnetic nanoparticles on these surfaces. Higher concentration of precursors can promote a higher concentration of

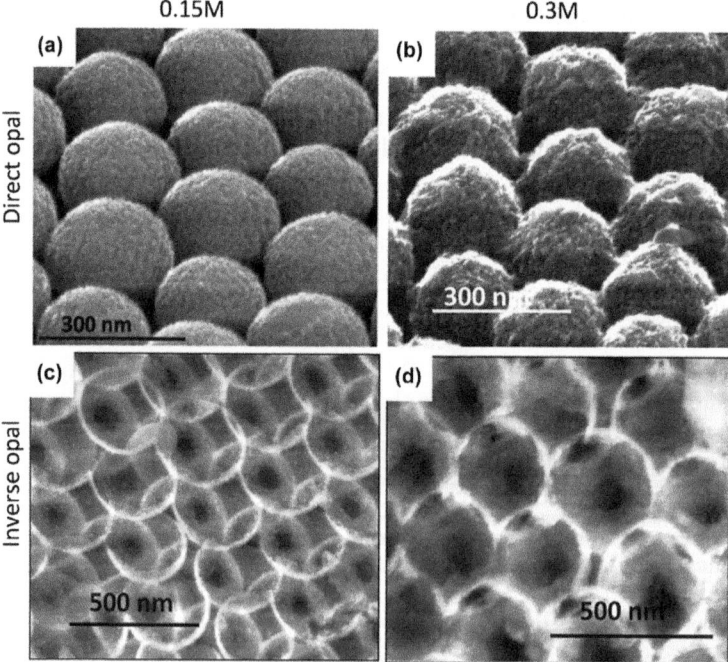

Figure 10.12   Influence of different precursors concentration over magnetic loading with $MnFe_2O_4$ NPs for direct and inverse opals. (a) MD.7; (b) MD.6; (c) MI.3 and (d) MI.2.

8 min (MD.1a)   13 min (MD.2)   21 min (MD.11)

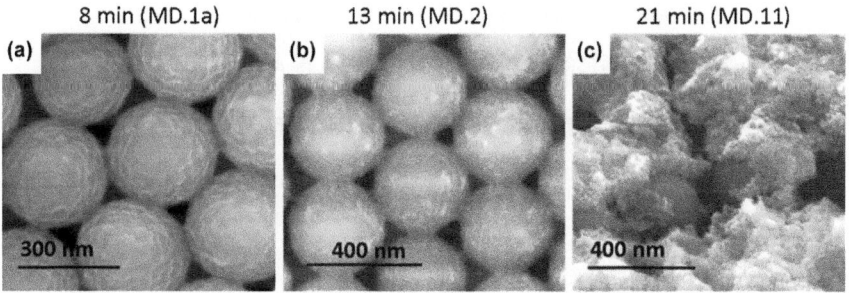

Figure 10.13 Influence of deposition time (t_{irrad}) on $MnFe_2O_4$ NPs coverage degree.

nuclei and subsequently a higher adsorption of them on the silica or alumina surface resulting in a higher coverage degree.

10.3.2.4 Influence of Deposition Time (t_{irrad})

Another influential factor on the coverage degree is the total irradiation time and, implicitly, the reaction time. By employing longer deposition time (t_{irrad}) the thickness of the conformal NPs layer can be increased to such an extent that the template voids are completely filled (Figure 10.13), losing the photonic structure. The same effect was observed in the case of 2D-MPCs using the microwave-assisted deposition approach.[84]

10.3.2.5 Influence of Amine Functionalization on the Magnetic Coverage

Based on the previous knowledge about microwave-assisted nanoparticles synthesis[85] we know that: i) the growth of the NPs is driven by the chemical reaction of condensation and ii) this reaction can be activated or inhibited by using different types of surfactants. Based on these facts, we investigated the effect of chemical functionalization of the opal surfaces on the magnetic coating. For this purpose trietylamine, known as a catalyst for the condensation reaction, was chosen.[81] Pristine opals (SiO_2 direct opal and Al_2O_3 inverse opals) were vertically immersed in trietylamine solution and kept for 30 min before the microwave-assisted infiltration process. In both types of opals, the magnetic loading corresponding to the functionalized opal was higher than in pristine opal, as expected (see Figure 10.14). A much more uniform conformal coverage with the magnetic layer was obtained, either at the surface or deep in the layers when the opal spheres were functionalized with trietylamine. For instance, it was observed that the chemical functionalization on the surface of direct opals promoted a more uniform coverage as comparing the pristine opal surface (see Figures 10.14(a) and (b)). Similarly, in the case of the inverse opals, trietylamine promotes the formation of magnetic nanoparticles in the voids of the alumina inverse opal. For instance, using a 0.15 M precursor concentration

Direct opal nonfunctionalized   Direct opal functionalized

Inverse opal nonfunctionalized   Inverse opal functionalized

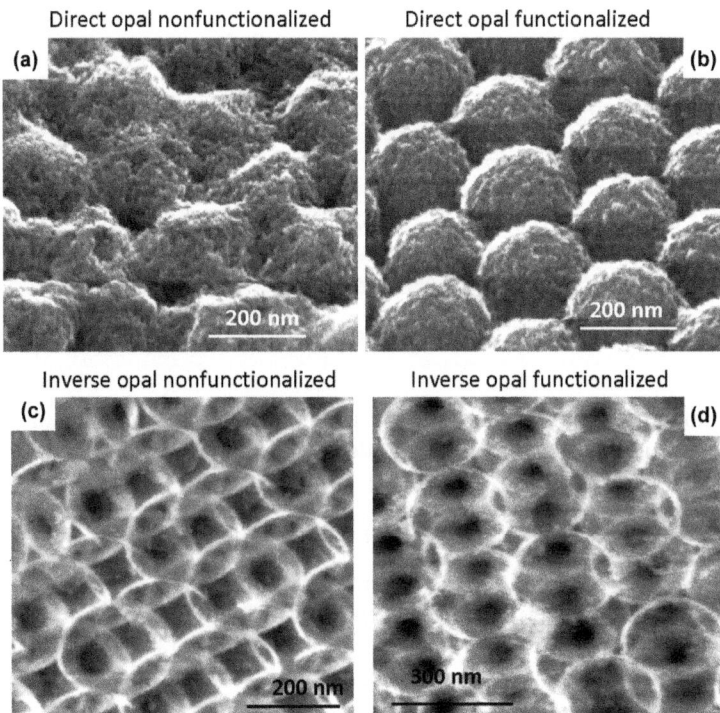

Figure 10.14   SEM micrographs showing the morphology after microwave deposition
of magnetic coatings over opal surfaces functionalized with triethylamine – MD.8 (b), and MI.7 (d) and without functionalization – MD.9
(a) and MI.6 (c).

and similar irradiation times (around 8 min), the magnetic loading was three
times more intense in the presence of trietylamine (see Figure 10.14(d)) than
without functionalization (see Figure 10.14(c)).

10.3.3   Magnetophotonic Crystals: Functionalities (Optical and Magneto-Optical) of Direct and Inverse Magnetic Opals. Enhanced Magneto-Optical Effects

The excellent quality of the 3D-MPCs obtained by microwave-assisted sol-gel
chemistry and the optimal magnetic layer coating in the central part of the opal
revealed by the SEM and TEM images anticipates an optimal magnetophotonic response. To assess their functional properties, we have performed
optical transmittance and MCD spectroscopy experiments in a range of
wavelengths $\lambda = 400$–800 nm. The transmittance experiments were undertaken
to determine the stop-bands of the photonic crystals (λ_B) and to assess the
influence of the magnetic coverage fraction on the optical properties.

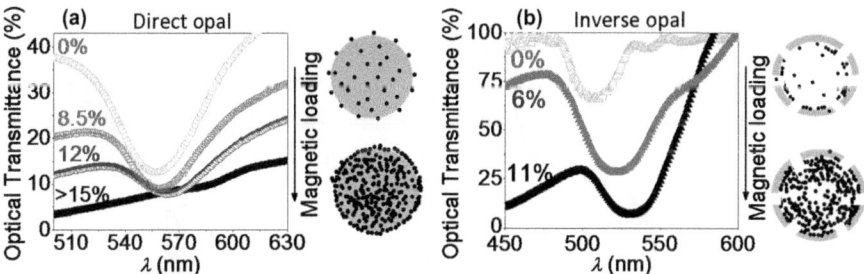

Figure 10.15 Optical Transmittance (%) spectra of (a) direct (0% – opal before deposition, 8.5%-MD.7-1, 12%-MD.7-2 and >15%-MD.7-4, respectively) and (b) inverse magnetic opals corresponding to different magnetic loadings (0% – opal before infiltration, 6%-MI.3, and 11%-MI.2, respectively).

10.3.3.1 Optical Properties of Direct and Inverse MPCs

We have exploited our control on the coverage degree of the opal structures to analyze its effect on the optical transmission and the Bragg peak position. As clearly observed in the optical spectra (Figure 10.15), the increase of the magnetic coating thickness, either on the outer surface or in the inner wall, resulted in the shift of Bragg peak to longer wavelengths; at the same time the intensity of the Bragg peak decreases to almost its disappearance. The decrease of the Bragg peak intensity could be originated by different mechanisms, either from the loss of ordered structures, some absorption of light by the magnetic material,[45,49] or by a decrease of the refractive-index contrast.[30] In view of the high quality of the crystals after the magnetic coating – corroborated by the SEM images showing the preservation of the ordered structure all along the crystals – the most probable cause of decreasing Bragg peak intensity is the reduction of the refractive-index contrast, which is in good agreement with the simulations made for different infiltration degrees of a silica direct opal with maghemite.[32]

10.3.3.2 Magneto-Optical Properties of Direct and Inverse MPCs and Enhanced Magneto-Optical Effects

As with the previous materials, we recorded the whole series of full MCD hysteresis loops for wavelengths in the visible range (see Figure 10.16) and obtained the MCD spectral response of opals infiltrated by the microwave-assisted chemical route. Figure 10.16 displays the loops at different wavelengths of two direct magnetophotonic crystals, loaded with 10% $MnFe_2O_4$ NPs (MD.1a) (see Figure 10.16(a) and 30% (MD.1b) (see Figure 10.16(b)).

Plotting the MCD spectral response using the maximum amplitude signal we observed that an intricate structure appeared in the MCD spectra around the stop-band regions for both direct (Figure 10.17(a)) and inverse opals (Figures 10.17(b) and (c)) indicating that a magnetophotonic response develops

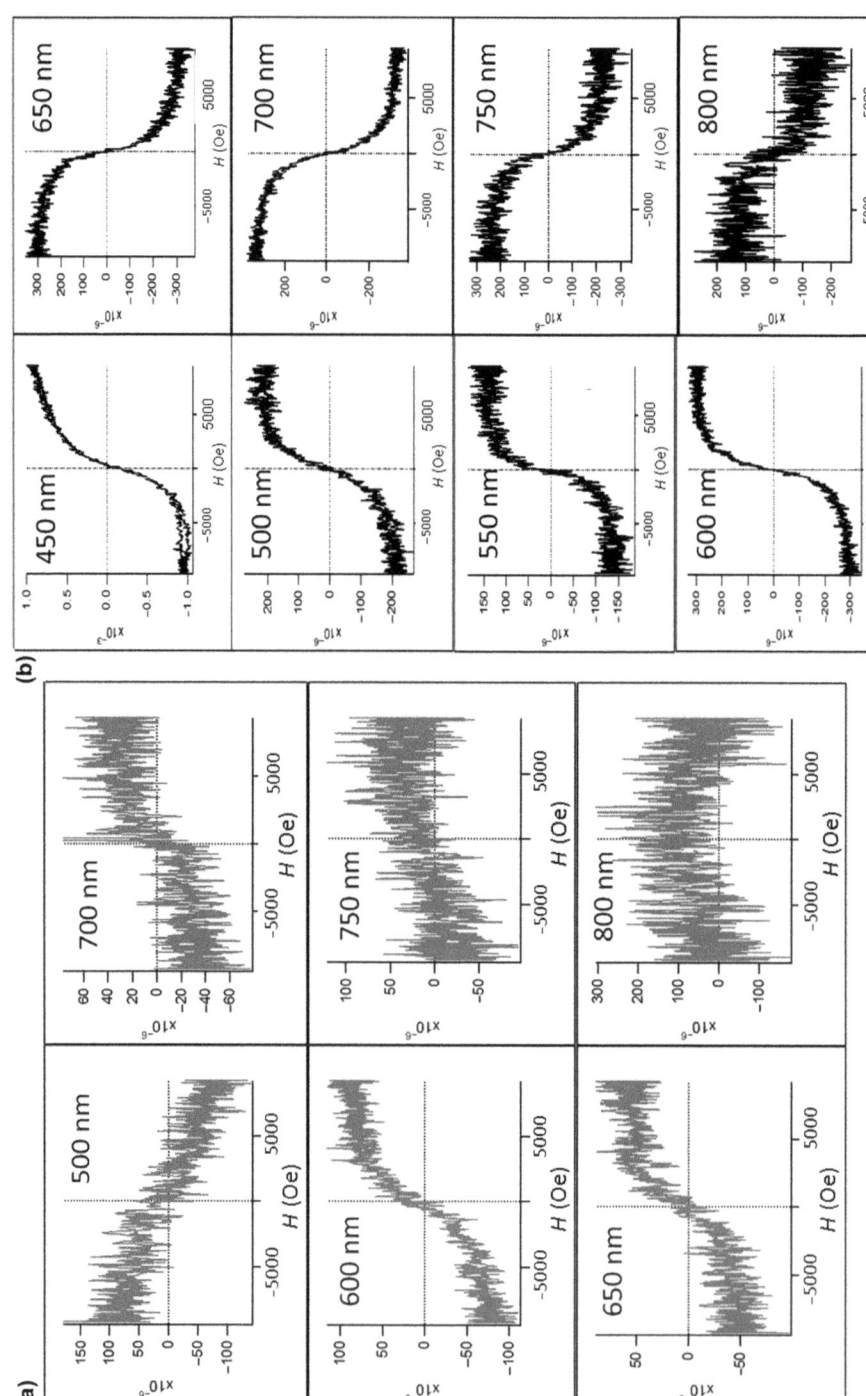

Figure 10.16 MCD hysteresis loops (MCD signal vs. H) recorder at different wavelengths of direct opal loaded with 10% (MD.1a) (a) and 30% (MD.1b) (b) magnetic material (MnFe₂O₄ NPs).

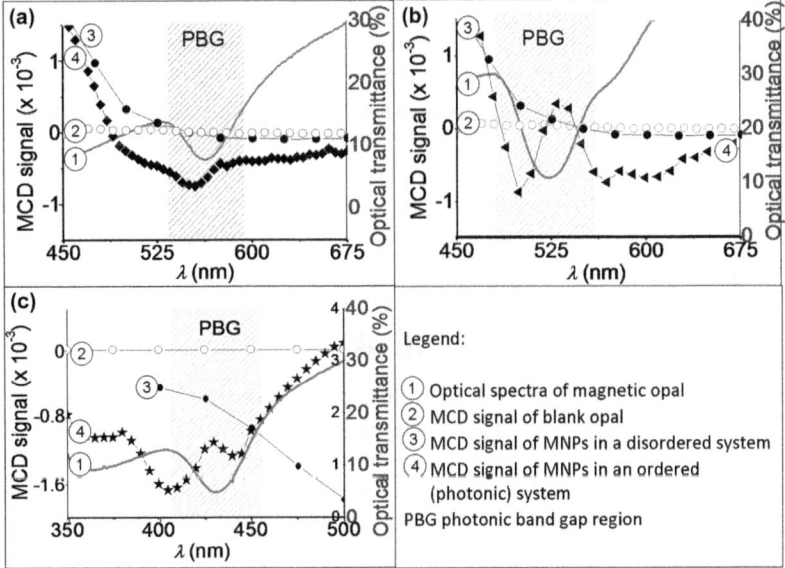

Figure 10.17 MCD and optical transmission spectra obtained in a direct opal (*MD.7*) (a) and two inverse opals with different periodicity without (*MI.3*) (b) and with (*MI.7*) (c) surface functionalization. Panels (a–c) also include as references the MCD spectra of a blank opal without magnetic loading and magnetic nanoparticles in a disordered media.

due to the interplay of the photonic and magneto-optical properties. To better understand the relationship between the photonic band structure and magneto-optics, we compare the magnetophotonic response of magnetic direct and inverse opals, taking into account the zero MCD signal of the blank opal.

Figure 10.17 shows the MCD spectra (right axis) of a direct SiO_2 opal of 366 nm periodicity and 10.8% filling factor (*MD.7*, Figure 10.17(a)) and two Al_2O_3 inverse opals of different periodicity, 465 nm with 6.4% filling factor (*MI.3*, Figure 10.17(b)) and 276 nm with 6.2% (*MI.7*, Figure 10.17(c)). The figure also includes the transmission spectra (pattern 1) in all three spectra as well as the MCD spectrum of magnetic nanoparticles distributed randomly on a glass substrate. Finally, note that all the materials comprising the photonic structures here described are dielectric and, thus, such structures do not support the propagation of surface plasmon polaritons. Thus, the enhanced magneto-optical response is the result of the strong interaction of light with the structure close to the photonic band frequencies (where the Bragg reflection condition is fulfilled), rather than coupling to plasmons, as recently reported for other systems combining magnetic materials and noble metals.[86]

A common feature observed for all the magnetic opals is that the MCD response is deeply modified with respect to that of the randomly distributed nanoparticles, indicating that the photonic band structure of the crystals radically influences the magneto-optical properties of the manganese ferrite nanoparticles. Interestingly, we observe that the magneto-optical signal is enhanced at the stop-band edges (even for low filling factor, e.g. MI.6 of 2.5%). This signal enhancement is particularly large in the inverse opals, and more specifically for the opal previously functionalized with amine (see Figure 10.17(c) – spectrum 4; MI.7) where two prominent shoulders are easily visible, indicating that in the vicinity of the band-edge frequencies the influence of the photonic structure on the magneto-optics is very intense. An important observation is that the optical transmittance of the inverse opals is larger than that of the direct opal.

10.4   Conclusions and Outlook

By infiltrating direct and inverse opals with magnetic nanoparticles, 3D-MPCs of high quality, showing a large enhancement of the magneto-optical signal at frequencies around the stop-band edges of the photonic crystals were fabricated. The resulting materials exhibited unambiguously enhanced magnetophotonic response at band-edge frequencies and a magneto-optical response tunable by magnetic fields. Two alternative routes for the magnetic infiltration were essayed. On the one hand, self-assembled Al_2O_3 inverse opal structures were impregnated with monodispersed nickel nanoparticles by vertical dip coating. Results of the optical and magneto-optical characterization are consistent with a homogeneous magnetic infiltration of these opal structures, although the magnetic loading was limited to volume fractions below 10%. Alternatively, in order to increase significantly the efficiency of the magnetic impregnation, we have also exploited a microwave-assisted route that allowed us to grow, in just a few minutes, stoichiometric and homogeneous conformal nanometric coatings with high magnetic load and covering extensive areas, of the order of cm^2. We have demonstrated the feasibility of this methodology by coating opals with superparamagnetic $MnFe_2O_4$ and iron-oxide nanoparticles, not requiring any postsynthesis material processing. Although previous studies[47,67] have also reported fairly high magnetic loads of nanoparticles embedded in direct opals, the homogeneity of the infiltration was not fully addressed and long processing times (>24 h) were required. In contrast, our 3D-MPCs obtained *via* the microwave-assisted chemical route have comparable volume fractions of the magnetic material – up to $\sim 30\%$ for direct and around 15% for inverse opals, processed at a much faster speed — typically less than 30 min – and forming a quasi-ideal conforming coating that extends across the opal depth. Moreover, it is shown that the magnetic coating coverage can be controlled by different factors, including the precursor concentration, irradiation time and surface functionalization of the opal.

We emphasize that the large versatility of the microwave-assisted chemical route, in terms of available materials and fast processing times, allows its easy extension to other areas beyond the photonics applications considered here. Interestingly, our results demonstrate the viability of this method to achieve ultrathin coatings of materials with complex stoichiometry over large areas of 3D intricate structures. Thus, this methodology could be extrapolated to applications in different fields, such as photonics, phononics, metamaterials or data storage, in which a large variety of materials and nanostructures may be involved. Additionally, other perspectives may open in other disciplines, such as materials for energy, sensors, or in catalysis where a full coverage with functional materials on intricate topologies is crucial to boost the device performance based on a significant increase of the total effective active area.

The work presented here shows the feasibility of exploiting photonic effects at visible wavelengths to enhance the magneto-optic activity of magnetic crystals. Interestingly, this is achieved with reasonably high optical transmittance – in most cases above 20% – which is much higher than the transmittance values observed in garnets at visible frequencies and similar thicknesses.[87] Although this is an important advance, the specific magneto-optic activity still requires to be further enhanced in the MPCs. One promising route towards larger enhanced optical responses is to excite photonic modes with higher energies, which usually exhibit flat bands over larger regions of the reciprocal space.[88,89] The excitation of these high-energy modes should contrive to make more intensive the light–matter interaction and the increase of the effective optical path, thus promoting a larger enhancement of the optical responses. Therefore, the propagation of visible electromagnetic radiation through photonic crystals with significantly reduced lattice parameter – let us say below 100 nm – could represent a prospective strategy to achieve still larger magneto-optic responses.

Acknowledgements

The authors warmly acknowledge the contributions of the coauthors to the several articles that constitute the main core of the works presented here, namely, J.M. Caicedo, M. Gich and J. Fontcuberta from ICMAB-CSIC and M. López-García, V. Canalejas, A. Blanco and C. López from ICMM-CSIC. Funding sources are also acknowledged, the Spanish Government (CONSOLIDER Nanoselect-CSD2007-00041, MAT2011-29269-C03-01, MAT2012-35324, MAT2009-06885) and the Generalitat de Catalunya (2009SGR-376, 2009SGR-203).

References

1. K. Tatsuno in *"Future Challenges in the Photonics Market" – Target at the Green Digital Economy-*, Optoelectronic Industry & Technology Development Association, Brussels, 2008, p. 17.
2. S. Kaierle *in Photonics Research in Europe. The European Technology Platform Photonics21*, European Laser Institute ELI, Fraunhofer, 2011.
3. S. A. Maier in *Plasmonics: Fundamentals and Applications*, Springer-Verlag, New York, 2007, **Vol. XXV**, p. 223.
4. J.-M. Lourtioz, *C. R. Phys.*, 2008, **9**, 4.
5. E. Yablonovitch in *Photonic Crystals Semiconductors of Light*, Scientific American, Inc., 2001, **Vol. Dec**, 47.

6. P. Russell, *Science*, 2003, **299**, 358.
7. J. F. Galisteo López, in *An Optical Study of Opals Based Photonic Crystals*, Thesis, Universidad Autónoma de Madrid, Madrid, 2005, p. 146.
8. B. E. A. Saleh and M. C. Teich in *Fundamentals of Photonics*, ed. B. E. A. Saleh, Wiley Interscience, New Jersey, 2nd edn, 2007, p. 1177.
9. H. Kato, T. Matsushita, A. Takayama, M. Egawa, K. Nishimura and M. Inoue, *IEEE Trans. Magn.*, 2002, **38**, 3246.
10. M. Inoue, R. Fujikawa, A. Baryshev, A. Khanikaev, P. B. Lim, H. Uchida, O. Aktsipetrov, A. Fedyanin, T. Murzina and A. Granovsky, *J. Phys. D: Appl. Phys.*, 2006, **39**, R151.
11. M. R. Mason in *A Practical Guide to Magnetic Circular Dichroism Spectroscopy*, John Wiley & Sons, Inc., Hoboken (New Jersey), 2007, p. 218.
12. *Optical isolators,* http://www.thorlabs.de/navigation.cfm?guide_ID=2015.
13. *Faraday Optical Isolators*, http://www.leysop.com/faraday.htm.
14. M. C. Sekhar, M. R. Singh, S. Basu and S. Pinnepalli, *Opt. Exp*, 2012, **20**, 9624.
15. C. Koerdt, G. L. J. A. Rikken and E. P. Petrov, *Appl. Phys. Lett.*, 2003, **82**, 1538.
16. V. M. N. Passaro, B. Troia, M. La Notte, and F. De Leonardis in *Chemical Sensors Based on Photonic Structures, Advances in Chemical Sensors*, ed. W. Wang, InTech, 2012, Chapter 5, p. 358.
17. *Light Propagation Controlled in Photonic Chips: Major Breakthrough in Telecommunications Field*, Science Daily, Columbia University, 2011, http://www.sciencedaily.com/releases/2011/2007/110710132825.htm.
18. M. Inoue, A. V. Baryshev, A. B. Khanikaev, M. E. Dokukin, K. Chung, J. Heo, H. Takagi, H. Uchida, P. B. Lim, and J. Kim, *IEICE Trans. Electron.*, 2008, **E91-C**, 1630.
19. P. Viktorovich, E. Drouard, M. Garrigues, J. L. Leclercq, X. Letartre, P. R. Romeo and C. Seassal, *C. R. Phys.*, 2007, **8**, 253.
20. A. K. Zvezdin and V. I. Belotelov, *Eur. Phys. J.*, 2004, **37**, 479.
21. A. A. Fedyanin, O. A. Aktsipetrov, D. Kobayashi, K. Nishimura, H. Uchida and M. Inoue, *J. Magn. Magn. Mater.*, 2004, **282**, 256.
22. M. Ghanaatshoar, M. Zamani and H. Alisafaee, *Opt. Commun.*, 2011, **284**, 3635.
23. I. E. Razdolski, T. V. Murzina, S. I. Khartsev, A. M. Grishin and O. A. Aktsipetrov, *Thin Solid Films*, 2011, **519**, 5600.
24. H.-X. Da, Z.-Q. Huang and Z.-Y. Li, *J. Appl. Phys.*, 2010, **108**, 063505.
25. M. Inoue, H. Uchida, K. Nishimura and P. B. Lim, *J. Mater. Chem.*, 2006, **16**, 678.
26. J. F. Torrado, J. B. Gonzalez-Díaz, G. Armelles, A. García-Martín, A. Altube, M. Lopez-Garcia, J. F. Galisteo-Lopez, A. Blanco and C. Lopez, *Appl. Phys. Lett.*, 2011, **99**, 193109.
27. A. Levy, H. Yang, M. Steel and J. Fujita, *J., Lightwave Technol.*, 2000, **119**, 1964.
28. K. Takahashi, F. Kawanishi, S. Mito, H. Takagi, K. H. Shin, J. Kim, P. B. Lim, H. Uchida and M. Inoue, *J. Appl. Phys.*, 2008, **103**, 07B331.

29. A. Mekis, J. C. Chen, I. Kurland, S. Fan, P. R. Villeneuve and J. D. Joannopoulos, *Phys. Rev. Lett.*, 1996, **77**, 3787.
30. S. Noda, K. Tomoda and N. Yamamoto, *Science*, 2000, **289**, 604.
31. N. Liu, H. Guo, L. Fu, S. Kaiser, H. Schweizer and H. Giessen, *Nature Mater*, 2008, **7**, 31.
32. M. Fang, T. T. Volotinen, S. K. Kulkarni, L. Belova and K. V. Rao, *J. Appl. Phys.*, 2010, **108**, 103501.
33. J. A. Lee, S. T. Ha, H. K. Choi, D. O. Shin, S. O. Kim, S. H. Im and O. O. Park, *Small*, 2011, **7**, 2581.
34. T. Kodama, K. Nishimura, A. V. Baryshev, H. Uchida and M. Inoue, *Phys. Status Solidi B*, 2004, **241**, 1597.
35. K. Napolskii, N. Sapoletova, A. Eliseev, G. Tsirlina, A. Rubacheva, E. Gan'shina, M. Kuznetsov, M. Ivanov, V. Valdner, E. Mishina, A. van Etteger and T. Rasing, *J. Magn. Magn. Mater.*, 2009, **321**, 833.
36. J. Sabataitytea, I. Simkienea, A. Rezaa, G.-J. Babonasa, R. Vaisnorasb, L. Rastenieneb, D. Kurdyukovc and V. Golubevc, *Superlattices Microstruct*, 2008, **44**, 664.
37. N. Sapoletova, T. Makarevich, K. Napolskii, E. Mishina, A. Eliseev, A. van Etteger, T. Rasing and G. Tsirlina, *Phys. Chem. Chem. Phys.*, 2010, **12**, 15414.
38. V. V. Pavlov, P. A. Usachev, R. V. Pisarev, D. A. Kurdyukov, S. F. Kaplan, A. V. Kimel, A. Kirilyuk and T. Rasing, *Appl. Phys. Lett.*, 2008, **93**, 072502.
39. J. M. Caicedo, E. Taboada, D. Hrabovský, M. López-García, G. Herranz, A. Roig, A. Blanco, C. López and J. Fontcuberta, *J. Magn. Magn. Mater.*, 2010, **322**, 1494.
40. O. Pascu, J. M. Caicedo, J. Fontcuberta, G. Herranz and A. Roig, *Langmuir*, 2010, **26**, 12548.
41. X. Xu, G. Friedman, K. D. Humfeld, S. A. Majetich and S. A. Asher, *Chem. Mater.*, 2002, **14**, 1249.
42. J. Ge, Y. Hu and Y. Yin, *Angew. Chem. Int. Ed.*, 2007, **46**, 7428.
43. T. Ding, K. Song, K. Clays and C.-H. Tung, *Adv. Mater.*, 2009, **21**, 1936.
44. J. Ge, L. He, J. Goebl and Y. Yin, *J. Am. Chem. Soc.*, 2009, **131**, 3484.
45. W. Libaers, B. Kolaric, R. A. L. Vallée, J. E. Wong, J. Wouters, V. K. Valev, T. Verbiest and K. Clays, *Colloids Surf. A*, 2009, **339**, 13.
46. W. Peng, S. Zhu, W. Wang, W. Zhang, J. Gu, X. Hu, D. Zhang and Z. Chen, *Adv. Funct. Mater.*, 2012, **22**, 2072.
47. X. Yu, Y. J. Lee, R. Furstenberg, J. O. White and P. V. Braun, *Adv. Mater.*, 2007, **19**, 1689.
48. J. Liu, Y. Cai, Y. Deng, Z. Sun, D. Gu, B. Tu and D. Zhao, *Microporous Mesoporous Mater.*, 2010, **130**, 26.
49. S. A. Grudinkin, S. F. Kaplan, N. F. Kartenko, D. A. Kurdyukov and V. G. Golubev, *J. Phys. Chem. C*, 2008, **112**, 17855.
50. M. Sadakane, T. Horiuchi, N. Kato, C. Takahashi and W. Ueda, *Chem. Mater.*, 2007, **19**, 5779.

51. A. V. Baryshev, T. Kodama, K. Nishimura, H. Uchida and M. Inoue, *J. Appl. Phys.*, 2004, **95**, 7336.
52. K. F. Cedric Yiu, C. H. Yu, H. Tang, H. He, S. C. Tsang and K. Y. Tam, *J. Phys. Chem. C*, 2008, **112**, 7599.
53. L. He, Y. Hu, X. Han, Y. Lu, Z. Lu and Y. Yin, *Langmuir*, 2011, **27**, 13444.
54. K. Nishimura, A. V. Baryshev, T. Kodama, H. Uchida and M. Inoue, *J. Appl. Phys.*, 2004, **95**, 6633.
55. K. Cheng, Q. Chen, Z. Wu, M. Wang and H. Wang, *CrystEngComm*, 2011, **13**, 5394.
56. C. Zhu, L. Chen, H. Xu and Z. Gu, *Macromol. Rapid Commun.*, 2009, **30**, 1945.
57. F. Caruso, M. Spasova, A. Susha, M. Giersig and R. A. Caruso, *Chem. Mater.*, 2001, **13**, 109.
58. J. Ge and Y. Yin, *Adv. Mater.*, 2008, **20**, 3485.
59. R. Xuan, Q. Wu, Y. Yin and J. Ge, *J. Mater. Chem.*, 2011, **21**, 3672.
60. H. Wang, Y.-B. Sun, Q.-W. Chen, Y.-F. Yu and K. Cheng, *Dalton Trans.*, 2010, **39**, 9565.
61. S. Y. Yang, H. E. Hornga, Y. T. Shiaoa, C.-Y. Hongb and H. C. Yangc, *J. Magn. Magn. Mater.*, 2006, **307**, 43.
62. Y. Saado, M. Golosovsky, D. Davidov and A. Frenkel, *Phys. Rev. B*, 2002, **66**, 195108.
63. L. He, Y. Hu, H. Kim, J. Ge, S. Kwon and Y. Yin, *Nano Lett.*, 2010, **10**, 4708.
64. S. Pu and M. Liu, *J. Alloys Compd.*, 2009, **481**, 85.1.
65. C. Zhu, W. Xu, L. Chen, W. Zhang, H. Xu and Z.-Z. Gu, *Adv. Funct. Mater.*, 2011, **21**, 2043.
66. C. B. Murray, S. Sun, W. Gaschler, H. Doyle, T. A. Betley and C. R. Kagan, *IBM J. Res. Dev.*, 2001, **45**, 47.
67. Y. Chen, D.-L. Peng, D. Lin and X. Luo, *Nanotechnology*, 2007, **18**, 505703.
68. H. Amekura, Y. Takeda and N. Kishimoto, *Thin Solid Films*, 2004, **464–465**, 268.
69. Š. Višňovský, V. Pařízek, M. Nývlt, P. Kielar, V. Prosser and R. Krishnan, *J. Magn. Magn. Mater.*, 1993, **127**, 135.
70. P. D. García, A. Blanco, A. Shavel, N. Gaponik, A. Eychmüller, B. Rodríguez-González, L. M. Liz-Marzán and C. López, *Adv. Mater.*, 2006, **18**, 2768.
71. H. Minguez, C. López, F. Meseguer, A. Blanco, L. Vázquez, R. Mayoral, M. Ocaña, V. Fornés and A. Mifsud, *Appl. Phys. Lett.*, 1997, **71**, 1148.
72. J.-M. Lourtioz, H. Benisty, V. Berger, J.-M. Gerard, D. Maystre, and A. Tchelnokov in *Photonic Crystals. Toward Nanoscale Photonic Devices*, ed. J.-M. Lourtioz, Springer, 2008, p. 426.
73. J. M. Caicedo, O. Pascu, M. López-García, V. Canalejas, A. Blanco, C. López, J. Fontcuberta, A. Roig and G. Herranz, *ACS Nano*, 2011, **5**, 2957.

74. V. Polshettiwar and R. S. Varma in *Aqueous Microwave Assisted Chemistry: Synthesis and Catalysis*, ed V. Polshettiwar and R. S. Varma, Royal Society of Chemistry, 2010, p. 228.
75. M. Baghbanzadeh, L. Carbone, P. D. Cozzoli and C. O. Kappe, *Angew. Chem. Int. Ed.*, 2011, **50**, 11312.
76. I. Bilecka and M. Niederberger, *Nanoscale*, 2010, **2**, 1358.
77. I. Bilecka, M. Kubli, E. Amstad and M. Niederberger, *J. Sol-gel Sci. Technol.*, 2011, **57**, 313.
78. O. Pascu, E. Carenza, M. Gich, S. Estradé, F. Peiró, G. Herranz and A. Roig, *J. Phys. Chem. C*, 2012, **116**, 15108.
79. O. Pascu, J. M. Caicedo, M. Lopez-Garcia, V. Canalejas, A. Blanco, C. Lopez, J. Arbiol, J. Fontcuberta, A. Roig and G. Herranz, *Nanoscale*, 2011, **3**, 4811.
80. E. P. Wohffarth in *Ferromagnetic Materials - A Handbook on the Properties and Magnetically Ordered Substances*, ed. E. P. Wohffarth, Elsevier, 1986, **Vol. 2**, p. 604.
81. K. L. Sorgi in *Triethylamine*, ed. L. A. Paquette, D. Crich, P. L. Funchs, and G. Molander), John Wiley & Sons, New York, 2010, *Vol. 14* , p. 12094.
82. S. J. Vallee and W. C. Conner, *J. Phys. Chem. B*, 2006, **110**, 15459.
83. W. C. Conner and G. A. Tompsett, *J. Phys. Chem. B*, 2008, **112**, 2110.
84. O. Pascu, M. Gich, G. Herranz and A. Roig, *Eur. J. Inorg. Chem.*, 2012, 2656.
85. O. Pascu in *Synthesis of Magnetic Nanoparticles and Strategies towards Magneto-Photonic Materials*, Thesis, Universitat Autonoma de Barcelona, 2012, p. 284.
86. V. I. Belotelov, I. A. Akimov, M. Pohl, V. A. Kotov, S. Kasture, A. S. Vengurlekar, A. V. Gopal, D. R. Yakovlev, A. K. Zvezdin and M. Bayer, *Nature Nano*, 2011, **6**, 370.
87. S. Kang in *Advanced Magneto-Optical Materials and Devices,* Thesis, Pennsylvania State University, 2007, p.161, https://etda.libraries.psu.edu/paper/7686/2980.
88. M. Botey, M. Maymó, A. Molinos-Gómez, L. Dorado, R. A. Depine, G. Lozano, A. Mihi, H. Míguez and J. Martorell, *Opt. Exp.*, 2009, **17**, 12210.
89. K. Sakoda, "Optical Properties of Photonic Crystals," Springer-Verlag, Berlin (2005).

CHAPTER 11

Polymer Nanocomposites: Conductivity, Deformations and Photoactuation

JEAN E. MARSHALL, YAN Y. HUANG AND
EUGENE M. TERENTJEV*

Cavendish Laboratory, JJ Thomson Avenue, Cambridge, CB3 0HE,
*Email: emt1000@cam.ac.uk

11.1 Introduction

Nanostructures comprise solid objects that are smaller than 100 nm in at least one dimension. They have been the focus of intense research interest over the last few decades, largely due to the high surface to volume ratio and quantum-mechanical effects that become observable at this length scale. Properties of nanoscale structures can be very different from those of the corresponding bulk materials; for example, the absorption spectrum, electrical properties, catalytic behavior, photonic properties, and mechanical strength of metal nanoparticles can be completely unlike those of the bulk metals.[1,2] A further aspect of this broad field is the dispersion of such nanostructures into a surrounding material in order to dramatically alter its properties. The formation of the resulting "nanocomposites" therefore has the potential to create new functional materials. However, creating a material of specified characteristics by the addition of a nanoscale filler is not a trivial matter because the properties of the composite depend not only on the components used but also on geometry and the method used for dispersion in the material.

RSC Smart Materials No. 5
Responsive Photonic Nanostructures: Smart Nanoscale Optical Materials
Edited by Yadong Yin
© The Royal Society of Chemistry 2013
Published by the Royal Society of Chemistry, www.rsc.org

In this review, we focus on nanostructures embedded in a soft, polymeric matrix. The added nanostructures can be used simply to impart greater stiffness to the polymer,[3] but we will explore cases where they endow the composite material with new optical, electronic or mechanical actuating behavior. The dispersion of nanostructures into many types of thermoset and thermoplastic polymers has been reviewed extensively elsewhere;[4] the focus of this work is on nanostructures embedded within an elastomeric matrix. As we discuss in detail, the elastic behavior of these materials combined with the optical and electrical effects introduced by embedded nanoparticles can lead to novel mechanical actuating behavior.

The broad definition of nanostructures encompasses nanosheets (which are smaller than 100 nm in one dimension only), nanotubes or filaments (smaller than 100 nm in two dimensions) and nanoparticles (smaller than 100 nm in all three dimensions). We can further divide the available nanostructures into those that are formed purely from carbon, and those that incorporate other elements. Nanocomposites containing carbon nanoparticles and nanotubes have been most prevalent so far in the scientific literature, largely because of their commercial availability, but this situation is changing as a wider variety of nanostructures have been successfully synthesized.

Some of the earliest attempts to incorporate carbon nanostructures into polymers focused on the use of carbon black, a particulate form of carbon formed by the partial combustion of oil. The typical particle size of carbon black can be in the tens or hundreds of nanometers[5,6] and its incorporation into polymers can have a dramatic effect on the light absorption, electrical resistivity and mechanical toughness of the resulting composite materials.[7] Other methods of producing carbon nanoparticles are available (for example, by laser ablation[8]) but carbon black is a desirable filler due to its commercial availability.

Extending this work on isotropic structureless carbon nanoparticles, research interest has turned to carbon nanostructures that are smaller than 100 nm in two dimensions; these can be orders of magnitude larger in their third dimension. The most studied system of this type is the carbon nanotube (CNT), which consists of one or more sheets of graphitic carbon "rolled" to form cylinders. The existence of CNTs has been known for a long time, but they have been far more intensively studied since high-quality CNTs were observed and publicised by Iijima in 1991.[9] CNTs' covalently bonded, graphitic structure results in excellent mechanical properties (with a Young modulus of the order of TPa) as well as unusual electrical, thermal and optical characteristics.[10,11] CNTs are therefore regarded as highly interesting nanostructures for the formation of nanocomposite materials. Even more recently, the discovery of a practical method for the production of graphene (single layers of covalently bonded sp²-hybridized carbon atoms)[12,13] has opened up promising avenues for research into graphene nanocomposites. This field is still in its infancy, but given the extremely low resistivity and high mechanical strength of graphene, it seems likely that this type of material will be the focus of new research for some time.

There are also a vast range of noncarbon nanostructures that are interesting candidates for the formation of composite materials; this is especially true of particles that are magnetically or electrically active. As in the case of carbon nanostructures, most filler materials are available in the form of nanoparticles (most often metal and metal-oxide nanoparticles) and filaments (such as metal-oxide nanowires). Many of these nanostructures must be synthesized specially for the purpose, unlike the carbon nanostructures that are now largely commercially available. For instance, a noncarbon "sheet", structurally similar to (and isoelectronic with) graphene, has been synthesized by the reaction of borazine with a metal surface to form a boron nitride (BN) "mesh";[14] composite materials have been formed from such nanosheets, but scaling up their production remains a significant challenge.[15,16]

As we will discuss below, the properties of the final nanocomposite material depend on the selection of elastomeric polymer used, the type and concentration of nanostructures embedded within the material, the homogeneity of the nanostructure dispersion, and the physical/chemical interaction between the nanostructure/polymer interface. We will discuss methods for creating a homogeneous dispersion of nanostructures through such polymer systems, beginning with a general discussion of nanostructures embedded in elastomers and progressing to the more specific case of elastomer materials containing liquid-crystal components (which, in themselves, can impart novel properties to the material). Nanostructure/polymer combinations, with which good dispersion has been achieved, will be examined and their actuation behavior in response to electrical and optical stimuli will be interpreted.

11.2   Dispersion of Nanostructures into Elastomers

In practice, elastomers can be formed from thermoset or thermoplastic polymers, and usually consist of lightly crosslinked polymer networks. Embedding nanostructures within polymeric systems can be a complex procedure, due to the natural tendency of nanostructures to aggregate (and hence to phase-separate from the polymeric or organic solvent medium, in which one attempts to disperse them). The individual nanostructures must therefore be separated from each other, and their surface may be chemically modified in order to increase their solubility in the polymer matrix. Here, we use carbon nanotubes (CNTs) as examples. CNTs can be classed as single-walled nanotubes (SWNTs) when they are formed from a single rolled layer of graphite, or as multiwalled nanotubes (MWNTs) when they are formed from multiple concentric graphitic cylinders. Both SWNTs and MWNTs are now commercially available, and considerable literature currently exists on the subject of their incorporation into composite materials.[17–19] Dispersion involves separation and then stabilization of CNTs in a medium. For the best decision on the choice of technique for a particular system, it is essential to distinguish and study these two processes individually. We will begin with the general topic of how CNTs are effectively dispersed into solvents, stabilized and subsequently incorporated into polymer materials, and will then progress toward the specific topic of nanostructures embedded in a polymer matrix containing liquid-crystal components.

11.2.1 Nanotube Separation and Dispersion

Strong interfacial energies due to van der Waals attraction are present among nanoparticulate structures. When being mixed in a solution, filler aggregates are subjected to shear stresses imparted from the medium (*e.g.*, solvent or polymer melt). Therefore, the flow of the medium in response to an external force (*e.g.*, through the rotation of a mixer blade, or cavitation in ultra-sonication) generates the local shear stresses that are ultimately responsible for dispersion. A mixing process can be interpreted as the delivery of mechanical energy into the solution to separate the CNT aggregates. Considering the interfacial binding energy and the shear energy input, one can establish the mixing criteria for effective aggregate separation as: the supplied energy (to be more precise, the local energy density) from the chosen mixing technique to be greater than the binding energy of the CNT aggregates (to be more precise, the energy per local volume of the contact). On the other hand, to retain the morphology of individual CNTs, the supplied energy should also be lower than the amount required to fracture a filament. Hence, an ideal aggregate separation technique should supply an energy density between the binding energy of the aggregates (lower limit), and the fracture resistance of individual nanotubes (upper limit).

Binding Energy

Using a continuum model by integration over the tubular surfaces of two parallel (10,10) SWNTs, Girifalco *et al.* [20] calculated the cohesion energy per unit length between the SWNT pair to be approximately –0.095 eV/Å; the associated energy is –0.36 eV/Å for a bundle formed by the same SWNTs. Since the pair cohesion energy per unit length is much greater in magnitude than the room temperature thermal energy ($\sim$0.025 eV), "unzipping" of a SWNT from an existing bundle is unlikely, despite the fact that tubes in a separated state have higher configuration entropy. It is noted that the van der Waals effect alone is not sufficient to account for the different bundling and clustering morphologies present. To form tight bundles, the neighboring nanotubes have to be in close proximity (< 1 nm spacing) to each other, and align parallel with respect to the bundle axis.[21] This means that nanotubes in a bundle will need to "adjust" their conformations to confront the packing requirements cooperatively. Inevitably, there is an associated strain energy of deforming (bending) a nanotube to follow the contour of its neighbors; the magnitude of this strain energy scales with the tube diameter to the fourth power, within a classical model of beam bending. Hence for MWNTs, the energy cost that results from bending may greatly exceed the energy saving from forming parallel bundles for thick carbon nanotubes. Although most MWNTs are not closely packed in order to retain their original contour shapes to minimize the strain energy, they are formed into networks and are subjected to van der Waals binding at the contact junctions. For a tube diameter of 10 nm, this was calculated to be approximately −15 eV per contact.[22] The number of contacts that can form between neighboring CNT pairs increases dramatically with the length of the nanotubes. Therefore, the binding of a clustered network of long MWNTs is extremely strong, where each contact acts effectively as a physical crosslink fixing the network.

To convert the values of these "contact" energies into the more macroscopically meaningful quantity of energy density (ε), the following analysis is applied. For SWNTs, the energy density ε required to separate a pair of parallel tubes of length L and diameter d, bound by an attractive potential energy V_{tot}, is $\varepsilon \approx |V_{tot}|/(Ld)^2$. Using $L = 1\,\mu m$, $d = 1\,nm$ and $V_{tot} = V_{//} \approx -10^3\,eV$ (here, the van der Waals binding energy per unit length is taken as −0.095 eV/Å), $\varepsilon \approx 100\,MPa$ is obtained. For MWNT networks, one needs to have additional information about the number density of nanotube contacts in the network, c_t (in units of number per cubic meter). Then, the energy density required to separate the raw MWNT network is $\varepsilon \approx c_t|V_{tot}|$. One can estimate c_t by looking at the spacing ζ between the direct crossing junctions of the neighboring nanotubes visible on the surface of a cluster (*e.g.* in a scanning electron micrograph). Then, $c_t \sim 1/\zeta^3$. Taking the spacing between MWNT junctions to be $\zeta \sim 100\,nm$ (estimated from the SEM image of MWNTs with diameters of 60–100 nm in Fig. 1 of ref. 23) and $V_{tot} \approx -100\,eV$ for a 80-nm diameter MWNT,[22] we obtain: $\varepsilon \sim 16\,kPa$. The above two examples demonstrate the vastly different energy levels, required to separate the CNTs of various as-grown morphologies.

Energy Density Delivered from Mixing

Separation of nanotubes is usually performed in a solution phase using shear mixing,[24–26] or ultrasonication.[27–30] These processes are both governed by the transfer of local shear stress that breaks down the aggregates. It is therefore intuitive to suggest that complete separation of the nanotubes would require the shear energy densities delivered to the bundle/cluster to exceed the binding energies of the system (as estimated above). Although the exfoliation state achieved may only be a temporary one, it greatly assists the surface adsorption of interfacial molecules (such as surfactants and compatible solvent molecules) which may subsequently stabilize the dispersion of CNTs.[31] During mixing, the level of energy density delivered into a solution is equivalent to the shear stress attainable. Shear stress (σ_s) is defined as the product of fluid viscosity (η) and fluid strain rate ($\dot{\gamma}$), *i.e.* $\sigma_s = \eta\dot{\gamma}$. We next consider the magnitude of the energy density that can be delivered by a mechanical mixing or a sonication technique.

Mechanical shear mixing through stirring can be performed in both low-viscosity solvents (*e.g.*, water or organic solvents with or without dissolved polymers), or highly viscous polymer melts. Hence, the η values employed in shear mixing can span from 0.01 Pa s to 10 Pa s. The fluid strain rate ($\dot{\gamma}$) for the common melt shear mixing is dependent on the rotational speed of the mixer blade (ω in units of rad/s), and the geometry of the mixer and the container. For a typical Couette (concentric cylinder) shear-mixing geometry, $\dot{\gamma} = R\omega / h$, with R the radius of the container, and h the spacing between the leading edge of the mixer blade and the inner wall of the container. The standard Couette mixing conditions[23] have yielded a strain rate of $500\,\text{s}^{-1}$; experimental setups of different geometries could obtain a fluid strain rate as high as $4000\,\text{s}^{-1}$ (see ref. 32). Therefore, using a viscous polymer melt such as uncrosslinked polydimethylsiloxane (PDMS, $\eta = 5.6\,\text{Pa s}$), the shear stress imparted by the mixing medium is below 20 kPa. Note that if one wishes to shear mix in a low-viscosity solvent (such as water, toluene or chloroform), the shear stress delivered to the CNT clusters will drop to below 50 Pa, offering very little hope of achieving full dispersion. The straightforward conclusion drawn from this analysis is that shear mixing is only suitable for dispersion of MWNT clusters in high-viscosity polymer melts.

Fracture of CNTs during Mixing

Compared to mechanical shear mixing, ultrasonication uses a very different mechanism for delivering the shear stress for dispersing aggregates. Cavitation occurs in a low-viscosity fluid above a certain ultrasonic intensity in the low-pressure regions of the traveling wave. Once created, the cavitation bubbles collapse causing an extremely high strain rate in the fluid in the proximate regions of bubble implosion. A strain rate of up to $10^9\,\text{s}^{-1}$ is thus produced.[33,34] The distribution of cavities is controlled by the geometry of the sonicator and the sonication settings, and is inhomogeneous throughout the solution.[35] Taking a typical low viscosity solvent of 0.1 Pa s, the local shear stress σ_{son} imparted in the vicinity of an imploding bubble can approach 100 MPa.

Shear stress acting on the surface of a CNT can induce a pulling effect (a tensile force) on the nanotube. As a result, dispersion methods supplying high energy input can also induce fracture of CNTs. Mixing-induced fracture is of particular relevance in ultrasonication, where literature studies have confirmed the breakage of CNTs during sonication, with an example illustrated in (Figure 11.1). Huang *et al.*[28] employed a potential-flow description of bubble

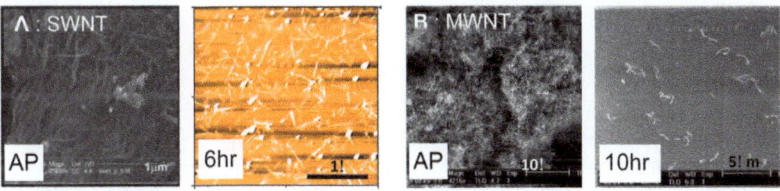

Figure 11.1 Effect of ultrasonication dispersion on the morphologies of CNTs. (A) Scanning electron image of as-produced SWNTs bundles and atomic force microscopy image of the SWNTs sonicated for 6 h. (B) Scanning electron images of as-produced MWNTs cluster and those which were sonicated for 10 h.

implosion dynamics to investigate the shear forces imposed on CNTs during sonication. Based on this model, an important filament fracture resistance parameter $0.5\sigma^*(d/L)^2$ was obtained, where σ^* is the breaking stress (ultimate tensile strength) of the filament. This parameter gives guidance on the morphology of nanofilaments achieved after mixing treatment. When $0.5\sigma^*(d/L)^2 > \sigma_{son}$, the filament will not fracture during sonication; on the other hand, when $0.5\sigma^*(d/L)^2 < \sigma_{son}$, fracture occurs, and L will decrease until $L = L_{lim}$, a limiting length below which fracture no longer takes place. This is approximated to be: $L_{lim} = \sqrt{d^2\sigma^*/2\eta(\dot{R}_i/R_i)}$. Therefore, filaments with higher strength and lower aspect ratio are more resistant to fracture during sonication, as one would expect.

Effects of Shear Mixing and Sonication

Combining the above analysis, an energy-density diagram is presented in (Figure 11.2), which compares the theoretical values of energy density input, filament fracture resistance, and binding energy of the CNT aggregates. The lower and upper binding-energy limits for MWNTs are calculated based on a tube diameter $d = 80$ nm, for nanotube junctions spaced by distances of four diameters, and one diameter apart, respectively. For SWNTs, the ranges of binding energies shown are determined from the energy balance between the bending strain and the van der Waals attraction. The span of values of the filament fracture resistance parameter $0.5\sigma^*(d/L)^2$ arises from the different ranges of strengths reported for different CNTs, such that $\sigma^*_{MWNT} \sim 3$–100 GPa, and $\sigma^*_{SWNT} \sim 10$–100 GPa.[36–40] To understand the

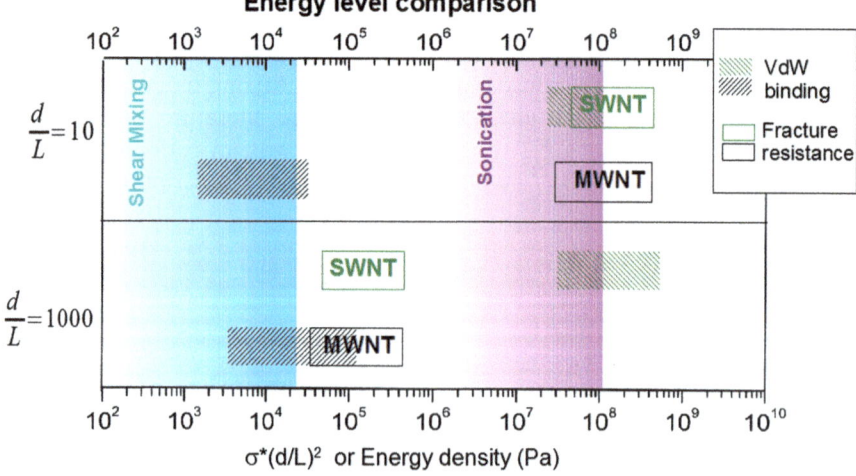

Figure 11.2   An energy diagram showing the theoretical capabilities and limitations of shear mixing and ultrasonication for CNT dispersion, using filament aspect ratios L/d of 10 and 1000 as examples.

diagram, we first focus on the mixing conditions. High-speed shear mixing in a high-viscosity polymer melt can deliver an energy density reaching 10 kPa. However, this energy density is still orders of magnitude lower than an ultrasonic cavitation event. For an aspect ratio $L/d = 10$, the fracture resistance of SWNTs is of a slightly higher value than the binding energy of their aggregates. Ultrasonication just delivers a sufficient stress level to separate the SWNT bundles without causing much fracture to individual nanotubes. For MWNTs, one finds that high-viscosity shear mixing is able to separate the network aggregates. Shear mixing is a better dispersion method because it can effectively separate the MWNTs without causing damage to the filament. For an aspect ratio $L/d = 1000$, which is relevant to the as-produced CNT sources, the fracture resistance of both SWNTs and MWNTs are lower than the stress-input level delivered by sonication. On the other hand, shear mixing is not able to achieve a local stress matching the binding-energy density. This creates the dilemma that for complete separation of long CNTs, the tubes will also be broken during the process.

Based on the above, one can predict the morphology of the CNTs immediately after mixing. It is important to emphasize that the findings stated above only hold when there exists a steady-state condition during mixing/sonication. For mechanical shear mixing, it is useful to determine the time t^*, required to achieve such a steady-state condition. A rheological prototype has been developed to determine t^*,[23] where the evolution of the solution micro-structure for $t < t^*$ was also studied. The viscous medium also provides a means to temporarily stabilize the as-produced dispersion state for subsequent composite fabrication. On the other hand, since ultrasonication both disperses and cuts the CNTs,[41] finding a good "recipe" for sonication dispersion is system dependent and requires significant trial and error experimentation. Since ultrasonication is performed in a low-viscosity solvent/solution (only where cavitation occurs), additional procedures are required to stabilize the as-dispersed state. Various stabilization methods of CNTs in a solution are discussed in the section below.

It is noted that the methods used for CNTs do not necessarily have universal applicability, however; for example, one group found that while PbTiO$_3$ nanoparticles were not well dispersed by ultrasonication, they could be effectively separated by mechanical mixing;[42] meanwhile, MoO$_{3-x}$ nanowires were adequately dispersed under ultrasonication.[83] At the moment, there exists no general 'recipe' for the effective dispersion of nanostructures in a solvent or polymer, but the method must be optimized for each new type of nanostructure and dispersing medium chosen. Nevertheless, one certainly can obtain valuable experience from investigating CNT systems.

11.2.2 Nanostructure Stabilization

In early studies, carbon black was initially incorporated into rubber materials by simple mechanical mixing of the carbon black with the polymer before a vulcanization step; researchers noted that the carbon-black particles remained highly aggregated even after long mixing times.[43] It was observed that the electrical resistivity of the material was dependent on the mixing time, but that the relationship between these variables was far from straightforward; this is a common theme in the study of nanocomposite materials, and the final properties of the material often depend largely on the conditions used for dispersion of nanostructures into the polymer matrix. Reasons for this complex behavior include the formation of aggregates of nanostructures, and variable interfacial interaction between the nanostructure surface and the polymer.

The high aspect ratio of CNTs means that composites containing them can potentially be made more conductive than carbon-black-containing materials, even at much lower loading concentration of the carbon nanoparticles. As discussed in the previous section, the separation of the nanotubes and their dispersion into polymer matrices pose significant challenges. Van der Waals interactions between CNTs are significant due to their high aspect ratio and high polarizability.[44,45] Nanotubes have a natural tendency to reaggregate even after initial separation using shear mixing or ultrasonication. In order to harness useful properties of CNTs in nanocomposites, methods must be found for stabilizing them in solution, and increasing adhesion at the CNT/polymer interface – this may involve chemical modification of the CNT surface. One method for achieving this is to add monomers to the liquid that the CNTs are sonicated in, in order to directly polymerize the monomers around the dispersed CNTs.[46] Alternatively, the surface of the CNTs may be chemically modified in order to promote greater interaction between the CNTs and the solvent, and to improve adhesion between the CNTs' surface and the polymer matrix.[47] This can be achieved by covalently binding organic molecules to the CNTs' surface,[48–51] but since such bonding must necessarily convert a proportion of the carbon atoms at the CNTs' surface from sp^2 to sp^3 hybridization, it is likely that some of the properties that make CNTs interesting for composites (*e.g.* high electrical conductivity) would be degraded in this process. Earlier attempts to covalently functionalize nanotubes tended to rely on harsh chemical conditions, including strong acids, to functionalize the nanotubes; these conditions can also lead to defects in the nanotubes. More recently, procedures allowing for gentler reaction conditions have been suggested based on cycloaddition reactions;[52] however, these methods still rely on disrupting the CNTs' π system by introducing covalent bonds at the surface.

A method for the solubilization of CNTs without disruption of their structure involves the addition of a suitable "surfactant" molecule, incorporating a chemical group that will interact noncovalently with the π system at the nanotube surface, increasing interfacial adhesion between the nanotube and the medium in which it is suspended.[53] Ionic and nonionic small molecules, as well as chemically functionalized polymers, have been used for this purpose. Small-molecule ionic surfactants have commonly been used to suspend SWNTs and MWNTs in water;[54–58] however, nonionic surfactants with a high molecular weight have been shown to be capable of suspending a higher concentration of CNTs; this is likely to be due to steric stabilization effects.[59] O'Connell *et al.* showed that in an aqueous suspension, certain water-soluble polymers are capable of "wrapping" around the CNTs, and suggested that the driving force for this solubilization is thermodynamic, with the energy cost of forcing the polymer into a low-entropy (wrapped) state rather lower than the energy gained by shielding the hydrophobic CNT surface from the aqueous surroundings.[60] Aqueous suspensions of CNTs have been used to create CNT–polymer composite materials; by mixing a surfactant-stabilized suspension of CNTs with an aqueous polystyrene latex solution (generated by emulsion polymerization) a CNT–polystyrene composite was formed.[61]

However, since most polymers are not water soluble, it is useful to be able to suspend CNTs in organic solvents if one wishes to create CNT–polymer composites. Since the surface of the CNTs is hydrophobic, it may be expected that CNTs would be more soluble in organic solvents than in water; however, only a few solvents have been shown to interact sufficiently well with the CNTs to overcome the van der Waals interactions between nanotubes.[62,63] Noncovalent functionalization is therefore desirable in this case, as with aqueous suspensions of CNTs; the surfactants employed should be nonionic and, for optimum dispersion, preferably polymeric. Dispersion of CNTs in acetone was achieved using polyoxyethylene 8 lauryl as the surfactant, allowing the CNTs to be mixed with epoxide polymerization precursors.[64] However, the most effective polymeric surfactants tend to be high molecular-weight polymers; again, this is likely to be due to the effect of CNTs becoming "wrapped" in the polymer. It should be noted that many of these polymers – such as polyvinylpyrrolidone (PVP) – are able to solubilize CNTs in both water and organic solvents.[65] Conjugated polymers often work well to stabilize CNTs, perhaps due to efficient interaction between the nanotubes and the π systems in the polymer "surfactant".[66–68]

Block copolymers are also of interest in this context, because the copolymer can be designed so that one block adheres well to the carbon nanotube surface while the other block improves interaction with an organic solvent. For this system to work well, the organic solvent chosen should be a poor solvent for the block that is anchored to the CNT, and a good solvent for the other block. Block copolymer surfactants has been shown to stabilize SWNTs very effectively in both organic and aqueous solvents.[69] This idea was taken still further by Zou *et al.*, who used a block copolymer surfactant where the anchoring block comprised a conjugated polymer, poly(3-hexylthiophene).[70] The authors showed that this polymer could effectively stabilize SWNTs in chloroform and could allow the CNTs to be well dispersed in a polystyrene film.

All noncovalent surfactants for CNTs require chemical functionality that anchors them to the CNT surface; however, this does not have to be a polymer block. Chemical functional groups that interact strongly with graphite structures include pyrene[71] and porphyrin;[72] incorporating one of these groups as an end-capping functionality on a polymer chain should allow the pyrene/porphyrin group to anchor itself to a carbon nanotube while the polymer extends into the solvent, thus solubilizing the CNT. If the functionalized CNTs are subsequently mixed into a polymer, then interfacial adhesion between the

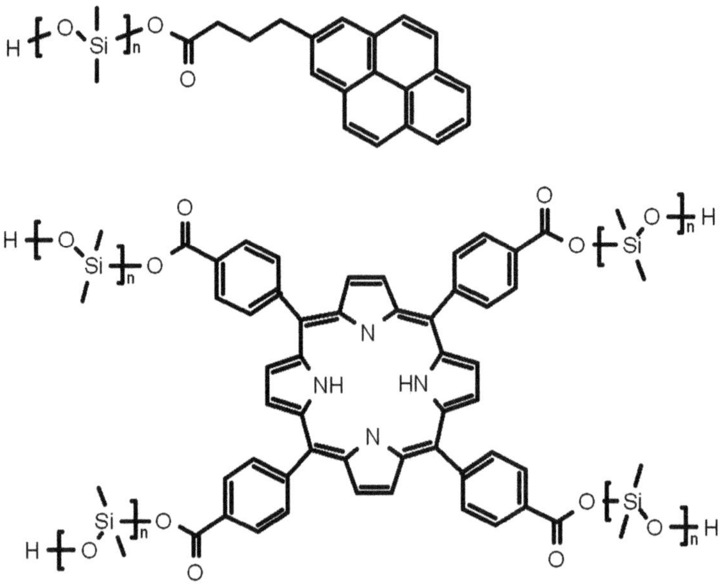

Figure 11.3   Chemical structures of "surfactants" used to solubilize carbon nanotubes in organic solvents and polymers. (a) a polysiloxane chain end-capped by a pyrene group; this surfactant more effectively stabilized MWNTs than SWNTs; (b) polysiloxane chains radiating from a central porphyrin group. This surfactant readily solubilized SWNTs.

polymer and the CNT will be promoted as long as the surfactant shows good compatibility with the polymer. In our group, we have shown that CNTs can be successfully dispersed in organic solvents and in siloxane polymers using this method.[73] The chemical structures of the pyrene/porphyrin functionalized siloxane polymers are shown in (Figure 11.3). When sonicated in petroleum ether in the presence of the surfactant, CNTs formed stable suspensions. We found that pyrene-modified siloxanes were good surfactants for MWNTs (solubilizing them in concentrations up to 10 mg/ml) but not compatible with SWNTs; meanwhile, the opposite situation applied for the porphyrin surfactants (SWNTs were made soluble up to 0.5 mg/ml, while MWNTs were not well dispersed). A possible explanation is that the rigid, planar pyrene group interacts better with MWNTs because the local curvature on the MWNTs' surface is less than that of the SWNTs; meanwhile the porphyrin group interacts with the CNTs preferably *via* the amine groups at its center, so that the surface curvature of the nanotubes become irrelevant. After dispersion in petroleum ether under sonication, our functionalized CNTs were effectively mixed with polydimethylsiloxane (PDMS); a widely used elastomeric material that exhibits desirable properties such as elasticity, chemical resistance and a low glass-transition temperature.

An increasing number of studies investigating the incorporation of 2D carbon nanostructures into polymer composites are now appearing. As in the case of CNTs, the main challenges in this area center around the need to separate the nanosheets and disperse them homogeneously throughout the polymer. Even prior to the discovery of graphene, it was known that in an aqueous medium graphitic oxide can decompose into layers whose thickness is of the order of nm.[74] Graphite oxide is an insulator, and is hence of lesser interest than graphene as a filler material; however, Stankovich and coworkers[75,76] recently demonstrated that flakes of graphite oxide can be reduced to form graphene layers. When this was done in the presence of polystyrene, composite materials were formed that had electrical properties comparable to, or better than, those of CNT–polymer composites.

Thus far, our discussion has been limited to the dispersion of carbon nanostructures in solvents and in polymers. Nanostructures made from metals and oxides are also of interest in terms of how they may modify the properties of a polymer matrix. Metal nanoparticles constitute a huge research field, dating back to the 1850s when Michael Faraday reduced a solution of gold chloride with phosphorus (and realized that the resulting red color was due to the small size of the particles). Several methods are available for the production of monodisperse gold nanoparticles, of which the best known is that invented by Turkevich *et al.*[77] and involves the reduction of chlorauric acid (H[AuCl₄]) with small amounts of sodium citrate. The citrate ions reduce the gold, but also "cap" the resulting gold particles, acting as a surfactant and solubilizing them in water. Alternative capping agents can be introduced, so as to solubilize the gold nanoparticles in organic solvents; in order to disperse them effectively in a polymeric matrix it is possible to make the capping agent itself polymeric.[78,79] Interest in nanoparticle-containing polymers has increased rapidly over the last few years, and methods have also been developed for polymer-coated nano-particles made from other metals, such as silver;[80] and for grafting polymers from oxide nanoparticles such as silica.[81] The incorporation of metal and oxide nanoparticles is of interest due to the possibility of making the materials optically, electrically or magnetically active. This field is not restricted to spherical nanoparticles; as in the case of carbon, it is possible to make high aspect ratio metal-oxide "tubes" that can be dispersed in a polymer.[82–84]

11.2.3 Dispersion of Nanoparticles in Liquid-Crystalline Elastomers

As discussed in Section 11.2.2, nanostructures such as nanoparticles and nanotubes can be dispersed in a wide variety of polymeric materials. We will now discuss the specific case of nanostructures dispersed in the class of materials termed "liquid-crystal elastomers" (LCEs). These weakly crosslinked anisotropic polymer networks are unique due to their ability to reversibly change shape in response to phase transitions; we are therefore interested in the question of whether this response can be enhanced by the dispersion of nanoparticles in the LCE matrix. Similar challenges must be met in the creation of LCE nanocomposites as in other polymer nanocomposites (in terms of dispersion and stabilization of the nanoparticles); however, an added complexity is introduced in that any stabilizing agent added to the reaction mixture must not disrupt the liquid-crystalline nature of the matrix polymer.

Liquid crystals (LCs) consist of rigid, anisotropic molecules. This anisotropy creates a macroscopic orientational order amongst them while they maintain the liquid-like ability to flow. The simplest LC phase is the nematic (in which the anisotropic liquid-crystal molecules, termed mesogens, have a preference to be aligned in a specific direction but have no positional order or regularity; the preferred alignment direction for the mesogens is termed the liquid-crystal director). Transitions between LC phases (induced by heating/cooling of the material) affect the transmission of light through the material, leading to novel optical behavior.

Early attempts to incorporate liquid-crystal mesogens into polymers commenced nine decades ago, and considerable synthetic control over the incorporation of liquid-crystal mesogens into polymers is now feasible;[85] the mesogens can form part of the polymer chains (main-chain liquid-crystal polymers) or be attached as pendant side-chain groups (side-chain liquid-crystal polymers). Useful liquid-crystal polymers have sufficient chain mobility that the mesogens can form liquid-crystal phases, as in the small-molecule system. When crosslinks are introduced between the LC polymer chains (to form a liquid-crystalline elastomer, or LCE), the properties of these systems become still more interesting.[86] The polymer network is now "topologically" fixed, at a determinate number of points; LC phase transformations undergone by the constituent mesogens can potentially therefore create dimensional changes in the bulk phase.

Large dimensional changes in nematic LCEs may only be seen in cases where the material is monodomain, *i.e.* the orientation of the liquid-crystal director is the same across the whole of the material. In a polydomain material, the direction of shape-change will differ between domains and the overall effect will be negligible. The creation of a bulk monodomain sample presented several problems, which were overcome by Finkelmann and coworkers using a two-step technique;[87,88] in this method, a preformed siloxane polymer partially reacts with a mesogen-containing group and a crosslinker, creating a weakly crosslinked gel. This gel is then mechanically stretched, which causes rotation of the LC director within individual domains, leading to an overall LC ordering across the bulk material. A subsequent crosslinking process fixes the aligned state of the LC mesogens. The chemistry of such a system is illustrated in Figure 11.4.

After the second crosslinking step, the LCEs appear transparent at room temperature. With increased temperature, the LC can switch to its isotropic state; in a well-aligned, monodomain sample this phase transformation can be accompanied by a substantial uniaxial contraction of the film, where the contraction direction is dictated by the LC director. Some of the largest contractions of LCEs observed so far have been achieved by incorporating "main-chain" LC polymers into the siloxane system; the polymer film can shrink by up to 500% in some cases,[89] exerting an actuation stress of several to

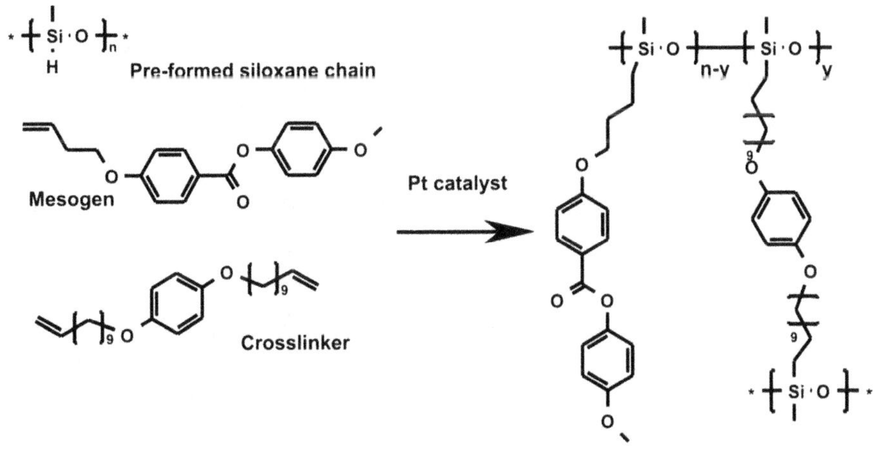

Figure 11.4 Chemical formulae of the siloxane polymer backbone, mesogen, and crosslinker used in a typical synthesis of a "side-chain" LCE.[88] In the presence of a Pt catalyst, a weakly crosslinked gel containing side-grafted mesogens is formed.

tens of kPa. This feature of LCEs' behavior makes them extremely interesting in the field of "artificial muscles", where strain changes of tens or hundreds of per cent are desirable. The actuating behavior of LCEs will be discussed further in Section 11.4.

Nanocomposites made from LCE materials offer new possibilities for enhanced material functionality. The nanostructures employed as LCE filler materials can be designed to have desirable optical, magnetic or electrical properties, and must be well dispersed within the polymer matrix. The primary motivation for this work is to make LCE materials that are responsive to nonthermal stimuli; this will extend the possible range of practical applications for LCEs. In particular, optical and electrical stimuli are more effective than thermal switching, because the stimulus can be highly localized and rapidly switched on and off.

Light-sensitive LCEs have in the past been created by the incorporation of light-sensitive chemical groups,[90–96] while attempts to sensitize LCEs to electrical stimuli have been made using resistive joule heating techniques[97] or the incorporation of ferromagnetic chemical groups into the LCEs.[98–100] Each of these techniques have disadvantages in terms of the response speed, strength of stimulus required, or mechanical contraction generated, but the dispersion of suitable nanostructures in the LCE polymer has been suggested as an approach that could overcome these problems.[101]

In the case of the siloxane LCE polymer depicted in Figure 11.5, all of the polymer precursors are dissolved in toluene before polymerization takes place. We might therefore expect that it is sufficient to disperse nanoparticles well in toluene before simply mixing them with the prepolymerization mixture. However, in the case of nanostructures dispersed in a surfactant, this is not

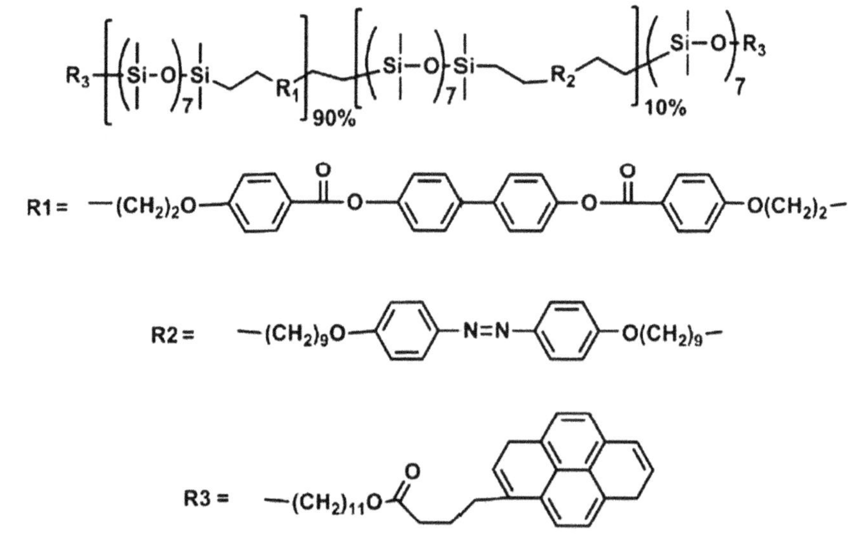

Figure 11.5   Chemical structure of a pyrene end-capped liquid-crystalline polymer that can be used to solubilize multiwalled CNTs in organic solvents and disperse them in liquid-crystalline polymers.

necessarily ideal; if the surfactant is polymeric and is not itself liquid crystalline, then the presence of too much surfactant may disrupt the liquid-crystal order in the LCE. Small-molecule surfactants could be used, but may lead to lesser interfacial adhesion than is possible with a polymeric surfactant. In order to overcome this problem, we designed a liquid-crystalline surfactant, comprising a liquid-crystalline siloxane polymer capped by pyrene groups.[102] The structure of this polymer is shown in Figure 11.5.

After sonication in toluene in the presence of this liquid-crystalline surfactant, MWNTs were well-dispersed in the organic solvent. This suspension was subsequently mixed with the LCE precursors and a catalyst in order to create composite films upon heating. Films with a MWNT loading of up to 3 wt% were fabricated (though calorimetry indicated that a certain amount of phase separation occurs at a loading above 1%). Monodomain LCE films were created by mechanical stretching of the material in between the crosslinking stages, as per Finkelmann's method.

The degree of LC ordering present in LCE-CNT samples of varying CNT concentrations was characterized by X-ray scattering (see Figure 11.6).[103] The scattering patterns observed by X-rays are typical of monodomain LCEs.[104] Polydomain LCEs would generate a uniform ring pattern at 0.5–0.6 nm (directed by the packing of the LC mesogens), but the alignment of the LC mesogens parallel to the director means that there is little scattering in that direction, and therefore there is a nonuniform intensity distribution around the ring. It is clear from these data that the addition of small concentrations of MWNTs does not significantly affect the internal structure of the material, and

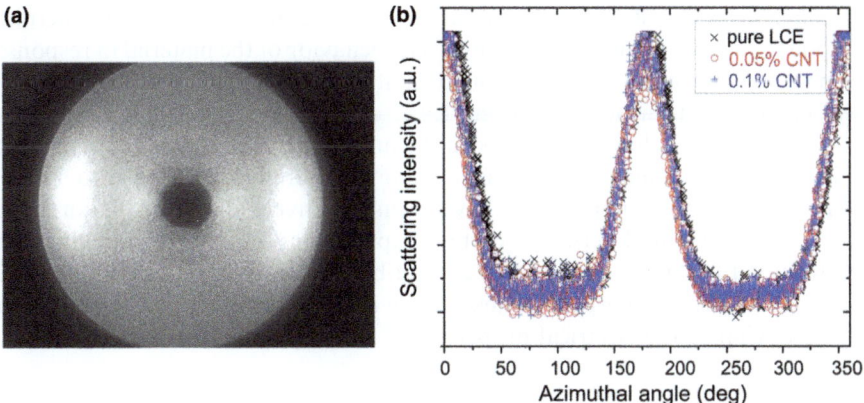

Figure 11.6 (a) A typical X-ray diffractogram produced from nematic LCEs. (b) A comparison of the scattering intensity *vs.* azimuthal angle for LCE films containing different CNT concentrations.

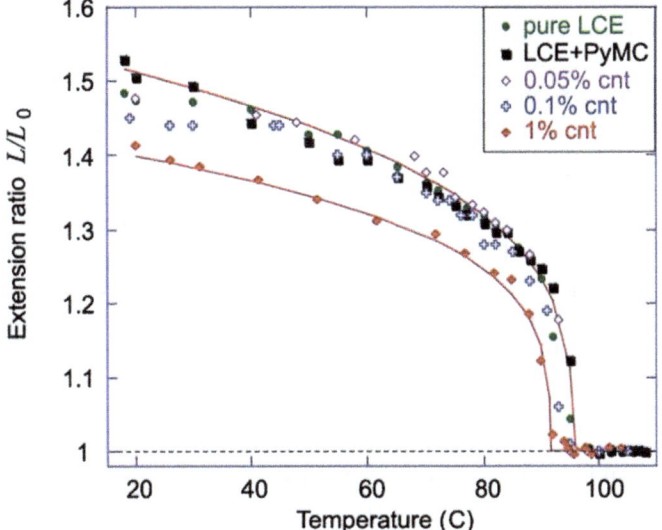

Figure 11.7 The equilibrium mechanical response (change of natural length) of various LCE samples to increasing temperature.

that we may therefore usefully compare the actuation response of these films of composite material.

The response of strips of this composite material to heating is illustrated in Figure 11.7 . The length of each strip decreases with increasing temperature, up to a transition point at approximately 90 °C; at this point, the material becomes fully isotropic and there is no further decrease in the LC order parameter. One would expect that at higher concentrations of embedded CNTs, the stiffness of the material would increase, thus altering the extent of contraction with heating; however, it is clear that at low filler concentrations, the thermal response does not significantly change. The behavior of the material in response to nonthermal stimuli can change significantly, however, and that phenomenon will be discussed further in Section 11.4.

Although CNTs have perhaps been the most widely studied filler particles in this area, other nanostructures have been successfully dispersed into LCEs. Spherical carbon nanoparticles have been effectively dispersed in an LCE matrix by a method termed "carbon reprocessing",[105,106] *i.e.* swelling the preformed LCE in a suspension of carbon-black nanoparticles. Attempts have also been made to incorporate metal-oxide nanowires into the LCE matrix, in order to change the electrical properties of the polymer.[83]

11.3 Nanocomposite Conductivity

Most polymers are insulating in Nature. Incorporating conducting nanoparticles such as carbon nanotubes in a polymer, may make them responsive to electrical stimuli but also allow them to retain the chemical and physical properties of the embedded polymer. These composite systems are promising candidates for a wide range of applications from plastic electronics to biosensing devices. Since the concentration and the conformation/distribution of nanoparticles control the formation of conductive network inside the polymer matrix, modifying one or both of these factors can lead to dramatically different electrical properties of the composite. In this section, we aim to understand the conductivity of polymer/nanoparticle composites from two different angles. First, how microstructures affect the composite conductivity. The observed phenomena can be described using a model based on the random walk of electrons. The second angle is to take a completely experimental approach, to correlate the electrical properties of the composite to the systematic change in one or more fabrication conditions. To begin our discussion, we first define conductivity following the concepts of complex linear response, *i.e.* $\sigma(\omega) = \text{Re}\,\sigma(\omega) + \text{Im}\,\sigma(\omega)$. This means that the frequency-dependent conductivity $\sigma(\omega)$ has a real part $\text{Re}\,\sigma(\omega)$ which is in phase with the applied AC voltage, and an imaginary part $\text{Im}\,\sigma(\omega)$, which is 90 degrees out of phase with voltage. $\text{Im}\,\sigma(\omega)$ is related to the dielectric constant $\varepsilon(\omega)$, *i.e.* $\text{Im}\,\sigma(\omega) = -i\omega\varepsilon(\omega)\,/\,4\pi$. When the frequency ω approaches 0, the imaginary part $\text{Im}\,\sigma(\omega)$ becomes negligible; thus $\sigma(\omega = 0)$ is the DC conductivity. Intrinsically, $\sigma(\omega)$ of a composite is a function of nanoparticle concentration. It is strongly affected by the morphology of the nanoparticles, such as size and aspect ratio (L/d). Carbon nanotubes will continue being our model nanoparticles here to illustrate the dependence of composite $\sigma(\omega)$ on the concentration and the various structural factors of the nanoparticles.

11.3.1 Nanotube Network Coating

First, we focus on the network formed by carbon nanotubes. For a composite to be conductive, the CNTs embedded must form a percolating network.

The CNT network system provides a simpler picture, which excludes the effect of the polymer matrix but still illustrates the key mechanisms that affect the electron transport process through the network formed by the CNTs. Experimentally, these pure CNT networks are usually made by depositing CNTs onto a planar surface, forming a film like structure. Due to the high electrical conductivity of CNTs, one can characterize the electrical property of the network by only considering $\sigma_0 = \sigma(\omega = 0)$, the DC conductivity. To describe the conductivity of a film-like structure in the planar direction, it is more convenient to introduce the concept of surface conductance S, where $S = \sigma_0 \cdot H$, with H being the thickness of the film (see Figure 11.8(A) for schematic representation of S and H). The film thickness H therefore scales with the number of CNTs per unit planar area. S can be measured conveniently using a 4-probe method. H can be evaluated from the light transmission (T) through the film. For thin films, H scales with the film absorption, *i.e.* $H \propto (1 - T)$. Therefore, the dependence of S *vs.* H is equivalent to that of S *vs.* (1-T). Figure 11.8(B) shows plots of normalized surface conductance (S/S^*) against $(1 - T)$, for conducting networks composed of different types of CNTs and graphene, deposited using different methods.[107–116] For a material with constant conductivity σ_0, the slope associated with a plot would be 1, because $S \propto \sigma_0.(1 - T)$. However, this is not the case for the majority of the data presented. One can identify the values of the slopes, which can be between 1 to 3, and change as the film thickness, or $(1 - T)$ increases. In other words, σ_0 is not a constant but depends on the film thickness, which is then reflected by its dependence on the measured $(1 - T)$ values. We therefore define:

$$S = H^* \sigma_0^* (1 - T)^a (1 - T), \quad i.e. \ S = S^*.(1 - T)^\gamma.$$

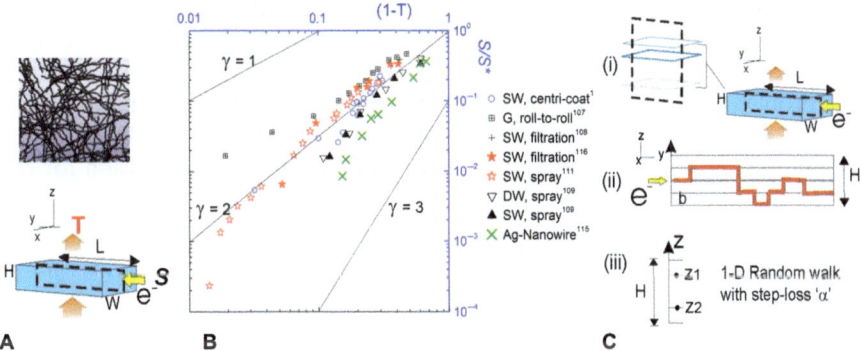

A **B** **C**

Figure 11.8 (A) Scheme of a transparent conductive coating with its coating thickness H and transparency T illustrated. (B) Log–log plots of scaled sheet conductance (S/S^*) against coating absorption (1–T) for conductive nanostructured coatings fabricated by different methods, as detailed in the corresponding references. (C) Overall scheme of the "constrained random walk model": (i) pseudoconduction planes for the drifting electrons; (ii) a probable flow path for an electron drifting along the z-axis when encountering "obstacles"; (iii) the final 1D random walk model that was employed in our calculations.

In the above, we have set a thickness-independent surface conductance through $S^* = H^* \sigma_0^*$, where H^* is a constant prefactor of length scale, σ_0^* is the material conductivity invariant of film thickness, the significance of H^* and σ_0^* will be discussed later. The dependence of σ_0 on film thickness is expressed through $(1 - T)^a$. The second equation is a simplified form of the first equation. The exponent $\gamma = 1 + a$, satisfies that when $\gamma = 1$ (or $a = 0$), the film has a conductivity invariant of film thickness. How can one explain the observed $1 \leq \gamma \leq 3$ (or, $0 \leq a \leq 2$) and the transition in the γ values as indicated in Figure 11.8(B)?.

Electron Transport through Random Paths

To address the above question, we will now illustrate an electron random walk model established in ref. 117. The model is schematically illustrated in Figure 11.8(C), with the objective to evaluate the film resistance in the plane perpendicular to its thickness (z-axis) along the entire length direction. The film morphology is assumed to be homogeneous along the plane, and the width (W) and the length (L) of the film are large. One can imagine the film to be formed by pseudolayers of conducting CNT planes, on which electrons can travel freely in the y-direction under an applied electric field. The total resistance is determined by the resistance of the film per unit width, and hence one only needs to consider the electron transport along the y-direction in a selected $y - z$ plane, Figure 11.8(C) (ii). On the other hand, electrons are also allowed to jump across to a neighboring CNT layer along the z-axis, the probability of which can be assumed to be Gaussian, associated with an energy barrier E_j for such a jump. In effect, this is a semiconducting property of a junction between two CNTs in lateral contact, and we shall discuss it in a different (3D bulk) setting in the next section. Each step is therefore associated with a finite reduction in conductance, expressed by a relative factor α ($\alpha \geq 1$), which is a function of E_j. The problem can be simplified to a 1D random walk of steps b along the z-axis, for any arbitrary starting position z_1 and ending position z_2, confined by the overall thickness of the coating, H.[118] Its solution gives the expression for the total amount of conduction:

$$C_z = \frac{c_0 H}{\ln \alpha} \left[1 - \frac{1}{\sqrt{1.5 \ln \alpha}(H/b)} \tanh\left(\sqrt{1.5 \ln \alpha}(H/b)\right) \right] \qquad (11.1)$$

where $c_0 H$ is the bare conductance of an average CNT path. We clearly see the dependence of C_z on H, the thickness of the coating. When $H \gg b/\sqrt{\ln \alpha}$, eqn (11.1) is simply reduced to $C_z = c_0 H/\sqrt{\ln \alpha}$. This corresponds to the constant-conductivity behavior: a classical Ohmic case, which predicts that the surface conductance is linearly scaled with cross-sectional film thickness (or $\gamma = 1$ in Figure 11.8). On the other hand, how does the relative crosslayer reduction factor α affect the dependence of C_z-H for intermediate values of H? Two limiting cases are relevant. First, when α is close to 1 (*i.e.* the barrier for the electron jumping across the CNT junction is very low), transition between layers is favorable and the conducting path therefore has a high roughness. In this case, $C_z = \pi^2 c_0 H^3 / (12b^2)$, which predicts the surface conductance to scale with H^3 (or $\gamma = 3$). On the other hand, when $\alpha \gg 1$, the z-jumps across the CNT junctions of the conducting path are strongly prohibited and the charge carriers move more or less within plane layers. In this case one again obtains $C_z = c_0 H / \sqrt{\ln \alpha}$, *i.e.* the total conductance is simply a sum from each flat path. One can plot the scaled $C_z / (c_0 b / \ln \alpha)$ *vs.* H / b, which allows us to super-impose all available experimental data by selecting suitable values of α, thus confirming the universality of eqn (11.1). This new analysis is plotted in Figure 11.9.

Comparing the theoretical and experimental results, one finds a high degree of matching for the various CNT coatings. In particular, one expects in this case the step length b to be of the order of the filament (*e.g.* nanotube) diameter, and indeed the thinnest coating of the dataset[116] is at $H = b$, that is, practically a monolayer of CNT. Based on the above analysis, one can deduce that the commonly asserted $S = S^* . (1 - T)^\gamma$ scaling behavior with the exponent $1 \leq \gamma \leq 3$ is due to the transition from a thin film to a bulk structure of a multilayered network. The exact form of correlation is significantly affected by the magnitude of α, the measure of the barrier for charge carriers to transverse the layers. If we return to the beginning of the section relating to σ^* and H^*, we see that σ^* is in fact the constant bulk conductivity, and H^* marks the thickness at which transition from fractual to the bulk phase occurs.

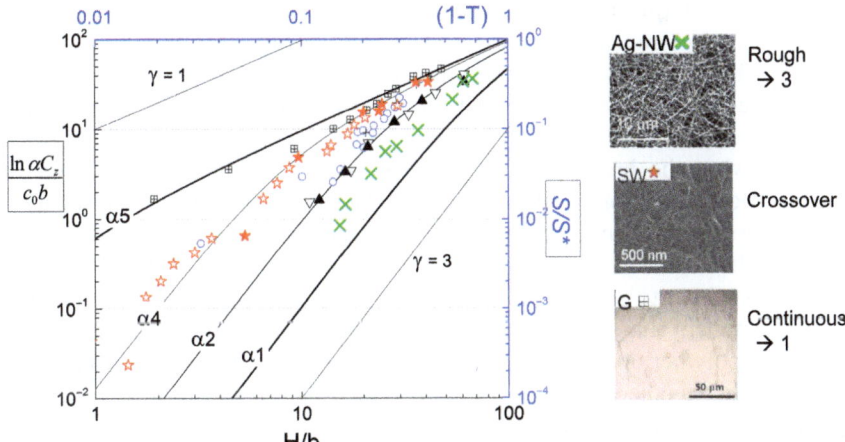

Figure 11.9 Comparison between the modelled data, $\ln(aC_z/c_0b)$ *vs.* H/b, and experimental data, for various nanofilament and graphene coatings. Here, $\alpha1 = 1{:}00008$, $\alpha2 = 1.0008$, $\alpha3 = 1.0025$, $\alpha4 = 1.1$, $\alpha5 = 5$. The labels for the experimental data are the same as those shown in Figure 11.8(B). Microscopic images are shown alongside to support the validity of the theoretical modelling.

Network Microstructure and Conductivity

We will now relate the α values to the associated coating microstructure. A relatively large magnitude of α indicates the preference of electrons to drift parallel to the applied electric field without diversion to neighboring layers. For this to take place, there should be a good continuity of conduction path bridging the start and the end probes. In contrast, smaller α indicates greater roughness of conducting paths. For films formed by highly conductive nano-filaments like CNTs, the length of the filament is a key parameter controlling the thickness-dependent behavior: once in a filament, an electron can flow relatively freely until it reaches the end of the filament. CNTs usually have high aspect ratios. Data from silver nanowires (Ag-NWs) acts as a good comparison to the results from SWNTs. Having a much smaller aspect ratio (filament diameter $\sim 80\,nm$, length $\sim 6.5\,\mu m$) than SWNTs, Ag-NWs show a much stronger dependence of surface conductance on the coating thickness, *i.e.* the electrons are having to jump across layers making very rough paths resulting in a low α value. Together with microscopic images shown in Figure 11.9, one again confirms that a film with higher α value is associated with a micro-structure of better uniformity. Accordingly, the above theory also predicts that a nonunit exponent would always exist for nanofilament coatings because of the finite filament lengths. This is also what is observed in practice. α is an important microstructure parameter that includes information about the filament aspect ratio, and the connectivity of the network.

Remark on Junction Resistance

Investigation of various nanostructured coatings has revealed a very different sheet resistance–transmittance $S = S^{*}.(1 - T)^{\gamma}$ correlation that cannot be described by the classical thin-film model, making them a special category of conductive coating. The theoretical model is based on counting the random paths of charge carriers through a multiply connected system, which allows us to relate what the broad literature perceives as the "exponent" γ, and the optoelectrical behavior of the coatings. The general equation allows us to predict the system parameters for ultimate coating performance. For rough film coatings, the exponent γ can implicate the gradual transition of the film from 2-dimension ($\gamma = 3$) to 3-dimension transition ($\gamma = 1$). Such a continuous change in the exponent has also been observed for a network made of carbon nanotube–polymer composite fibers.[118] Therefore, the carbon-nanotube network discussed here represents a general model for bare conducting network systems, as well as networks in a matrix.

11.3.2 Bulk Nanotube Composite

Studying the conductivity of a pure CNT network shows that conduction is a process of electrons flowing through the paths established by CNTs that connect the start and the end probes. Electrons "crossing" through the CNT contacts cost energy, and this is the main contributor to network resistance. The above study reveals the key roles played by the aspect ratio of individual conducting filaments, and the connectivity of the filaments, in addition to the filament concentration. The same factors also influence the resulting electrical properties of the polymer/CNT composites, which is the topic of the present section. It is noted that the electrical properties of composites are complicated by the dispersion quality of CNTs, which leads to varying spatial distribution of CNTs inside the matrix. In this section, we focus on experimental evaluation on how the variation in a fabrication process would affect the mechanism of conductivity and thus the final electrical properties of the carbon nanotube–polymer composite. Since the polymer matrix present a dielectric surrounding to the CNT network, we discuss the full form of complex conductivity $\sigma(\omega) = \text{Re}\,\sigma(\omega) + \text{Im}\,\sigma(\omega)$. It is hoped that the following section can provide a systematic guide to selecting and designing the downstream processing of carbon nanocomposites.

Processing Parameters

Polydimethylsiloxane (PDMS) and carbon nanotube composites are used as the model system. Debundling and dispersion of nanotubes were achieved by either ultrasonication or mechanical shear-mixing (see Section 11.2). The notation of MW_u-, MW_m-, and SW_u- are introduced. The first two letters of each term denote the type of CNTs being dispersed, and the subscript denotes the debundling method, which corresponds to ultrasonication (subscript u) or mechanical shear mixing (subscript m).

In order to reduce the uncertainty in the tube length distribution as a result of sonication scission, all the MW_u- samples were sonicated for 8 h (unless otherwise stated) in a PDMS-solvent solution, ensuring the majority of CNTs have reached their limiting length, L_{lim}.[28] The MW_m- samples were prepared by melt shear-mixing following the procedures outlined in ref. 23. Unless otherwise stated, they were all mixed for times well beyond the t^*, the critical time to achieve a steady rheological signature of the dispersion. Polysiloxane surfactant $mPSi_{70}$[73] was also applied in some cases (corresponding samples labeled as MW_u-PSi, and MW_m-PSi) to make systematic comparison with pristine CNT-PDMS composites. To test the effect of nondispersed CNT clusters on the composite conductivity, samples based on the 2 wt% pure MWNT-PDMS composition mechanically mixed for times much smaller than the critical mixing time were also fabricated. These samples are labeled as $MW_m(2)(t)$. Finally, as reference, a 0.1 wt% surfactant-free SWNT-PDMS sample by sonication: $SW_u(0.1)$, and a 5 wt% pristine carbon black sample by mechanical mixing: $CB_m(5)$, were also shown.

DC Response

First, we consider the DC responses of these composites, *i.e.* $\sigma_0 = \sigma(\omega = 0)$ through the DC voltage, current response. A number of representative current density *vs.* electric field (*J–E*) curves (essentially analogous to current-voltage, *I–V*, but independent on the measurement geometry) are plotted on one master plot in Figure 11.10. One observes a clear separation in the DC response, making three different groups as labeled on the plot. Crossing the percolation threshold, *i.e.* moving from Group C to Group B, one finds a jump in the current density of around 6 orders of magnitude for the same applied field. However, the composites in Group A, with a really high conductivity, have the current density another 2–3 orders of magnitude higher. Based on this master plot, we will investigate the *I–V* characteristics of each conductivity group, and then analyze how each fabrication method has affected the DC response of the CNT-PDMS composites.

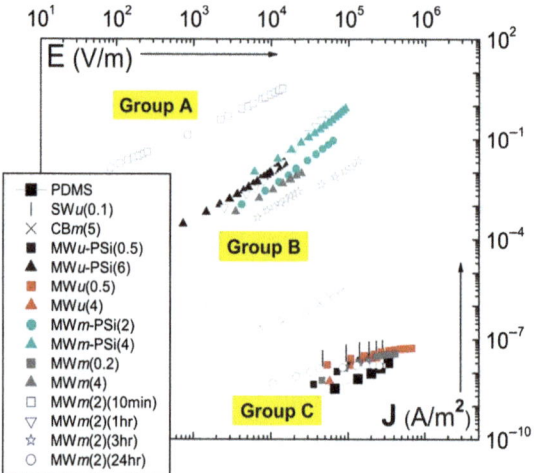

Figure 11.10   Master plot of current density *vs.* applied field *J–E* for a range of carbon polymer composites (this is analogous to *I–V*, but presents the relevant intensive parameters). The details of fabrication process corresponding to each sample are illustrated in Table 11.1.

Table 11.1   List of the fabrication conditions represented by each sample label in Figure 11.10.

Notation	Filler addition	Processing method	Mixing time
MWu (wt%)	MWNTs: 0.5 wt%, 4 wt%	ultrasonication	8 h
MWu -PSi(wt%)	MWNTs: 0.5 wt% (filtered) 6 wt% (filtered)	Psi assisted dispersion ultrasonication	8 h
MWm (wt%)	MWNTs: 0.2 wt% to 4 wt%	Shear mixing	24 h
MWm -PSi(wt%)	MWNTs: 2 wt%, 4 wt%	PSi assisted dispersion, Shear mixing	24 h
MWm (2)(t)	MWNTs: 2 wt%	Shear mixing	*t*: 10 min; 1 h; 3 h; 24 h
SWu (0.1)	SWNTs: 0.1 wt%	ultrasonication	40 min
CBm (5)	Carbon Black: 5 wt%	Shear mixing	24 h

As illustrated in Figure 11.10, the DC *J–E* profiles of composites lying in different conductivity regimes are distinct. For the conducting Group A one finds a linear (Ohmic) *I–V* dependence. For Group B, an apparent increase in conductivity with voltage resembles the semiconducting response. The conductivity value between 10^5 to 10^7 S/m is at the lower end of typical semiconductors as well. Similar *I–V* profiles have been reported by other authors,[119–121] who attributed this to a Schottky contact formation between the metal electrodes and the semiconducting composite. Another interpretation based on CNT contact resistance was suggested by Bryning *et al.*[122] Finally, for the insulating Group C the opposite effect is found: the current saturates at higher voltages. The conductivity of composites in Group C, in the region of 10^{12} to 10^{13} S/m, is similar to pure PDMS. For PDMS, however, one does not find the current saturation within the field range investigated: it shows a linear *I–V* relation as expected for a plain dielectric. The leakage-current saturation at intermediate field strength implies an apparent increase in resistivity with applied field, which closely resembles an effect observed in dielectric liquids when residue charges are present.[123]

Next, let us turn our focus to how the variation in fabrication could ultimately change the associated conduction mechanism and thus the electrical properties of the composites. First, comparing the ultrasonicated samples (MW_u-) with their mechanically mixed counterparts (MW_m-) reveals a strong dependence of composite conductivity on the aspect ratio of tubes. It has been determined previously in ref. 28 that an 8 h sonication will reduce the length of the present MWNT type to 1/3 of their original lengths (which still gives an aspect ratio of ~ 100). Therefore, despite the high MWNTs content, $MW_u(4)$ exhibits a much lower conductivity than even the carbon-black shear-mixed composite *CBm*(5). This result suggests that long-time sonication has not only resulted in significant tube scission, but also caused detrimental amorphization of the graphene walls, making the $MW_u(4)$ composite effectively an insulator. For the conductive composites, the MW_u-PSi(6)(filtered) composite also has a conductivity value lower than the MW_m-PSi(4) sample, see Figure 11.10.

Now we can look at the effect of surfactants, in this case mPSi70. Since MWNTs were used, mPSi70 has induced an electrical percolation in the 2 wt% MWNT-PDMS composite that is otherwise insulating in the pure form (compare MW_m-PSi(2) and MW_m(2)(24 h) in Figure 11.10). For concentrations well above percolation, the addition of mPSi70 also leads to approximately an order of magnitude increase in conductivity for the same MWNT addition (compare MW_m-PSi(4) and MW_m(4). A possible explanation for the improvement in conductivity is that mPSi is in a liquid state at room temperatures; thin layers of mPSi surrounding the CNTs promote connection of the loose dangling ends/loops of CNT in the matrix. Electron transport through the contact junctions of CNTs has been shown to be the dominant mechanism for macroscopic conductivity.[124,125] This is unlike surfactants such as SDS (sodium dodecyl sulfate),[126] which form a dense shell layer surrounding the CNTs, increasing the tunneling barrier for electron transport across the CNT junctions. Surfactants studied elsewhere (*e.g.*, Mitchell and

Krishnamoorti;[127] Zou et al.[70]) also appear to have a similar effect of improving the electrical percolation behavior of CNTs in polymer matrices.

From Figure 11.10, the most interesting result comes from comparing the pure (surfactant-free) MWNT-PDMS mixture subjected to an increasing time of shear mixing, i.e. the $MW_m(2)(t)$ series. These materials belong to the three distinct groups (A, B and C) in spite of their identical chemical composition. At this concentration, the composites are fully mixed when the mixing times are greater than 5 h. For the short-time mixed samples, the nondispersed CNT clusters are inevitably present in the matrix, leading to strong inhomogeneity of the composite. With increasing mixing time, a dramatic decrease in conductivity is found, and finally the well-mixed sample has become effectively insulating. This is equivalent to a change in conductivity of 9 orders of magnitude at the same concentration. Nevertheless, the above results seem to suggest that it is the nondispersed clusters that form a conductive network spanning between the top and bottom electrodes, which promote the conductivity of the composite. Any thin coating of polymer on the nanotube surface can significantly reduce the rate of charge transport,[125,128,129] while physical entanglement between the neighboring tubes can still remain.

AC Response

The J–E master plot in Figure 11.10 clearly shows a strong dependence of DC response on the microstructure of the composites, with conductivity dominated largely by the nature of contact between nanotubes. Next, we will look at how the AC response differs between composites processed by different methods for Group B and Group C. AC impedance was measured at a voltage amplitude of 0.1 V (root mean squared amplitude), where the modulus of complex conductivity $|\sigma(\omega)^*|$, and the loss factor $\tan(\delta)$, are plotted against frequency in Figures 11.11(A) and (B). Although the composites in Group B all demonstrate similar I–V profiles, their AC responses appear to be much more varied in the frequency window ($f = \omega/2\pi$) of between 100 Hz to 2 MHz. One would expect a low-frequency $|\sigma(\omega)^*|$ plateau for the Group B composites since they all show a reasonable DC conductivity. However, there are no signs of such a plateau developing for either $MW_m(4)$ and $MW_m(2)(3$ h) composites. In fact, $MW_m(2)(3$ h) shows an incredibly low loss factor of ~ 0.03 throughout the frequency range, comparable with the experimentally determined $\tan(\delta) \sim 10^3$ for pure PDMS. Clearly, the semiconducting values of DC response seen in Figure 11.11(A) require a much longer relaxation time than 10^{-2} s at the lowest frequency measured. Due to their significant DC conductivity these composites may not hold potential for practical applications as low-frequency dielectrics, however, they may be considered as high-permittivity dielectrics in the high-frequency range, above MHz. For the complementary $\tan(\delta)$–f plots, maxima in $\tan(\delta)$, are observed for the more conducting composites, indicating the relaxation process moving into the observed frequency range. This frequency range would translate into length scales of 100 to 10^6 m, unlikely to be caused by the actual size of any physical objects present. Nevertheless, there is a

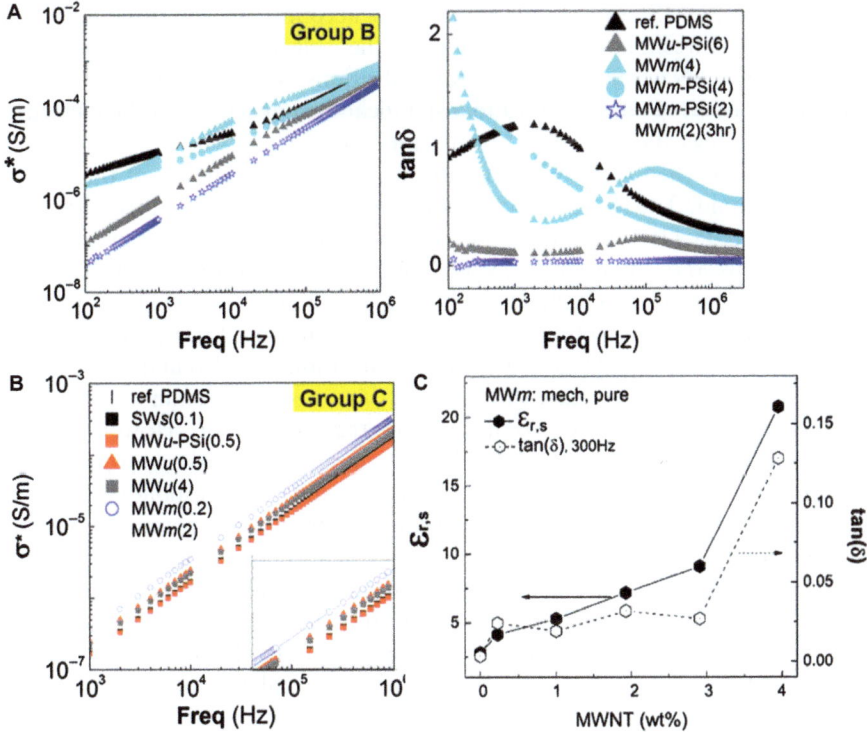

Figure 11.11 (A) Plot of complex modulus of conductivity (left) and loss factor (right) against frequency for various Group B composites; (B) Plot of complex modulus of conductivity against frequency for various Group C composites; the insert shows a zoomed-in profile for clearer representation of the data order. (C) Plot of $\varepsilon_{r,s}$ and $\tan\delta$ (at 300 Hz) *vs.* the CNT concentration in the well-mixed, surfactant-free CNT–PDMS system.

correlation between the relaxation frequency and the mesh size of the homogenous MWNT-networks (related to the CNT concentration in the matrix and the tube length). The $\tan(\delta)$, maxima shift to higher frequencies for increased doping of longer tubes.

For the insulating samples of Group C one uniformly finds a very low loss factor ($\tan(\delta) < 0.01$); therefore only the $|\sigma(\omega)^*|$-f data is presented in Figure 11.11(B). All the Group C materials behave close to an ideal dielectric filling the parallel-plate capacitor, as does the pure PDMS, showing $|\sigma(\omega)^*| \sim f^1$. Therefore, one can extrapolate the static dielectric constant ($\varepsilon_{r,s}$) from the relationship of a classical capacitor $\varepsilon_{r,s} \sim \sigma/(2\pi f \varepsilon_0)$. Values of $\varepsilon_{r,s}$ for this group of samples are estimated to be between $\sim$2.8–7.8, where a more depressed $|\sigma(\omega)^*|$-f line corresponds to the higher $\varepsilon_{r,s}$. SWNTs appear to give more effective enhancement in permittivity than the MWNTs doping. The higher $\varepsilon_{r,s}$ of MWm(2) also supports the picture of physically interacting tubes separated by layers of PDMS coating, the model that accounts for the discrepancy in percolation thresholds derived from the electrical and rheological measurements. Nevertheless, with concentrations much below percolation, addition of MWNTs did not lead to a significant enhancement in the dielectric permittivity, regardless of the fabrication method.

What about when the CNT concentration is at the boundary of the percolation threshold? The static dielectric constants were extrapolated from the AC impedance data (100 Hz–2 MHz) for the pure mechanically mixed samples: homogeneous MW_m composites, with MWNTs concentrations from 0.2 wt% to 4 wt%. Figure 11.11(C) illustrates how $\varepsilon_{r,s}$ and tan(δ) (at 300 Hz) vary with the concentration of nanotubes. The percolation threshold of this system is at $\sim$3 wt% MWNTs in PDMS, at which concentration one also finds the growth in the loss factor. Below this threshold, $\varepsilon_{r,s}$ increases approximately linearly with concentration without compromising the loss factor. However, such an increment is not very significant. As the concentration crosses over 4 wt%, one immediately observes a step rise in $\varepsilon_{r,s}$, and a corresponding rise in dissipative current reflected by tan(δ). Therefore, the studied composite system did not reproduce the nondissipative, divergent permittivity predicted by percolation theory.[130] Careful analysis of available literature reveals that CNT/ polymer composites with a claim for the significant enhancement in relative permittivities were similarly accompanied by high dielectric losses (tanδ > 0.1).[131–133] It seems that significant dielectric enhancement is difficult to obtain by simple mixing of a highly conductive filler of high aspect ratio (like CNTs) in a dielectric matrix. In order to prevent the intertube contact, which contributes to the dissipation current, systems based on surfactant-coated CNTs[102,134] have been investigated elsewhere.

11.4   Light-Induced Actuation

As we have shown, the inclusion of nanostructures into a polymer matrix may simply be used to alter the properties of the material, such as conductivity. This topic is of particular interest when the composite material is actuating, *i.e.* capable of producing a mechanical response to an applied stimulus (such as heat, light, or an electromagnetic field).[17,135,136] We will herein discuss systems where a polymer is sensitized to light by the addition of functional nanoparticles. This can be achieved using polymers that have little intrinsic actuating ability (beyond the thermal expansion/contraction exhibited by all polymers under changing temperature) but gain responsivity when nanoparticles are added; alternatively, it is possible to begin with a responsive polymer (such as an LCE) and enhance its properties by the inclusion of embedded nanoparticles.

11.4.1   Sensitizing Polymers to Light

Thus far, most photoactuating nanocomposites have been based on polymers containing CNTs. This is largely due to the optical behavior of CNTs, which absorb light effectively over a range of wavelengths. MWNTs' light absorption is relatively constant over a wide range of wavelengths,[137] while SWNTs can show specific absorption peaks.[138,139] The absorption spectra of SWNTs do, however, depend on their dispersion and alignment; "forests" of vertically aligned SWNTs have been observed to almost perfectly absorb all incident light over a wavelength range of 500–1200 nm (behaving almost as a perfect blackbody absorber).[140] An absorber in thermal equilibrium with its surroundings must re-emit all absorbed light as heat; indeed SWNTs, when exposed to a photographic flash, can emit so much heat that they spontaneously ignite.[141] Given that there are already a wide range of polymers that are responsive to heat, this property of CNTs is therefore an obvious way in which to sensitize them to light; CNTs dispersed in a polymer can emit heat, thus triggering dimensional changes in the heat-sensitive matrix. It has also been suggested that CNTs themselves can contract under light irradiation;[142] this effect could also aid CNT–polymer composites to produce useful dimensional changes in response to an optical stimulus, but is yet to be fully understood.

Most materials expand isotropically when heated, but if alignment is induced in a series of crosslinked polymer chains (*e.g.* by mechanical stretching) then increased temperature causes contraction of the material along the direction of alignment, and this contraction is driven by entropic effects. When CNTs are added to an aligned polymer of this type, there are two possible light-sensitizing effects to consider. The CNTs can absorb light and re-emit this as heat, thus driving the chain contraction in the same fashion as direct heating; also, contraction of the nanotubes themselves under light irradiation may create a strain difference between the matrix and the CNTs, thus inducing contraction of the bulk material.

It has been shown that a common commercial elastomer, polydimethylsiloxane (PDMS), can become responsive to light when CNTs are well dispersed within it.[143] The strength and direction of the response is dependent on the initial strain applied to the material; as shown in Figure 11.12, at low values of prestrain the composite material expands when irradiated with infrared light, whereas when the material is highly strained it shows a contraction (and there is a crossover point, at roughly 8–10% strain, at which the material shows little expansion or contraction). This effect has been shown to be similar for embedded MWNTs, SWNTs[144] and graphene.[145]

11.4.2 Enhancement of Shape-Memory Properties

The observed contraction of aligned polymers when heated is not the only type of directional stimulus response available. The rapidly growing field of "shape-memory polymers" (SMPs) deals with macromolecular materials that have been deformed into a "programmed" shape at an elevated temperature, before having the programmed shape "frozen in" during a rapid cooling process. These materials are capable of re-forming their original shape upon phase transitions triggered by heat.[146] The phase transition can simply be the

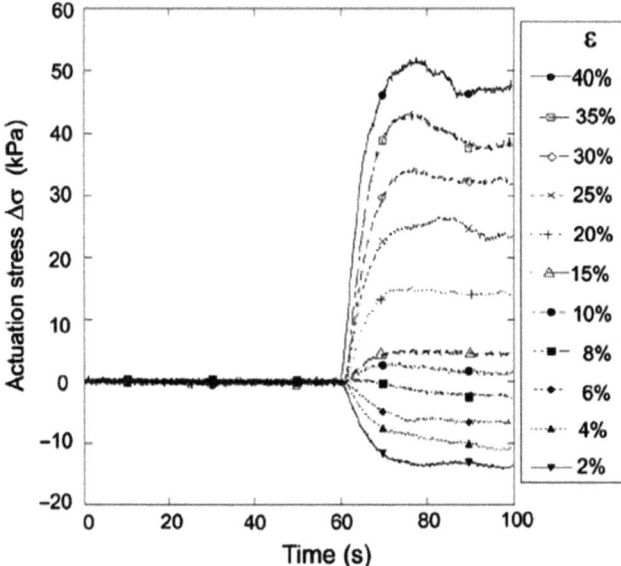

Figure 11.12   The speed of actuation response, illustrated by plotting the actuation stress of a fixed-length PDMS-CNT composite as a function of time for different prestrain values. Increasing prestrain leads to the increasing alignment of CNT segments; the positive actuation stress means the sample contracts on illumination, while the negative $\Delta\sigma$ means the sample increasing its natural length.

glass transition of an elastomer; when deformed at a temperature above the T_g, and subsequently rapidly cooled, the polymer retains the "new" shape, but recovers the original shape when reheated above the T_g. Alternatively, a shape-memory polymer can be composed of a block copolymer whose component blocks have different melting temperatures, and the shape-memory transition can occur during melting of one of the blocks.

In order to sensitize these materials to light rather than heat, it is necessary to either modify the chemical structure of the polymer or to include a light-absorbing component (as we described previously for the PDMS-CNT composites). Polymers containing chemical functional groups that can be crosslinked under UV light have been demonstrated; these can be deformed into the required shape, "fixed" by crosslinking under UV light, and subsequently recover their original shape by photocleaving of chemical bonds under UV light of a different wavelength.[147] The inclusion of embedded light-absorbing nanoparticles, however, has the advantage that this process can be made more flexible, with actuation occurring over a much wider range of light wavelengths (including visible and infrared light). In this case the shape-memory effect is triggered by heating across a phase transition, but the heat is provided by the light-absorbing filler material. The light-absorbing particles used can be CNTs[148] or other light-absorbing nanoparticles such as carbon black.[149]

11.4.3   Actuation of Aligned LCE Nanocomposites

The shape-memory materials described in Section 11.4.2 can recover their original (undeformed) state upon heating, but this is a one-way shape-memory effect; after recovery, the material will not spontaneously regain its deformed state on cooling. The ability to switch repeatedly between two states is a major advantage of liquid-crystal elastomers, which we briefly described in Section 11.2.2. These materials could be considered as a different type of shape-memory material, but in this case the mechanism for actuation is not simple stress

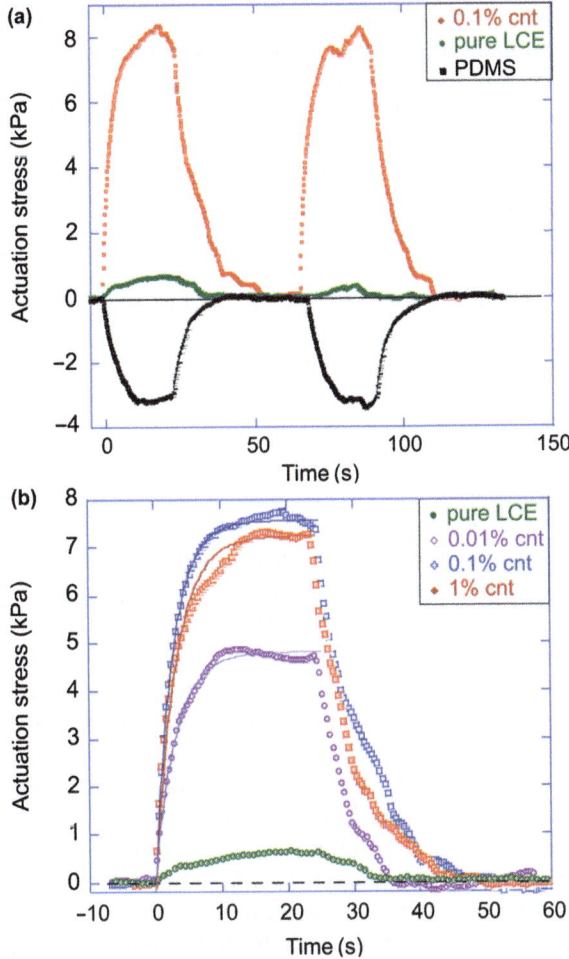

Figure 11.13   (a) Stress response to light irradiation for 3 different materials; a "pure" LCE shows little response to the applied light, while a "nonaligned" CNT-PDMS composite expands under the incident light. An aligned, monodomain LCE-CNT composite, however, contracts under the same light source. (b) Stress response data showing a nonlinear dependence on CNT concentration.

release (as in the case of the shape-memory polymers) but is linked to the LC phase changes undergone by the covalently bound mesogens that form part of the polymer.

As in the case of shape-memory polymers, monodomain LCEs undergo phase transformations (and thus bulk deformations) in response to heating. The material can be sensitized to incident light by the inclusion of light-absorbing nanoparticles, as we have already observed for the shape-memory polymer nanocomposites described in Section 11.4.2; embedded nanostructures convert the incident light into heat, thus altering the LC order parameter and triggering the phase change.[103,102]

The actuation response of a light-sensitive material of this type can be measured by clamping one end of a film to a dynamometer (while the other end is fixed) and measuring the force output as the film is irradiated with an incident light source. As shown in Figure 11.13(a), a "pure" LCE produces little response to an incident white-light source, while an LCE-CNT composite contracts when the light is switched on (producing an actuation stress of *ca.* 6 kPa). This is in contrast to an "unstretched" PDMS material containing embedded CNTs (since the material is not stretched, the polymer chains in the PDMS matrix will not be aligned), which (unsurprisingly) expands due to heating under irradiation. Figure 11.13(b) shows that the response to the incident light is not linearly dependent on CNT concentration. This may be because higher concentrations of CNTs decrease the penetration depth of light into the material, or because the increased concentration of embedded CNTs increases the stiffness of the LCE, thereby suppressing the actuation response.

11.5   Conclusions

We have shown that nanostructured fillers can be embedded in polymers to make novel composite materials with interesting conductive and actuating properties. The selection of both the polymer and nanostructures used will direct the final properties of the material, and a wide variety of possible combinations exist; however, the processing techniques used in order to disperse and stabilize nanostructures in a polymer matrix is also crucial. The most widely used techniques for the dispersion of nanostructures (in a solvent or a polymer) are mechanical shear mixing and ultrasonication; however, the optimum conditions for dispersion must be optimized for each polymer/ nanostructure system chosen. In order to prevent reaggregation of the nanos-tructures after dispersion, it is possible to add extra stabilizing agents; this has been effectively demonstrated using polymeric "surfactants" that incorporate chemical functionality compatible with both the nanotube surface and the dispersing medium. The strong dependence of the nanocomposite material properties on processing conditions is evidenced by the large differences in electrical conductivity shown by polymer–CNT composites that have simply been shear mixed for varying lengths of time.

The optical absorption properties of certain types of nanostructures, particularly CNTs, gives rise to their ability to sensitize a polymer to light.

This can be achieved in a general way, using CNTs embedded in stretched elastomers (where incident light causes the CNTs to generate heat, causing a contraction in the polymer to maximize entropy), but also to enhance the actuating properties of LCEs (where heat generated by the CNTs causes a contraction due to a phase transformation). The use of light as an actuation trigger could be revolutionary for a number of practical applications, because the actuation stimulus is localized, safe, and not in contact with the actuating component. Actuation stresses of several kPa have been observed, making these actuating materials comparable to muscle. In future work, it should be possible to further optimize the materials' response, for example by adding organic light-absorbing components.

References

1. D. L. Feldheim and A. F. Colby, *Metal Nanoparticles- Synthesis, Characterization and Applications*, CRC Press, New York, 2001.
2. J. H. Fendler, ed., *Nanoparticles and Nanostructured Films*, Wiley-VCH, Weinheim, 1998.
3. P. Podsiadlo, A. K. Kaushik, E. M. Arruda, A. M. Waas, B. S. Shim, J. Xu, H. Nandivada, B. G. Pumplin, J. Lahann, A. Ramamoorthy and N. A. Kotov, *Science*, 2007, **318**, 80–83.
4. S. S. Ray and M. Bousmina, *Polymer Nanocomposites and Their Applications*, American Scientific Publishers, Valencia, CA, 2006.
5. G. Kraus, *Rubber Chem. Technol.*, 1978, **51**, 297.
6. G. Kraus, *Angew. Makromol. Chem.*, 1976, **60**, 215.
7. O. A. Al-Hartomy, F. Al-Solamy, A. Al-Ghamdi, N. Dishovsky, V. Iliev and F. El-Tantawy, *Int. J. Polym. Sci.*, 2011, Article ID 837803.
8. Y. Suda, T. Ono, M. Akazawa, Y. Sakai, J. Tsujino and N. Homma, *Thin Solid Films*, 2002, **415**, 15–20.
9. S. Iijima, *Nature*, 1991, **354**, 56–8.
10. M. S. Dresselhaus, G. Dresselhaus, J. C. Charlier and E. Hernández, *Philos. Trans. R. Soc. Lond. A*, 2004, **362**, 2065–2098.
11. R. Khare and S. Bose., *J. Min. Mater. Charact. Eng.*, 2005, **4**, 31–46.
12. K. S. Novoselov, A. K. Geim, S. V. Morozov, D. Jiang, Y. Zhang, S. V. Dubonos, I. V. Grigorieva and A. A. Firsov, *Science*, 2004, **306**, 666–669.
13. K. S. Novoselov, D. Jiang, F. Schedin, T. J. Booth, V. V. Khotkevich, S. V. Morozov and A. K. Geim, *Proc. Natl. Acad. Sci. USA*, 2005, **102**, 10451–10453.
14. M. Corso, W. Auwärter, M. Muntwiler, A. Tamai, T. Greber and J. Osterwalder, *Science*, 2004, **303**, 217.
15. C. Zhi, Y. Bando, C. Tang, H. Kuwahara and D. Golberg, *Adv. Mater.*, 2009, **21**, 2889–2893.
16. D. Golberg, Y. Bando, Y. Huang, T. Terao, M. Mitome, C. Tang and C. Zhi, *ACS Nano*, 2010, **4**, 2979–2993.

17. S. V. Ahir, Y. Y. Huang and E. M. Terentjev, *Polymer*, 2008, **49**, 3841–3854.
18. M. Moniruzzaman and K. I. Winey, *Macromolecules*, 2006, **39**, 5194–5205.
19. R. Verdejo, M. A. Lopez-Manchado, L. Valentini, and J. M. Kenny, 'Carbon Nanotube Reinforced Rubber Composites' in: *Rubber Nanocomposites: Preparation, Properties and Applications*, T. Sabu and S. Ranimol ed., Wiley, Singapore, 2009.
20. L. A. Girifalco, M. Hodak and R. S. Lee, *Phys. Rev. B.*, 2000, **62**, 13104–13110.
21. Y. Y. Huang and E. M. Terentjev, *Polymer*, 2012, **4**, 275.
22. A. I. Zhbanov, E. G. Pogorelov and Y. C. Chang, *ACS Nano*, 2010, **4**, 5937–5945.
23. Y. Y. Huang, S. V. Ahir and E. M. Terentjev, *Phys. Rev. B*, 2006, **73**, 125422.
24. R. Andrews, D. Jacques, M. Minot and T. Rantell, *Macromol. Mater. Eng.*, 2002, **287**, 395–403.
25. J. H. Park, P. S. Alegaonkar, S. Y. Jeon and J. B. Yoo, *Compos. Sci. Technol.*, 2008, **68**, 753–759.
26. M. H. G. Wichmann, J. Sumeth, B. Fiedler, F. H. Gojny and K. Schulte, *Mech. Comput. Mater.*, 2006, **42**, 395–406.
27. Y. Y. Huang and E. M. Terentjev, *Int. J. Mater. Form.*, 2008, **1**, 63–74.
28. Y. Y. Huang, T. P. J. Knowles and E. M. Terentjev, *Adv. Mater.*, 2009, **21**, 3945–3948.
29. Y. Yamamoto, Y. Miyauchi, J. Motoyanagi, T. Fukushima, T. Aida, M. Kato and S. Maruyama, *Jpn. J. Appl. Phys.*, 2008, **47**, 2000–2004.
30. R. Ramasubramaniama and J. Chen, *Appl. Phys. Lett.*, 2003, **83**, 2928–2930.
31. M. Strano, V. C. Moore, M. K. Miller, M. Allen, E. Haroz, C. Kittrell, R. H. Hauge and R. E. J. Smalley, *Nanosci. Nanotechnol.*, 2003, **3**, 81–86.
32. G. X. Chen, Y. J. Li and H. Shimizu, *Carbon*, 2007, **45**, 2334–2340.
33. T. Q. Nguyen, Q. Z. Liang and H. H. Kausch, *Polymer*, 1997, **38**, 3783–3793.
34. D. Lohse, *Nature*, 2005, **434**, 33–34.
35. A. Gedanken, *Ultrason. Sonochem.*, 2004, **11**, 47–55.
36. E. W. Wong, P. E. Sheehan and C. M. Lieber, *Science*, 1997, **277**, 1971–1975.
37. S. S. Xie, W. Z. Li, Z.W. Pan, B. H. Chang and L. F. Sun, *J. Phys. Chem. Solid.*, 2000, **61**, 1153–1158.
38. W. Ding, L. Calabri, K. M. Kohlhaas, X. Chen, D. A. Dikin and R. S. Ruoff, *Exp. Mech.*, 2007, **47**, 25–36.
39. B. Peng, M. Locascio, P. Zapol, S. Li, S. L. Mielke, G. C. Schatz and H. D. Espinosa, *Nature Nanotechnol.*, 2008, **3**, 626–631.
40. A. H. Barber, R. Andrews, L. S. Schadler and H. D. Wagner, *Appl. Phys. Lett.*, 2005, **87**, 203106.

41. C. J. Kerr, Y. Y. Huang, J. E. Marshall and E. M. Terentjev, *J. Appl. Phys.*, 2011, **109**, 094109.
42. V. Domenici, B. Zupančič, V. V. Laguta, A. G. Belous, M. O. I. V'yunov, Remškar and B. Zalar, *J. Phys. Chem. C*, 2010, **114**, 10782–1078.
43. E. M. Dannenberg, *Ind. Eng. Chem.*, 1952, **44**, 813–818.
44. A. I. Zhbanov, E. G. Pogorelov and Y.-C. Chang, *ACS Nano*, 2010, **4**, 5937–5945.
45. R. S. Ruoff, J. Tersoff, D. C. Lorents, S. Subramoney and B. Chan, *Nature*, 1993, **364**, 514–516.
46. C. Park, Z. Ounaies, K. A. Watson, R. E. Crooks, J. Smith Jr., S. E. Lowther, J. W. Connell, E. J. Siochi, J. S. Harrison and T. L. St. Clair, *Chem. Phys. Lett.*, 2002, **364**, 303–308.
47. C. Velasco-Santos, A. L. Martinez-Hernandez and V. M. Castano, *Compos. Interf.*, 2005, **11**, 567–586.
48. M. J. Park, J. K. Lee, B. S. Lee, Y.-W. Lee, I. S. Choi and S. Lee, *Chem. Mater.*, 2006, **18**, 1546–1551.
49. D. R. Shobha Jeykumari and S. Sriman Narayanan, *Nanotechnology*, 2007, **18**, 125501.
50. P. Abiman, G. G. Wildgoose and R. G. Compton, *Int. J. Electrochem. Sci.*, 2008, **3**, 104–117.
51. H. Kong, C. Gao and D. Yan, *J. Am. Chem. Soc.*, 2004, **126**, 412–413.
52. I. Kumar, S. Rana and J. W. Cho, *Chem. Eur. J.*, 2011, **17**, 11092–11101.
53. L. Vaisman, H. D. Wagner and G. Marom, *Adv. Colloid Interf. Sci.*, 2006, **128–130**, 37–46.
54. M. J. O'Connell, S. M. Bachilo, C. B. Huffman, V. C. Moore, M. S. Strano, E. H. Haroz, K. L. Rialon, P. J. Boul, W. H. Noon, C. Kittrell, J. Ma, R. H. Hauge, R. B. Weisman and R. E. Smalley, *Science*, 2002, **297**, 593–596.
55. L. Jiang, L. Gao and J. Sun, *J. Colloid Interf. Sci.*, 2003, **260**, 89.
56. J. Steinmetz, M. Glerupa, M. Paillet, P. Bernier and M. Holzinger, *Carbon*, 2005, **43**, 2397–2400.
57. Y. Tan and D. E. Resasco, *J. Phys. Chem. B*, 2005, **109**, 14454–14460.
58. O. Matarredona, H. Rhoads, Z. Li, J. H. Harwell, L. Balzano and D. E. Resasco, *J. Phys. Chem. B*, 2003, **107**, 13357–13367.
59. V. C. Moore, M. S. Strano, E. H. Haroz, R. H. Hauge and R. E. Smalley, *Nano Lett.*, 2003, **3**, 1379–1382.
60. M. J. O'Connell, P. Boul, L. M. Ericson, C. Huffman, Y. Wang, E. Haroz, C. Kuper, J. Tour, K. D. Ausman and R. E. Smalley, *Chem. Phys. Lett.*, 2001, **342**, 265–271.
61. O. Regev, P. N. B. ElKati, J. Loos and C. E. Koning, *Adv. Mater.*, 2004, **16**, 248–251.
62. B. Kima, Y.-H. Leeb, J.-H. Ryub and K.-D. Suh, *Colloids Surf. A*, 2006, **273**, 161–164.
63. B. J. Landi, H. J. Ruf, J. J. Worman and R. P. Raffaelle, *J. Phys. Chem. B.*, 2004, **108**, 17089–17095.

64. X. Gong, J. Liu, S. Baskaran, R. D. Voise and J. S. Young, *Chem. Mater.*, 2000, **12**, 1049–1052.
65. S. Manivannan, Il Ok Jeong, Je Hwang Ryu, Chang Seok Lee, Ki Seo Kim, Jin Jang and Kyu Chang Park, *J. Mater. Sci.: Mater. Electron.*, 2009, **20**, 223–229.
66. A. Star, J. F. Stoddart, D. Steuerman, M. Diehl, A. Boukai, E. W. Wong, X. Yang, S.-W. Chung, H. Choi and J. R. Heath, *Angew. Chem. Int. Ed.*, 2001, **40**, 1721–1725.
67. A. B. Dalton, C. Stephan, J. N. Coleman, B. McCarthy, P. M. Ajayan, S. Lefrant, P. Bernier, W. J. Blau and H. J. Byrne, *J. Phys. Chem. B*, 2000, **104**, 10012.
68. A. Ikeda, K. Nobusawa, T. Hamano and J. Kikuchi, *Org. Lett.*, 2006, **8**, 5489.
69. R. Shvartzman-Cohen, Y. Levi-Kalisman, E. Nativ-Roth and R. Yerushalmi-Rozen, *Langmuir*, 2004, **20**, 6085–6088.
70. J. Zou, L. Liu, H. Chen, S. I. Khondaker, R. D. McCullough, Q. Huo and L. Zhai, *Adv. Mater.*, 2008, **20**, 2055–2060.
71. E. J. Katz, *Electroanal. Chem.*, 1994, **365**, 157.
72. H. Li, B. Zhou, Y. Lin, L. Gu, W. Wang, K. A. S. Fernando, S. Kumar, L. F. Allard and Y.-P. Sun, *J. Am. Chem. Soc.*, 2004, **126**, 1014–1015.
73. Y. Ji, Y. Y. Huang, A. R. Tajbakhsh and E. M. Terentjev, *Langmuir*, 2009, **25**, 12325–12331.
74. T. Hwa, E. Kokofuta and T. Tanaka, *Phys. Rev. A.*, 1991, **44**, R2235–R2238.
75. S. Stankovich, D. A. Dikin, G. H. B. Dommett, K. M. Kohlhaas, E. J. Zimney, E. A. Stach, R. D. Piner, S. T. Nguyen and R. S. Ruoff, *Nature*, 2006, **442**, 282–286.
76. T. Ramanathan, A. A. Abdala, S. Stankovich, D. A. Dikin, M. Herrera-Alonso, R. D. Piner, D. H. Adamson, H. C. Schniepp, X. Chen, R. S. Ruoff, S. T. Nguyen, I. A. Aksay, R. K. Prudhomme and L. C. Brinson, *Nature Nanotechnol.*, 2008, **3**, 327–331.
77. J. Turkevich, P. C. Stevenson and J. Hillier, *Discuss. Faraday. Soc.*, 1951, **11**, 55–75.
78. M. K. Corbierre, N. S. Cameron, M. Sutton, S. G. J. Mochrie, L. B. Lurio, A. Rühm and R. B. Lennox, *J. Am. Chem. Soc.*, 2001, **123**, 10411–10412.
79. M. Yoo, S. Kim, S. G. Jang, S.-H. Choi, H. Yang, E. J. Kramer, W. B. Lee, B. J. Kim and J. Bang, *Macromolecules*, 2011, **44**, 9356–9365.
80. S. Lin, Y. Cheng, J. Liu and M. R. Wiesner, *Langmuir*, 2012, **28**, 4178–4186.
81. C. Chevigny, F. Dalmas, E. Di Cola, D. Gigmes, D. Bertin, F. Boué and J. Jestin, *Macromolecules*, 2011, **44**, 122–133.
82. D. Fragouli, B. Torre, G. Bertoni, R. Buonsanti, R. Cingolani and A. Athanassiou, *Microsc. Res. Tech.*, 2010, **73**, 952–958.
83. V. Domenici, M. Conradi, M. Remškar, M. Viršek, B. Zupančič, A. Mrzel, M. Chambers and B. Zalar, *J. Mater. Sci.*, 2011, **46**, 3639–3645.

84. T. Nagai, N. Aoki, Y. Ochiai and K. Hoshino, *ACS Appl. Mater. Interf.*, 2011, **3**, 2341–2348.
85. A. M. Donald, A. H. Windle, and S. Hanna, *Liquid Crystalline Polymers*, Cambridge University Press, Cambridge, 2006.
86. M. Warner and E. M. Terentjev, *Liquid Crystal Elastomers*, Oxford University Press, Oxford, 2007.
87. J. Küpfer and H. Finkelmann, *Macromol. Chem. Phys.*, 1994, **159**, 1353–1367.
88. J. Küpfer and H. Finkelmann, *Makromol. Chem., Rapid Commun*, 1991, **12**, 717–726.
89. S. M. Clarke, A. Hotta, A. R. Tajbakhsh and E. M. Terentjev, *Phys. Rev. E.*, 2002, **65**, 021804.
90. H. Finkelmann, E. Nishikawa, G. G. Pereira and M. Warner, *Phys. Rev. Lett.*, 2001, **87**, 015501:1–015501:4.
91. M. Camacho-Lopez, H. Finkelmann, P. Palffy-Muhoray and M. Shelley, *Nature Mater.*, 2004, **3**, 307–310.
92. C. S. Li, C. W. Lo, D. F. Zhu, C. H. Li, Y. Liu and H. R. Jiang, *Macromol. Rapid Commun.*, 2009, **30**, 1928–1935.
93. Y. Yu, M. Nakano and T. Ikeda, *Nature*, 2003, **425**, 145.
94. J. Garcia-Amorós, A. Piñol, H. Finkelmann and D. Velasco, *Org. Lett.*, 2011, **13**, 2282–2285.
95. A. Sánchez-Ferrer, A. Merekalov and H. Finkelmann, *Macromol. Rapid Commun.*, 2011, **32**, 671–678.
96. C. L. Harvey and E. M. Terentjev, *Eur. Phys. J. E*, 2007, **23**, 185–189.
97. Y. Y. Huang, J. Biggins, Y. Ji and E. M. Terentjev, *J. Appl. Phys.*, 2010, **107**, 083515:1–083515:8.
98. W. Lehmann, H. Skupin, C. Tolksdorf, E. Gebhard, R. Zentel, P. Krüger, M. Löschel and F. Kremer, *Nature*, 2001, **410**, 447–450.
99. P. Papadopoulos, P. Heinze, H. Finkelmann and F. Kremer, *Macromolecules*, 2010, **43**, 6666–6670.
100. Y. H. Na, Y. Aburaya, H. Orihara and K. Hiraoka, *Phys. Rev. E*, 2011, **83**, 061709:1–061709:5.
101. Y. Ji, J. E. Marshall and E. M. Terentjev, *Polymers*, 2012, **4**, 316–340.
102. Y. Ji, Y. Y. Huang, R. Rungsawang and E. M. Terentjev, *Adv. Mater.*, 2010, **22**, 3436–3440.
103. J. E. Marshall, Y. Ji, N. Torras, K. Zinoviev and E. M. Terentjev, *Soft Matter*, 2012, **8**, 1570.
104. X. Ao, X. Wen and R. B. Meyer, *Physica A*, 1991, **176**, 63–71.
105. M. Chambers, B. Zalar, M. Remskar, S. Zumer and H. Finkelmann, *Appl. Phys. Lett.*, 2006, **89**, 243116:1–243116:3.
106. M. Chambers, B. Zalar, M. Remskar, J. Kovac, H. Finkelmann and S. Zumer, *Nanotechnology*, 2007, **18**, 415706.
107. S. Bae, H. Kim, Y. Lee, X. F. Xu, J. S. Park, Y. Zheng, J. Balakrishnan, T. Lei, H.R. Kim, Y. I. Song, Y.-J. Kim, K. S. Kim, B. Özyilmaz, J.-H. Ahn, B. H. Hong and S. Iijima, *Nature Nanotechnology*, 2010, **5**, 574–578.

108. H. Z. Geng, K.K. Kim, K.P. So, Y.S. Lee, Y. Chang and Y.H. Lee, *J. Am. Chem. Soc.*, 2007, **129**, 7758–7759.
109. Z.R. Li, H.R. Kandel, E. Dervishi, V. Saini, Y. Xu, A.R. Biris, D. Lupu, G. J. Salamo and A.S. Biris, *Langmuir*, 2008, **24**, 2655–2662.
110. S. De, T. M. Higgins, P. E. Lyons, E. M. Doherty, P. N. Nirmalraj, W. J. Blau, J. J. Boland and J. N. Coleman, *ACS Nano*, 2009, **3**, 1767–1774.
111. V. Scardaci, R. Coull and J.N. Coleman, *Appl. Phys. Lett.*, 2010, **97**, 023114.
112. H. A. Becerril, J. Mao, Z. Liu, R.M. Stoltenberg, Z. Bao and Y. Chen, *ACS Nano*, 2008, **2**, 463–470.
113. X. S. Li, Y. W. Zhu, W. W. Cai, M. Borysiak, B. Y. Han, D. Chen, R. D. Piner, L. Colombo and R.S. Ruoff, *Nano Lett.*, 2009, **9**, 4359–4363.
114. C. Mattevi, G. Eda, S. Agnoli, S. Miller, K. A. Mkhoyan, O. Celik, D. Mostrogiovanni, G. Granozzi, E. Garfunkel and M. Chhowalla, *Adv. Funct. Mater.*, 2009, **19**, 2577–2583.
115. S. De, P. J. King, M. Lotya, A. O'Neill, E. M. Doherty, Y. Hernandez, G. S. Duesberg and J. N. Coleman, *Small*, 2010, **6**, 458–464.
116. E. M. Doherty, S. De, P. E. Lyons, A. Shmeliov, P. N. Nirmalraj, V. Scardaci, J. Joimel, W. J. Blau, J. J. Boland and J. N. Coleman, *Carbon*, 2009, **47**, 2466–2473.
117. Y.Y. Huang and E. M. Terentjev, *ACS Nano*, 2011, **5**, 2082.
118. F.M. Blighe, Y. R. Hernandez, W. J. Blau and J. N. Coleman., *Adv. Mater.*, 2007, **19**, 4443–4447.
119. C. H. Hu, C. H. Liu, L. Z. Chen and S. S. Fan, *Appl. Phys. Lett.*, 2009, **95**, 103103.
120. C. H. Liu and S. S. Fan, *Appl. Phys. Lett.*, 2007, **90**, 041905.
121. P. C. Ramamurthya, A. M. Malsheb, W. R. Harrell, R. V. Gregorya, K. McGuirec and A. M. Rao, *Solid State Electron.*, 2004, **48**, 2019.
122. M. B. Bryning, M. F. Islam, J. M. Kikkawa and A. G. Yodh, *Adv. Mater.*, 2005, **17**, 1186.
123. I. Adamczewski, *Ionization, Conductivity and Breakdown in Dielectric Liquids*, Taylor & Francis Ltd, London, 1969.
124. S. Cui, R. Canet, A. Derre, M. Couzi and P. Delhaes, *Carbon*, 2003, **41**, 797–809.
125. A.V. Kyrylyuk and P. van der Schoot, *Proc. Nature Acad. Sci.*, 2008, **105**, 82218226.
126. Q. Zhang, S. Rastogi, D. Chen, D. Lippits and P. J. Lemstra, *Carbon*, 2006, **44**, 778–785.
127. C. Mitchell and R. Krishnamoorti, *Macromolecules*, 2007, **40**, 1538.
128. S. Cui, R. Caneta, A. Derrea, M. Couzib and P. Delhaes, *Carbon*, 2003, **41**, 797.
129. B. Kilbride, J. Coleman, J. Fraysse, P. Fournet, M. Cadek, A. Drury, S. Hutzler, S. Roth and W. Blau, *J. Appl. Phys.*, 2002, **92**, 4024.
130. D. Grannan, J. Garland and D. Tanner, *Phys. Rev. Lett.*, 1981, **46**, 375.

131. C. Grimes, C. Mungle, D. Kouzoudis, S. Fang and P. Eklund, *Chem. Phys. Lett.*, 2000, **319**, 460.
132. B. Kim, J. Lee and I. Yu, *J. Appl. Phys.*, 2003, **94**, 6724.
133. L. Wang and Z. Dang, *Appl. Phys. Lett.*, 2005, **87**, 042903.
134. R. Kohlmeyer, A. Javadi, B. Pradhan, S. Pilla, K. Setyowati, J. Chen and S. Gong, *J. Phys. Chem. C*, 2009, **113**, 17626.
135. L. Hsu, C. Weder and S. J. Rowan, *J. Mater. Chem.*, 2011, **21**, 2812–2822.
136. C. Li, E. T. Thostenson and T.-W. Chou, *Compos. Sci. Technol.*, 2008, **68**, 1227–1249.
137. L. Liu, S. Zhang, T. Hu, Z.-X. Guo, C. Ye, L. Dai and D. Zhu, *Chem. Phys. Lett.*, 2002, **359**, 191–195.
138. M.A. Hamon, M.E. Itkis, S. Niyogi, T. Alvaraez, C. Kuper, M. Menon and R.C. Haddon, *J. Am. Chem. Soc.*, 2001, **123**, 11292–11293.
139. H. Kataura, Y. Kumazawa, Y. Maniwa, I. Umezu, S. Suzuki, Y. Ohtsuka and Y. Achiba, *Synth. Met.*, 1999, **103**, 2555–2558.
140. K. Mizuno, J. Ishii, H. Kishida, Y. Hayamizu, S. Yasuda, D. N. Futaba, M. Yumura and K. Hata, *Proc. Natl. Acad. Sci. USA*, 2009, **106**, 6044–6047.
141. P. M. Ajayan, M. Terrones, A. de la Guardia, V. Huc, N. Grobert, B.-Q. Wei, H. Lezec, G. Ramanath and T.W. Ebbesen, *Science*, 2002, **296**, 705.
142. S. Lu, S. V. Ahir, E. M. Terentjev and B. Panchapakesan, *Appl, Phys. Lett.*, 2007, **91**, 103106.
143. S. V. Ahir and E. M. Terentjev, *Nature Mater.*, 2005, **4**, 491–495.
144. S. Lu and B. Panchapakesan, *Nanotechnology*, 2007, **18**, 305502.
145. J. Loomis, B. King, T. Burkhead, P. Xu, N. Bessler, E. M. Terentjev and B. Panchapakesan, *Nanotechnology*, 2012, **23**, 045501.
146. A. Lendlein and S. Kelch, *Angew. Chem. Int. Ed.*, 2002, **41**, 2034–2057.
147. A. Lendlein, H. Jiang, O. Jünger and R. Langer, *Nature*, 2005, **434**, 879–882.
148. H. Koerner, G. Price, N. A. Pearce, M. Alexander and R. A. Vaia, *Nature Mater.*, 2004, **3**, 115–120.
149. J. S. Leng, X. L. Wu and Y. J. Liu, *Appl. Polym. Sci.*, 2009, **114**, 2455–60.

Subject Index